Progress in Optimization

Applied Optimization

Volume 39

The titles published in this series are listed at the end of this volume.

Progress in Optimization

Contributions from Australasia

edited by

Xiaoqi Yang

The Hong Kong Polytechnic University,
Hong Kong

Alistair I. Mees

Mike Fisher

and

Les Jennings

The University of Western Australia,
Perth, Australia

KLUWER ACADEMIC PUBLISHERS

DORDRECHT / BOSTON / LONDON

A C.I.P. Catalogue record for this book is available from the Library of Congress.

ISBN-13:978-0-7923-6286-9 e-ISBN-13:978-1-4613-0301-5
DOI: 10.1007/978-1-4613-0301-5

Published by Kluwer Academic Publishers,
P.O. Box 17, 3300 AA Dordrecht, The Netherlands.

Sold and distributed in North, Central and South America
by Kluwer Academic Publishers,
101 Philip Drive, Norwell, MA 02061, U.S.A.

In all other countries, sold and distributed
by Kluwer Academic Publishers,
P.O. Box 322, 3300 AH Dordrecht, The Netherlands.

Printed on acid-free paper

Contents

Preface

'Optimization Day' (OD) has been a series of annual mini-conferences in Australia since 1994. The purpose of this series of events is to gather researchers in optimization and its related areas from Australia and their collaborators, in order to exchange new developments of optimization theories, methods and their applications. The first four OD mini-conferences were held in The University of Ballarat (1994), The University of New South Wales (1995), The University of Melbourne (1996) and Royal Melbourne Institute of Technology (1997), respectively. They were all on the eastern coast of Australia.

The fifth mini-conference Optimization Days was held at the Centre for Applied Dynamics and Optimization (CADO), Department of Mathematics and Statistics, The University of Western Australia, Perth, from 29 to 30 June 1998. This is the first time the OD mini-conference has been held at the western coast of Australia. This fifth OD preceded the International Conference on Optimization: Techniques and Applications (ICOTA) held at Curtin University of Technology. Many participants attended both events. There were 28 participants in this year's mini-conference and 22 presentations in the mini-conference.

The presentations in this volume are refereed contributions based on papers presented at the fifth Optimization Days mini-conference. The volume is divided into the following parts: Global Optimization, Nonsmooth Optimization, Optimization Methods and Applications.

The first part, *Global Optimization,* starts from two state-of-the-art review papers: one is a keynote lecture on location theory and distance geometry, where various global optimization methods for studying these problems were reviewed and another one is an invited paper on mixed integer linear programming problems. This is followed by a paper on separability of star-shaped sets where the separation is made by a collection of linear functions, a generalization of a well known separation theorem by a linear function in Convex Analysis. Next, two brief review papers summarize the development of newly-introduced nonlinear Lagrangian methods for nonconvex optimization. Their emphasis is different: the first one mainly discusses the zero duality gap and nonlinear penalty function method for nonconvex continuous optimization, while the second paper focuses on the application of nonlinear Lagrangian methods in constrained integer programs. The nonlinear penalty function method is further developed in the last chapter of this part where both classical penalty functions and nonlinear penalty functions are applied to the study of the binary multicommodity flow problem via simulated annealing.

The second part of the book is *Nonsmooth Optimization.* This starts from an invited paper on the study of a quadratic recourse representation of the two-stage stochastic linear problem where the quadratic recourse function is SC^1 (i.e., it is continuously differentiable and its derivative is semismooth). The paper on Lagrange multipliers for nonconvex optimization establishes Kuhn-Tucker type optimality conditions for problems where directional derivatives of the functions involved are not convex functions of direction. There then follow two papers on generalized convex functions. It is worth noting that the study of convex functions and their generalizations is very important as they play significant roles in the investigation of optimization theories and algorithms. One paper concentrates on the class-inclusion properties of generalized convex functions. The second paper considers locally convexifiable smooth vector functions on smooth manifolds. The complementarity and variational inequality problems are traditionally two very closely related topics. This book includes one paper on each of them. The paper on complementarity studies the connectedness of solution sets of the problem, while the paper on variational inequalities considers the generalization of scalar variational inequalities to the vector case.

The third part, *Optimization Methods,* includes two review papers as well as research papers on methods of solving optimization and optimal control problems. The first review paper is a keynote lecture on parameter estimation in dynamic systems: it presents numerical methods for identifying parameters in systems of ordinary differential equations. The review paper on feasible direction methods for solving constrained optimization problems makes a comparative study of their numerical efficiency. The optimal switching control problem presented in the next paper involves the determination of switching time points over the system's time horizon; numerical examples are also given to illustrate the proposed methods. Although the trajectory following method is a relatively old concept, one paper in this part shows that this method, in conjunction with a good "stiff' integration routine, can provide a tool that is easy to use for a wide range of optimization problems. The upwind finite difference method is investigated to obtain numerical approximations of viscosity solutions of a two dimensional Hamilton-Jacobi-Bellman (HJB) equation arising from a class of optimal control problems. Optimal control methods are further applied to solving a class of continuous linear programs which involves an integral objective function and a Volterra type inequality constraint with a convolution kernel.

The last part of the book, *Applications,* presents several practical applications of optimization and optimal control methods. These are (i) a transport planning process via a maximum likelihood model solved by a generic algorithm, (ii) optimal strategy for fuel consumption of train control within a fixed travel time using methods of optimal control theory, (iii) determination of optimal batch size for a manufacturing system solved by a heuristic method, and (iv) parameter estimation for substrate diffusion in a metabolically active cutaneous tissue via partial differential equations.

The collection of the presentations (including invited papers) in the fourth annual conference started a series of edited books. This is the second volume of this series and is a refereed edited book. All the papers were reviewed either by external examiners or by the editors. We are grateful to the following reviewers for reviewing some of the papers: S. Dragomir (Victoria University of Technology), A. Ernst (Division of Mathematical and Information Sciences, CSIRO), X. Huang (The University of Western Australia), J. Ong (The University of Western Australia), P.J. Pudney (The University of South Australia), V. Rehbock (Curtin University of Technology), Y. Xiao (The University of Melbourne), G.L. Xue (The University of Vermon), J.C. Yao (National Sun Yat-Sen University), Z. Zang (Curtin University of Technology), and X.Y. Zhou (The Chinese

University of Hong Kong). Finally we thank the partial financial support from the Department of Mathematics and Statistics, The University of Western Australia and Professor K.L. Teo and Dr. D. Ralph for their suggestions and help during the preparation of this book.

X.Q. YANG, A.I. MEES, M.E. FISHER AND L.S. JENNINGS

Participants

Miss Kirsty Campbell, Centre for Applied Dynamics and Optimization, Department of Mathematics and Statistics, The University of Western Australia, Nedlands,Western Australia 6907, Australia. Email: kirsty@maths.uwa.edu.au.

Dr. Bruce D. Craven, Department of Mathematics and Statistics, The University of Melbourne, Parkville, Victoria 3052, Australia.
Email: craven@mundoe.maths.mu.oz.au.

Dr. Andrew Eberhard, Department of Mathematics, Royal Melbourne Institue of Technology, 124 Latrobe St, Melbourne, Victoria. 3000, Australia. Email: andy.eb@rmit.edu.au.

Associate Professor Mike Fisher, Centre for Applied Dynamics and Optimization, Department of Mathematics and Statistics, The University of Western Australia, Nedlands, Western Australia 6907, Australia.
Email: fisher@maths.uwa.edu.au.

Dr. Valadimir Gershkovich, Department of Mathematics and Statistics, The University of Melbourne, Parkville, Victoria 3052, Australia

Dr. Barney M. Glover, School of Mathematics and Statistics, Curtin University of Technology, Perth, Western Australia 6845, Australia. Email: rgloverb@alpha2.curtin.edu.au.

Dr. Bean San Goh, Department of Mathematics and Statistics, The University of Western Australia, Nedlands, Western Australia 6907, Australia. Email: gohsan@maths.uwa.edu.au.

Mr. Renlong Han, Department of of Information Management and Marketing, The University of Western Australia, Nedlands, Western Australia 6907, Australia. Email: rlhan@ecel.uwa.edu.au.

Associate Professor Phil Howlett, Centre for Indutrial and Applicable Mathematics, The University of South Australia, The Levels, South Australia, Australia. Email: p.howlett@unisa.edu.au.

Professor George Isac, Department of Mathematics and Computer Science, Royal Military College of Canada, P.O.Box 17000, STN Forces, Kingston, Ont., Canada, K7K 7B4. Email: isac-g@rmc.ca.

Associate Professor Les Jennings, Centre for Applied Dynamics and Optimization, Department of Mathematics and Statistics, The University of Western Australia, Nedlands, Western Australia 6907, Australia.
Email: les@maths.uwa.edu.au

Mr. M. Kasiama, Department of Mathematics, The University of the Witwatersrand, South Africa.

Professor Michael M. Kostreva, Department of Mathematical Sciences, Clemson University, Clemson, SC 29634-1907, USA.
Email: flstgla@clemson.edu.

Associate Professor Duan Li, Department of Systems Engineering and Engineering Management, Chinese University of Hong Kong, Shatin, N.T., Hong Kong. Email: dli@se.cuhk.edu.hk.

Dr. Yanqun Liu, Department of Applied Mathematics, The Hong Kong Polytechnic University, Kowloon, Hong Kong. Email: mayliu@polyu.edu.hk.

Professor Alistair Mees, Centre for Applied Dynamics and Optimization, Department of Mathematics and Statistics, The University of Western Australia, Nedlands, Western Australia 6907, Australia.
Email: alistair@maths.uwa.edu.au.

Associate Professor Charles Pearce, Department of Applied Mathematics, The University of Adelaide, South Australia 5005, Australia.
Email: cpearce@maths.adeliade.edu.au.

Professor Alex Rubinov, School of Information Technology and Mathematical Sciences, The University of Ballarat, Victoria 3353, Australia. Email: amr@ballarat.edu.au.

Dr. Ruhul Sarker, School of Computer Science, University College, The University of New South Wales, ADFA, Northcott Drive, Canberra, ACT 2600, Australia. Email: r-sarker@adfa.oz.au.

Professor Klaus Schittkowski, Department of Mathmatics, The University of Bayreuth, 95440 Bayreuth, Germany.
Email: klaus.schittkowski@uni-bayreuth.de.

Associate Professor Xiaoling Sun, Department of Mathematics, Shanghai University, Shanghai, China. Email: xlsun@guomai.sh.cn.

Dr. Patrick Tobin, Department of Mathematics, Swinburne University of Technology, Australia. Email: ptobin@swin.edu.au.

Professor Hoang Tuy, Institute of Mathematics, P.O. Box 631, Bo Ho, Hanoi, Vietnam. Email: htuy@hn.vnn.vn.

Professor Thomas L. Vincent, Aerospace and Mechanical Engineering, University of Arizona, Tucson, AZ 85721, USA. Email: vincent@u.arizona.edu.

Dr. Song Wang, Centre for Applied Dynamics and Optimization, Department of Mathematics and Statistics, The University of Western Australia, Nedlands, Western Australia 6907, Australia.
Email: swang@maths.uwa.edu.au.

Dr. K. H. Wong, Department of Mathematics, The University of the Witwatersrand, South Africa. Email: wong@gauss.cam.wits.ac.za.

Dr. Xiao-qi Yang, Department of Applied Mathematics, The Hong Kong Polytechnic University, Kowloon, Hong Kong. Email: mayangxq@polyu.edu.hk.

Editors

Xiaoqi Yang is an assistant professor in Department of Applied Mathematics, the Hong Kong Polytechnic University. He was awarded a PhD in Applied Mathematics from the University of New South Wales in 1994. He worked as an Australian Research Council Postdoctoral Fellow and a Lecturer in the University of Western Australia. His current research interest is in nonsmooth optimization, global optimization and mathematical finance. He has published over 50 refereed international journal papers in optimization.

Alistair Mees graduated from Edinburgh University in Mathematical Physics and did a PhD in Control Theory at Cambridge University. He has worked in the Department of Pure Mathematics and Mathematical Statistics, Cambridge University, the Department of Electrical Engineering and Computer Science at The University of California, Berkeley. He is currently Professor of Applied Mathematics and director of the Centre for Applied Dynamics and Optimization at The University of Western Australia. His research interests include discrete optimization and applied operations research as well as nonlinear dynamics and control, with around 100 publications in these areas.

Michael E. Fisher obtained his B.Sc. degree (first class honours) in applied mathematics and M.Sc. and Ph.D. degrees from the University of Western Australia. He is currently an Associate Professor in the Department of Mathematics and Statistics at the University of Western Australia which he joined as a Tutor in 1970. His research interests include optimization and control theory, and their applications to a variety of areas including industrial problems and biomedicine.

Leslie S. Jennings was born in Adelaide, South Australia on August 25, 1947. He received his B. Sc. (Honours) from the University of Adelaide in 1969 and a Ph. D. in numerical analysis from the Australian National University in 1973. This was followed by a nine month post-doctoral appointment at Stanford University in the Computer Science Department. Since then he has held a Lecturer, Senior Lecturer, then Associate Professor position in the Department of Mathematics at the University of Western Australia, Perth. His interests lie in numerical analysis, and in the application of optimal control to human movement modelling, multi-body systems, chemical engineering and filter design. Currently he is working on the interface of optimal control, numerical analysis and software engineering.

I GLOBAL OPTIMIZATION

1 GLOBAL OPTIMIZATION METHODS FOR LOCATION AND DISTANCE GEOMETRY PROBLEMS

Hoang Tuy

Institute of Mathematics
P.O. Box 631, Bo Ho,
Hanoi, Vietnam

Abstract: A general problem in location theory and distance geometry is to find the configuration of p unknown points in $\mathbb{R}^n$ satisfying a number of constraints on their mutual distances and their distances to N fixed points, while minimizing a given function of these distances. Global optimization methods recently developed for studying different variants of this problem are reviewed.

Key words: location problems, distance geometry, global optimization.

X. Yang et al. (eds.), Progress in Optimization, 3–20.

1 INTRODUCTION

The general problem we will be considering in this paper is to find the configuration of p unknown points in $\mathbb{R}^n$ (most often $n \leq 3$) satisfying a number of constraints on their mutual distances and their distances to N fixed points, while minimizing a given function of these distances. Often the unknown points represent the locations of facilities to be constructed to serve the users located at the fixed points, so as to minimize a cost function (e.g. the total travel cost from the users to the facilities). In other cases, the unknown points represent the locations of a set of objects (atoms, particles) whose interactions depend upon their mutual distances. The objective function in these problems is then interpreted as a "potential energy function" which should attain a global minimum at the unknown configuration.

This general formulation subsumes a wide class of important problems of location theory and distance geometry encountered in economics, engineering, computer science, physics, chemistry, biology and other fields.

For many years, combinatorial geometric reasoning and nonlinear programming methods have been the basic tools in the study of these problems. However, since most nonconvex problems are characterized by multiextremality, i.e. the existence of many local non-global minimizers, other more suitable methods are necessary for efficiently coping with this difficulty.

Global optimization methods began to be introduced in location theory about a decade ago. Branch and bound methods, such as the Big Square Small Square method (Hansen et al. (1985)), Plastria (1992)) were first developed for certain constrained location problems. Subsequently, techniques of d.c. optimization (i.e. global optimization of problems involving differences of convex functions) were used for the study of facility location problems with nonconvex objective functions (Tuy et al. (1992)), Weber's problems with attraction and repulsion (Chen et al. (1992), Maranas and Floudas (1993), Tuy et al. (1994) and also multisource Weber's problems (Chen et al. (1994)). Most recently, distance geometry problems arising from molecular conformation and related problems in biology, physics, chemistry, etc. have been identified as d.c. optimization problems with the same structure as location problems, though generally involving a much larger number of nonconvex variables (Pardalos et al (1997) and references therein).

The primary purpose of the present paper is to review global optimization and especially d.c. optimization methods in the formulation, analysis and numerical study of this important class of nonconvex optimization problems.

In the next section we will first recall some fundamental results of global optimization, with an emphasis on basic techniques for problem formulation. In Section 3 we will present a generic branch and bound method for solving the prototype d.c. optimization problem which is also the general form of most problems to be discussed in the sequel. Sections 4 through 6 are devoted to the fundamental problem of location (generalized Weber's problem) together with its variations and further extensions. The success of application of d.c. optimization methods to these problems is partially due to the relatively small number of "complicating" variables involved in their formulation. Section 6 discusses some of the most challenging problems of location and distance geometry, characterized by their large dimensionality. Since the distance between two points x, y in $\mathbb{R}^n$ is a convex quadratic function of x, y the objective functions in these problems are often compositions of quadratic functions with d.c. functions.

2 GLOBAL OPTIMIZATION

A global optimization problem is a problem of the form

$$(GOP) \qquad \min\{f(x) |\ x \in \Omega \subset \mathbb{R}^n\}$$

where Ω is a closed set while $f(x)$ is a continuous function on Ω and we are interested in finding not just a local minimizer, but a global minimizer, i.e. a point $\bar{x} \in \Omega$ such that $f(\bar{x}) \leq f(x) \quad \forall x \in \Omega$.

A typical problem of this class is the concave minimization problem which corresponds to the case when $f(x)$ is concave while Ω is convex and closed. It is known to be NP-hard even when Ω is a polytope. Despite its difficulty, this problem began to be studied in the middle of the sixties (see e.g. Horst and Tuy (1996), Tuy (1998)). Since then, as optimization models become widely used in engineering, economics and other sciences, an increasing number of different nonconvex optimization problems are encountered that cannot be handled successfully by conventional methods of nonlinear programming. This situation, in conjunction with the spectacular advances of computer technology, has led to an intensive development of global optimization methods in recent years.

A theoretical result which gives much insight into the general mathematical structure of global optimization problems states that any closed set in $\mathbb{R}^n$ is the projection of a difference of two convex sets in $\mathbb{R}^{n+1}$. (Thach (1987,93)). As a consequence, any problem (GOP) can be converted to the form

$$\min\{cx |\ x \in D \setminus C\}$$

where D is convex closed and C is convex open. The prototype global optimization problem can thus be stated as

$$(P(X)) \qquad \min\{f_0(x) |\ f_i(x) \leq 0 (i = 1, \ldots, m), x \in X\}$$

where X is a closed convex set in $\mathbb{R}^n$ and each function $f_i, i = 0, 1 \ldots, m$ is a difference of two convex functions (briefly, a d.c. function):

$$f_i(x) = g_i(x) - h_i(x), \quad g_i, h_i \text{ convex.} \tag{2.1}$$

The following facts (whose proofs can be found e.g. in Konno et al (1997) or Tuy (1998)) are of frequent use for the formulation and analysis of global optimization problems.

Proposition 2.1 *The pointwise maximum or pointwise minimum of a finite family of d.c. functions is also a d.c. function. In particular, if* $f_i(x), i = 1, \ldots, m$ *are convex, then*

$$\min_{i=1,\ldots,m} f_i(x) = \sum_{i=1}^{m} f_i(x) - max_{j=1,\ldots,m} \sum_{i \neq j} f_i(x).$$

Proposition 2.2 *Let* $u(x), v(x)$ *be convex positive valued functions on a compact convex set* $M \subset \mathbb{R}^m$ *such that* $u(x) - v(x) \geq 0 \; \forall x \in M$. *If* $q : \mathbb{R}_+ \to \mathbb{R}$ *is a convex nonincreasing function such that* $q'_+(0) > -\infty$ *and* $K \geq |q'_+(0)|$ *then* $g(x) := q(u(x) - v(x)) + K[u(x) + v(x)]$ *is a convex function. Hence*

$$q(u(x) - v(x)) = g(x) - K[u(x) + v(x)].$$

More generally, under mild conditions, a convex (or concave) monotone function of a d.c. function over a compact convex set is a d.c. function whose explicit d.c. representation can easily be obtained.

3 A GENERIC BB METHOD

One of the most suitable methods for global optimization is branch and bound (BB). In this section we present a generic BB method for solving the prototype d.c. optimization problem $P(X)$ which is also the general form of most problems to be discussed below.

A BB algorithm involves two basic operations: *branching*, i.e. partitioning the space into polyhedral domains of the same kind; and *bounding*, i.e. for each partition set M estimate a lower bound for the objective function values over the feasible points in M.

• BOUNDING We shall use simplicial subdivision, so that a partition set is an n-simplex M with vertex set $V(M) = \{v^1, \ldots, v^{n+1}\}$. To estimate a lower bound for the subproblem:

$$(P(X \cap M)) \qquad \min\{f_0(x) | \; f_i(x) \leq 0 (i = 1, \ldots, m), \; x \in X \cap M\}$$

we define a convex relaxation of $P(X \cap M)$, i.e. a program

$$(RP(M)) \qquad \min\{\varphi_0^M(x) | \; \varphi_i^M(x) \leq 0 (i = 1, \ldots, m), \; x \in X \cap M\}$$

such that $\varphi_i^M(x)$ is a *tight convex minorant* of $f_i(x)$ on M $(i = 0, 1, \ldots, m)$. The latter means that $\varphi_i^M(x)$ is a convex function satisfying $\varphi_i^M(x) \leq f_i(x) \; \forall x \in M$ and $\varphi_i^M(v) = f_i(v)$ at every $v \in V(M)$. Specifically, since g_i, h_i in (2.1) are convex we can take

$$\varphi_i^M(x) = g_i(x) - \psi_i^M(x)$$

where $\psi_i^M(x)$ is the affine function satisfying $\psi_i^M(v) = h_i(v)$ for every $v \in V(M)$. Thus, $\beta(M)$ is taken to be the optimal value of the convex program

$$(RP(M)) \qquad \begin{array}{ll} \min & g_0(x) - \sum_{j=1}^{n+1} \lambda_j h_0(v^j) \\ \text{s.t.} & g_i(x) - \sum_{j=1}^{n+1} \lambda_j h_i(v^j) \leq 0 \quad i = 1, \ldots, m \\ & x = \sum_{j=1}^{n+1} \lambda_j v^j \in X, \; \sum_{j=1}^{n+1} \lambda_j = 1, \; \lambda_j \geq 0. \end{array}$$

Let $\omega(M)$ be an optimal solution of this convex program. Since every $\varphi_i^M(x)$ is a tight convex minorant of $f_i(x)$ it is immediate that whenever $\omega(M)$ coincides with an extreme point of M then $\omega(M)$ is feasible to $P(X \cap M)$ and hence, $\beta(M)$ is the exact minimum of problem $P(X \cap M)$. We express this fact by saying that the above bounding is *tight*.

Often one may prefer to define a linear rather than convex relaxation. To this end for each $i = 0, 1, \ldots, m$ and $v^j \in V(M)$ let $p_{ij} \in \partial g_i(v^j)$. Then a tight convex minorant of $f_i(x)$ on M is the convex piecewise linear function $\varphi_i(x) = \max_{j=1,\ldots,n+1}[\langle p_{ij}, x - v^j\rangle + g_i(v^j)] - \psi_i(x)$ so that $\beta(M)$ can be taken to be the optimal value in the linear program:

$$(LR(M)) \quad \begin{array}{ll} \min & t_0 - \sum_{j=1}^{n+1} \lambda_j h_0(v^j) \\ \text{s.t.} & t_i - \sum_{j=1}^{n+1} \lambda_j h_i(v^j) \leq 0 \qquad i = 1, \ldots, m; \\ & \langle p_{ij}, x - v^j\rangle + g_i(v^j) \leq t_i \quad i = 0, 1, \ldots, m; j = 1, \ldots, n+1 \\ & x = \sum_{j=1}^{n+1} \lambda_j v^j \in X, \quad \sum_{j=1}^{n+1} \lambda_j = 1, \ \lambda_j \geq 0. \end{array}$$

Note that by simple manipulations using Propositions 2.1 and 2.2 it is always possible to reduce the original d.c. optimization problem to an equivalent $P(X)$ with f_0 linear and $m = 1$. The above linear relaxation should be applied to this equivalent problem rather than to the original one, in order to avoid too numerous linear constraints in $LR(M)$. Sometimes, to get a tighter bound, one can take a point $v^0 \in M$, e.g. $v^0 = \frac{v^1 + \ldots + v^{n+1}}{n+1}$, and add to $LR(M)$ the linear constraints

$$\langle p_{i0}, x - v^0\rangle + g_i(v^0) \leq t_i \quad i = 0, 1, \ldots, m$$

where $p_{i0} \in \partial g_i(v^0)$. Also, when the functions $f_i(x)$ are quadratic, there are several different ways to construct linear or convex relaxations, including semidefinite programming relaxations (see e.g. Tuy (1998)).

• Branching At each iteration, a partition set M with lowest $\beta(M)$ is further subdivided. A common subdivision rule is bisection, i.e. the partition of M into two smaller simplices of equal volume by a hyperplane through the midpoint of the longest edge of M. This subdivision is *exhaustive* in the sense that any infinite nested sequence of simplices generated by a bisection process shrinks to a point. However, a more efficient subdivision rule which can be used jointly with a tight bounding is the following

Normal Subdivision Rule: Select an infinite sequence $\Delta \subset \{1, 2, \ldots\}$. Assign a generation index $\sigma(M)$ to every generated partition set M. If $\sigma(M) \in \Delta$ then bisect M; otherwise divide M into subsimplices via $\omega(M)$.

• Algorithm Using the above specified branching and bounding operations, a BB algorithm can be defined in the standard way, such that at any iteration k :

- If for some partition set M the optimal solution $\omega(M)$ of the bounding subproblem ($RP(M)$ or $LR(M)$) coincides with a vertex of M then, since the bound is tight, $\omega(M)$ is a feasible solution and can thus be used to update the incumbent (current best solution) $\bar{x}^k$.

- The candidate for branching is $M_k \in \text{argmin}\{\beta(M) | \ M \in \mathcal{R}_k\}$, where $\mathcal{R}_k$ denotes the collection of all partition sets still of interest at this iteration.

- The algorithm terminates when $\mathcal{R}_k = \emptyset$ (then $\bar{x}^k$ is a global optimal solution).

As proved in Tuy (1998) a normal subdivision rule ensures that for any nested sequence $\{M_k, k \in K\}$ generated by the algorithm, the sequence $\{\omega(M_k), k \in K\}$ contains at least one subsequence converging to a vertex of the simplex $M_\infty = \cap_{k \in K} M_k$. From this "semi-exhaustive" property it is easy to derive the following

Theorem 3.1 *A BB algorithm for $P(X)$ using a* tight *bounding and a* normal *branching rule either yields a global optimal solution after finitely many steps, or generates an infinite sequence $\{\bar{x}^k\}$ whose every accumulation point is a global optimal solution.*

4 THE GENERALIZED WEBER'S PROBLEM

The classical Weber's problem (sometimes called Fermat-Weber problem), is to locate a single facility in the plane that minimizes the sum of weighted Euclidean distances to the locations of N known users. This problem can be generalized to the case of p facilities as follows.

Suppose p facilities (providing the same service) are designed to serve N users located at points $a^1, \ldots, a^N$ in the plane. We wish to determine the locations $x^1, \ldots, x^p \in R^2$ of the facilities so as to minimize the total cost (travel time, transport cost for customers, etc.), or to maximize the total attraction (utility, number of customers, etc.). The cost or attraction is a given function of the distances from the users to the facilities.

Traditionally, the standard mathematical formulation of this problem is (Brimberg et al (1994)):

$$\begin{array}{ll} \min_{x,w} & \sum_{i=1}^p \sum_{j=1}^N w_{ij} d(x^i, a^j) \\ \text{s.t.} & \sum_{i=1}^p w_{ij} = r_j, j = 1, \ldots, N, \\ & w_{ij} \geq 0, i = 1, \ldots, p, j = 1, \ldots, N, \\ & x^i \in S, \quad i = 1, \ldots, p, \end{array} \tag{4.1}$$

where x^i is the unknown location of facility (factory) $i, i = 1, \ldots, p$; w_{ij} is an unknown weight representing the flow from facility i to the fixed point (warehouse) a^j, $j = 1, \ldots, N$; $d(x, a)$ is the distance function which estimates the travel distance between any two points $x, a \in R^2$, and S is a rectangular domain in R^2 where the facilities must be located.

A serious disadvantage of the above traditional formulation is that it involves a very large number of variables (pN variables w_{ij} and $2p$ variables x_1^i, x_2^i) and a highly nonconvex objective function. Because of that, the problem is very difficult to handle, even for small values of p and N. In fact, while nonlinear programming methods can be trapped at a Kunh-Tucker point which may even not be a local optimum, the difficulties considerably increase when the model is further extended to deal with complicated but more realistic objective functions (e.g. when the cost function is measured by a monotone rather than linear function of the distances, or when the facilities are attractive only to a number of users and repelling to others).

Using the d.c. approach it is possible to significantly reduce the number of variables, and express the objective function in a much simpler form. Noticing that in an optimal solution each user must be served by the closest facility and setting

$$h_j(x^1,\dots,x^p) = \min_{i=1,\dots,p} \|x^i - a^j\| \tag{4.2}$$

we can rewrite the problem as:

$$\min \sum_{j=1}^{N} r_j h_j(x^1,\dots,x^p) \quad \text{s.t.} \quad x^i \in S \subset R^2,\ i = 1,\dots,p. \tag{4.3}$$

Since by Proposition 2.1

$$h_j(x) = \sum_{i=1}^{p} \|x^i - a^j\| - \max_k \sum_{i \neq k} \|x^i - a^j\| \tag{4.4}$$

where the functions

$$\sum_{i=1}^{p} \|x^i - a^j\|, \quad \max_k \sum_{i \neq k} \|x^i - a^j\|$$

are convcx it follows that (4.3) is a *d.c. optimization problem*. One obvious advantage of this formulation is that it uses only $2p$ variables : $x_1^i, x_2^i, i = 1,\dots,p$, instead of $pN + 2p$ variables as in the standard formulation. Furthermore, the objective function being a d.c. function, the problem can be practically solved by currently available global optimization methods, especially by the BB algorithm described in Section 3. In Tuy et al (1995), Al-Khayyal et al (1997), Al-Khayyal et al (1997a) results of solving large-scale problems with $N = 100,000; p = 1,\ N = 10,000; p = 2,\ N = 1,000; p = 3$ are reported. Note that the traditional approach would have difficulty in solving even problem instances with $N = 500, p = 2$ (then (4.1) would be a nonlinear programming problem with more than 1000 variables).

5 VARIOUS OBJECTIVES IN FACILITY LOCATION

Facility with Attraction and Repulsion

In more realistic models, the attraction of facility i to user j at distance t away is measured by a convex decreasing function $q_j(t)$. Furthermore, for some users the attraction effect is positive, for others it is negative (which amounts to a repulsion). Let J_+ be the set of attraction points (attractors), J_- the set of repulsion points (repellers). The problem is to find the locations so as to maximize the total effect, i.e.

$$\max \sum_{j \in J_+} q_j(h_j(X)) - \sum_{j \in J_-} q_j(h_j(X)) \quad \text{s.t.} \quad X = (x^1,\dots,x^p) \in S^p \subset (R^2)^p.$$

where $h_j(X)$ is given by (4.2). Using Proposition 2.2 it is easy to find an explicit d.c. representation of the objective function. Hence, the problem can be solved by the same method as problem (4.3). For detail, we refer the interested reader to Chen et al (1992), Chen et al (1994), Tuy et al (1992), Tuy et al (1995) and also Maranas et al (1994) and Al-Khayyal et al (1997), Al-Khayyal et al (1997a), where results in solving problems with up to 100,000 attractors and repellers have been reported. A local approach to this problem under more restrictive assumptions ($J = \emptyset$) can be found in Idrissi et al (1988).

MAXIMIN LOCATION

The maximin location problem is to determine the locations of p facilities so as to maximize the minimum distance from a user to the closest facility. Examples are obnoxious facilities, such as nuclear plants, garbage dumps, sewage plants, etc. With $h_j(x^1, \ldots, x^p)$ defined by (4.2) the mathematical formulation of this problem is:

$$\max\{\min_{j=1,\ldots,N} h_j(x^1,\ldots,x^p)|\ x^i \in S,\ i=1,\ldots,p\}. \tag{5.1}$$

An analogous problem, when locating emergency facilities, such as fire stations, hospitals, patrol car centers for a security guard company, etc., consists in minimizing the maximum distance from a user to its nearest facility. With the same $h_j(x^1, \ldots, x^p)$ as above the latter problem can be formulated as

$$\min\{\max_{j=1,\ldots,N} h_j(x^1,\ldots,x^p)|\ x^i \in S, i=1,\ldots,p\}. \tag{5.2}$$

and is often referred to as the p-center problem.

Since $h_j(x^1, \ldots, x^p)$ is a d.c. function (see (4.4)) it follows from Proposition 2.1 that $\min_{j=1,\ldots,N} h_j(x^1, \ldots, x^p)$ and $\max_{j=1,\ldots,N} h_j(x^1, \ldots, x^p)$ are also d.c. functions with known explicit d.c. representations. Therefore, (5.1) as well as (5.2) are d.c. optimization problems which can be solved by the generic BB method presented in Section 3 (see Al-Khayyal et al (1997) for the case $p = 1$).

DESIGN CENTRING

When the users are not finite in number but form a closed set $S \subset R^2$ the maximin location problem becomes

$$\max\{\min_{y\in S} h_y(x^1,\ldots,x^p)|\ x^i \in S, i=1,\ldots,p\},\quad h_y(x^1,\ldots,x^p) = \min_{i=1,\ldots,p} \|x^i - y\|. \tag{5.3}$$

and is called the continuous maximin problem. This is for example the problem of locating p obnoxious facilities so as to maximize the minimum distance of a facility to the residential area S.

Another important interpretation of the maximin location problem is the so called *design centring* problem (see e.g. Vidigal and Director Vidigal et al (1982), Grouch et al. Groch et al (1985), also Thach (1988)). Suppose that in a manufacturing process the quality of a manufactured item is characterized by an n-dimensional parameter and an item is accepted when the parameter value lies in a given region of acceptability $M \subset \mathbb{R}^n$. If the nominal (designed) value of the parameter is x, then, because of unavoidable random fluctuations its actual value y usually deviates from x. For every fixed x the probability that the deviation $\|x-y\|$ does not exceed a given level r monotonically increases with r, so for a given nominal value of x the expected production yield can be measured by the value

$$r_M(x) = \begin{cases} \max\{r \mid B(x,r) \subset M\} & \text{if } x \in M \\ 0 & \text{if } x \notin M \end{cases}$$

where $B(x,r) := \{y : \|x-y\| \leq r\}$ is the ball of radius r around x. Under these conditions, to maximize the expected production yield one has to solve the problem

$$\max\{r_M(x) \mid x \in \mathbb{R}^n\}. \tag{5.4}$$

Noting that

$$r_M(x) = \inf\{\|x-y\| \mid y \in \mathbb{R}^n \setminus M\}, \tag{5.5}$$

we see that the problem is

$$\max\{\inf_{y \notin M} \|x-y\| \mid x \in \mathbb{R}^n\}$$

i.e. a maximin problem of the form (5.3) where $p = 1$. In most cases of interest, at least one of the following conditions holds:

(i) the norm $\|.\|$ is ellipsoidal, i.e. there exists a positive definite matrix Q such that $\|x-y\|^2 = (x-y)^T Q (x-y)$.

(ii) $M = \{x \in \mathbb{R}^n \mid \alpha_i \leq f_i(x) \leq \beta_i, \quad i = 1, \ldots, m\}$ where $f_i(x)$ are convex functions.

If (i) holds then

$$\begin{aligned} r_M^2(x) &= \inf\{(x-y)^T Q (x-y) \mid y \in \mathbb{R}^n \setminus M\} \\ &= \inf\{x^T Q x + y^T Q y - 2x^T Q y \mid y \in \mathbb{R}^n \setminus M\}, \end{aligned}$$

so $r_M^2(x) = x^T Q x - h(x)$ with $h(x) = \sup_{y \notin M}\{2x^T Q y - y^T Q y\}$. Since for each $y \notin M$ the function $x \mapsto 2x^T Q y - y^T Q y$ is affine, $h(x)$ is a convex function and therefore, $r_M^2(x)$ is a d.c. function (see, e.g., Thach (1988)). Thus under Assumption (i) the design centring problem reduces to the d.c. optimization problem

$$\max\{x^T Q x - h(x) \mid x \in \mathbb{R}^n\}, \quad h(x) = \sup_{x \notin M}\{2x^T Q y - y^T Q y\} \tag{5.6}$$

An outer approximation method for solving this problem has been proposed in Thach (1988) (see also Konno et al (1997)).

If (ii) holds then

$$M = C \cap (\bigcap_{i=1}^{m} D_i),$$

where C and $\mathbb{R}^n \setminus D_i$ are convex sets defined by

$$C = \{x \mid f_i(x) \leq \beta_i, \ i = 1, \ldots, m\}, \quad D_i = \{x \mid \alpha_i \leq f_i(x)\}, \ i = 1, \ldots, m.$$

It is easily verified that

$$r_M(x) = \min\{r_C(x), r_{D_1}(x), \ldots, r_{D_m}(x)\}.$$

Proposition 5.1 *We have $r_C(x) = \rho_C(x)\ \forall x \in C$, where $\rho_C(x)$ is a concave function on $\mathbb{R}^n$. The functions $r_{D_i}(x),\ i = 1, \ldots, m$, are convex on $\mathbb{R}^n$.*

Proof: By representing C as the intersection of all its supporting halfspaces we have $C = \{x|\ \langle a^k, x\rangle \geq \theta_k,\ k \in K\}$, where $\|a^k\| = 1$. Setting

$$\rho_C(x) = \inf_{k\in K}(\langle a^k, x\rangle - \theta_k).$$

we observe that

$$r_C(x) = \rho_C(x) \quad \forall x \in C$$

because the distance from a point $x \in C$ to the hyperplane $H_k = \{x|\ \langle a^k, x\rangle = \theta_k\}$ is precisely $\langle a^k, x\rangle - \theta_k$. The first part of the Proposition then follows since the concavity of $\rho_C(x)$ is obvious (pointwise minimum of a family of affine functions). Note that

$$r_C(x) = \max\{\rho_C(x), 0\} = \rho_C(x) - \min\{\rho_C(x), 0\} \quad \forall x \in \mathbb{R}^n,$$

so $r_C(x)$ is concave on C and d.c. on $\mathbb{R}^n$. Analogously we can show that for $C_i = \mathbb{R}^n \setminus \text{int}D_i$:

$$r_{D_i}(x) = \max\{-\rho_{C_i}(x), 0\}.$$

Therefore $r_{D_i}(x),\ i = 1, \ldots, m$ are convex on $\mathbb{R}^n$. □

Thus, under Assumption (ii) the design centring problem reduces to the d.c. optimization problem

$$\max_{x\in M} \min\{r_C(x), r_{C_1}(x), \ldots, r_{C_m}(x)\}. \tag{5.7}$$

For solving this problem a branch and bound method has been proposed in Al-Khayyal et al (1997) which is a modification of the BB procedure in Section 3 applied to problem (5.7).

Problems similar to (5.4) appear in engineering design, coordinate measurement technique etc. and can hence be handled by the same method.

6 CONSTRAINED LOCATION

We now show that constrained location problems can also be reformulated as d.c. optimization problems under d.c. constraints. To simplify the notation, we will consider only single facility problems but the reader should be aware that the approach applies as well to multisource and multifacility problems (Al-Khayyal et al (1997)).

LOCATION ON UNION OF CONVEX SETS

In the real world, human and natural factors may set restrictions to the facility sites, so that the facility can be located only in one of several given convex regions, i.e. the feasible domain is a union of several convex sets $C_i =$

$\{x : c_i(x) \leq 0\}, i = 1 \ldots, k$ where $c_i(x)$ are convex functions (Hansen et al (1982)). Using the results of the previous section, the constraint

$$x \in \cup_{i=1}^{k} C_i$$

can be expressed as

$$\min_{i=1,\ldots,k} c_i(x) \leq 0$$

or else $p(x) - q(x) \leq 0$, with $p(x) = \sum_{i=1}^{k} x_i(x)$, $q(x) = \max_{j=1,\ldots,k} \sum_{i \neq j} c_i(x)$. Therefore, such problems (with the objective function as discussed above) can be formulated as:

$$\max\{G(x) - H(x) : \ x \in M, \ p(x) - q(x) \leq 0\} \tag{6.1}$$

where $G(x), H(x), p(x), q(x)$ are convex functions on R^2, M is a convex polygon in R^2.

The problem considered in Hansen et al (1982) and Chen et al (1992) is a particular case when each C_i is a polygon, i.e. each $c_i(x)$ is a polyhedral function:

$$c_i(x) = \sup_{\nu \in N_i} l_{i\nu}(x),$$

with $|N_i| < +\infty$ and $l_{i\nu}(x)$ being affine functions.

Location on Area with Forbidden Regions

In other circumstances, the facility can be located only outside certain forbidden regions. When these are open convex sets $C_i^o = \{x : \ c_i(x) < 0\}$, where $c_i(x)$ are convex functions (see e.g. Aneja et al (1994)), the constraint $x \notin \cup_{i=1}^{k} C_i^o$ is equivalent to $\min_{i=1,\ldots,k} c_i(x) \geq 0$, which allows the problem to be written as

$$\max\{G(x) - H(x) : x \in M, \ p(x) - q(x) \geq 0\} \tag{6.2}$$

with $G(x), H(x), M, p(x), q(x)$ having the same meaning as previously. This is a problem of the same type as (6.1), but with the roles of $p(x)$ and $q(x)$ interchanged.

General Constrained Location Problem

The most general situation occurs when the constraint set is a closed, not necessarily convex, set. For a single source, we can formulate the problem:

$$\max\{G(x) - H(x) | \ x \in M, \ x \in D\}, \tag{6.3}$$

where $G(x), H(x), M$ are as previously and D is a closed subset of R^2. By introducing an additional variable t we can rewrite the problem as

$$\min\{H(x) - t | \ x \in M, \ G(x) \leq t, \ x \in D\}. \tag{6.4}$$

When an explicit d.c. reformulation of the constraints is not readily available, we can solve this problem by an adaptation of an earlier outer approximation method by Tuy-Thuong, called the visible point method (Tuy et al (1988), see also Konno et al (1997), Tuy (1998)).

7 CHALLENGING PROBLEMS

As shown in the preceding sections, the d.c. optimization approach has proved to be very useful for various location problems. There are, on the other hand, many distance geometry problems with a similar structure, whose practical solution still remains challenging due to their large dimensionality. Some progresses, though, have been achieved in the understanding of their special d.c. structure which results from the composition of d.c. functions with distance functions.

MULTIFACILITY LOCATION AND MOLECULAR CONFORMATION

In the case of $p \geq 2$ facilities providing different services, aside from the costs due to interactions between facilities and users, one may also consider the costs due to pairwise interactions between facilities. The latter costs can be expressed by functions of the form $\rho_{il}[h_{il}(x^i, x^l)]$, where again $h_{il}(x^i, x^l)$ are convex nonnegative valued functions and $\rho_{il}(t)$ are concave increasing functions on $[0, +\infty)$ with finite right derivatives at 0. The total cost which one would like to minimize is then

$$\sum_{i=1}^{p} F_i(x^i) + \sum_{i<l} \rho_{il}[h_{il}(x^i, x^l)], \tag{7.1}$$

where $F_i(x^i) = \sum_{j\in J_1} q_{ji}[h_j(x^i)] - \sum_{j\in J_2} q_{ji}[h_j(x^i)]$ and q_{ji}, h_j are as in minisum single facility problems. As we saw in Section 5, the functions $F_i(x^i)$ as well as $\rho_{il}[h_{il}(x^i, x^l)]$ are d.c., so (7.1) is again a d.c. function on $(R^2)^p$. In the special case when there are no repulsion points (every $F_i(x^i)$ is convex) and the pairwise interactions between facilities $\rho_{il}(t)$ are convex, this is simply a convex function. Also, in the absence of interactions between facilities ($\rho_{ij}(.) = 0\ \forall ij$), the minimization of the function (7.1) splits into p independent single facility minisum problems. However, in the general case, if p (the number of facilities) is not small, solving this d.c. optimization problem may not be an easy task.

A variant of multifacility problem which has risen to attract much research in recent years is the so called *molecular conformation* problem encountered in Computational Biology, Computational Chemistry and Protein Folding. This is the problem of determining ground states or stable states of certain classes of molecular clusters and proteins and can be stated as follows (Maranas et al (1994)). Given a cluster of p atoms (in the 3-dimensional space), we wish to locate their centres $x^1, \ldots, x^p$, so as to minimize the potential energy function

$$V_p(x^1, \ldots, x^p) = \sum_{1\leq i<j\leq p} v(\|x^i - x^j\|)$$

where $\|.\|$ is the euclidean norm and $v(r)$ is the interatomic pair potential. This can be viewed as a multifacility problem in which there are no user but many facilities (the number p may be rather large, see e.g. Maranas et al (1994)). In models used for computation, the pair potentials of interest include the following:

$$v(r) \quad = \quad r^{-12} - 2r^{-6} \text{ (Lennard – Jones)}$$

$$v(r) = [1 - e^{\alpha(1-r)}]^2 - 1 \text{ (Morse)}$$
$$v(r) = z^{\alpha} z^{\frac{\beta}{r}} + Ae^{\frac{-r}{\rho}} \text{ (Born – Meyer)}$$

For $v(r) = r^{-12} - 2r^{-6}$ (Lennard-Jones) G.L. Xue (1997) has shown that the global minimum does not exceed $-(p-1)$ and that an optimal solution must satisfy

$$\|x^i - x^j\| \geq 0.5 \quad i < j.$$

Therefore , one may add the above constraints to the problem and assume $r \geq 0.5$. Since for $r \geq 0.5$ the function $v(r) = r^{-12} - 2r^{-6} = t^2 - 2t, t = 1/r^6$, can be decomposed into a difference of two convex functions, the problem can be written as a d.c. optimization problem. The difficulty, however, is that the dimension of the problem in cases of practical interest may be too large for currently available algorithms.

A related problem which also has applications in molecular conformation, and other questions such as surveying and satellite ranging, data visualization and pattern recognition, etc, is the *multidimensional scaling problem* or *distance geometry problem.* It consists in finding p objects $x^1, \ldots, x^p$ in $\mathbb{R}^n$ such that the quantity

$$V_p(x^1, \ldots, x^p) = \sum_{i<j} w_{ij}(\delta_{ij}^2 - \|x^i - x^j\|^2)^2 \tag{7.2}$$

is smallest, where $\Delta = (\delta_{ij}), W = (w_{ij})$ are symmetric matrices of order p such that

$$\delta_{ij} = \delta_{ji} \geq 0, \quad w_{ij} = w_{ji} \geq 0 \quad (i < j); \qquad \delta_{ii} = w_{ii} = 0 \quad (i = 1, \ldots, p).$$

By writing this problem as

$$\begin{array}{ll} \min & \sum_{i<j} w_{ij}\|x^i - x^j\|^2 - 2\sum_{i<j} w_{ij}\delta_{ij}\|x^i - x^j\| \\ \text{s.t.} & x^i \in \mathbb{R}^n \ (i = 1, \ldots, p) \end{array} \tag{7.3}$$

or, alternatively, as

$$\min \sum_{i,j} w_{ij} t_{ij}^2 \left| \begin{array}{l} -t_{ij} \leq \delta_{ij}^2 - \|x^i - x^j\|^2 \leq t_{ij} \ (\forall i < j) \\ x^i \in \mathbb{R}^n, \quad i = 1, \ldots, p \end{array} \right. \tag{7.4}$$

we again obtain a d.c. optimization problem which can, in principle, be solved by the BB algorithm in Section 3, or by recently developed methods of quadratic programming (see e.g. Tuy (1998)). An interesting local optimization algorithm called DCA has also been developed in An et al (1998) for solving the multidimensional scaling problem on the basis of the d.c. formulation (7.3). Perhaps by combining global optimization methods with these local ones one should be able to solve much larger problems.

CONTINUOUS p-CENTER PROBLEM

The continuous p-center problem differs from the p-center problem only in that the set of users is a rectangle $S \subset R^2$. rather than a finite set of points. It consists of finding the locations $x^1, \dots, x^p$ of p facilities in S so as to minimize the maximal distance from a point $x \in S$ to the nearest facility, i.e.:

$$\min\{\max_{y \in S} h_y(x^1, \dots, x^p) |\, x^i \in S,\ i = 1, \dots, p\}, h_y(x^1, \dots, x^p) = \min_{i=1,\dots,p} \|x^i - y\|. \tag{7.5}$$

For each facility j define the Voronoi (Dirichlet) cell

$$D_j = \{x |\ \|x - x^j\| \leq \|x - x^i\|\ \forall i = 1, \dots, p\}.$$

Computational results of Suzuki-Okabe (1995) for $p = 30$ and $p = 49$ suggest the conjecture that for large values of p most Voronoi cells tend to be *hexagons*, only a few are *pentagons*.

DISTRIBUTION OF POINTS ON A SPHERE

We close this paper by mentioning a problem which has recently attracted the attention of mathematicians, biologists, chemists, and physicists, working in such fields as viral morphology, crystallography, molecular structure, and electrostatics. Like many famous difficult problems of mathematics, the formulation of this problem is rather simple: find the configuration of p points on a sphere in R^3 so that

$$\min V_p(x^1, \dots, x^p) := \sum_{i<j} \frac{1}{\|x^i - x^j\|} \quad \text{s.t.} \quad \|x^i\| = 1 \quad i = 1, \dots, p. \tag{7.6}$$

Physically, $V_p(x^1, \dots, x^p)$ represents the energy of p charged particles that repel each other according to Coulomb's law. An optimal solution $X = (x^1, \dots, x^p)$ of the problem (7.6) is called a p-tuple of elliptic Fekete points. Computational experiments have shown that for $32 \leq p \leq 200$ optimal configurations arrange points according to *hexagonal-pentagonal pattern* (i.e. all but exactly 12 Voronoi cells are hexagonal, the exceptional cells being pentagons).

A related problem is to find a configuration of p points on the sphere that maximizes the product

$$W_p(x^1, \dots, x^p) := \prod_{i<j} \|x^i - x^j\|. \tag{7.7}$$

The difficulty of problem (7.6) is that there are many local minima that are not global; in addition the local minimizers have objective function values very close to the global minimum, which make it very difficult to determine the exact value of the global minimum.

Observe that $\frac{1}{\|x^i - x^j\|} = q(\|x^i - x^j\|)$ with $q(t) = 1/t$ convex decreasing function for $t \geq \delta > 0$. Hence problem (7.6) can be rewritten as

$$\min \sum_{i<j} (g(x^i, x^j) - K\|x^i - x^j\|) \quad \text{s.t.} \quad \|x^i\| = 1 \quad i = 1, \dots, p. \tag{7.8}$$

where $g(t)$ is convex, $K > 0$. This is a d.c. minimization problem over a sphere, which should be solvable for small values of p. Analogously, the problem

$$\max \prod_{i<j} \|x^i - x^j\| \quad \text{s.t.} \quad \|x^i\| = 1 \quad i = 1, \dots, p \tag{7.9}$$

or equivalently

$$\max \sum_{i<j} \log(\|x^i - x^j\|) \quad \text{s.t.} \quad \|x^i\| = 1 \ i = 1, \ldots, p$$

can be written as a d.c. optimization problem over a sphere:

$$\max \sum_{i<j} (L\|x^i - x^j\| - h_{ij}(x^i, x^j)) \quad \text{s.t.} \quad \|x^i\| = 1 \ i = 1, \ldots, N \tag{7.10}$$

since $\log(\|x^i - x^j\|) = q(\|x^i - x^j\|)$ with $q(t) = \log t$ (concave increasing for $t \geq \eta > 0$).

8 CONCLUSIONS

Location problems and distance geometry problems have similar structures which allow them to be approached essentially by the same methods. However, while most location problems involve a relatively small number of nonconvex variables, this is not usually the case with many distance geometry problems of practical interest. The latter problems are thus generally more difficult and require for their practical solution perhaps a hybrid approach, combining deterministic global optimization methods with stochastic and heuristic ones.

References

Al-Khayyal, Faiz A., Tuy, H. and Zhou, F. (1997), Large-scale single facility continuous location by d.c. optimization, Preprint, School of Industrial and Systems Engineering, Georgia Institute of Technology, Atlanta.

Al-Khayyal, Faiz A., Tuy, H. and Zhou, F. (1997), D.C. optimization methods for multisource location problems, Preprint, School of Industrial and Systems Engineering, Georgia Institute of Technology, Atlanta.

Hoai-An, L.T., and Tao, P.D. (1998), D.C. programming approach and solution algorithm to the multidimensional scaling problem, Preprint, Laboratory of Mathematics, National Institute for Applied Sciences, Rouen, France.

Androulakis, I.P., Maranas, C.D. and Floudas, C.A. (1997), Prediction of oligopeptide conformations via deterministic global optimization, *Journal of Global Optimization*, Vol. 11, pp. 1-34.

Aneja, Y.P. and Parlar, M. (1994), Algorithms for Weber facility location in the presence of forbidden regions and/or barriers to travel, *Transportation Science*, Vol. 28, pp. 70-216.

Brimberg, J. and Love, R.F. (1994), A location problem with economies of scale, *Studies in LocationAnalysis*, Issue 7, pp. 9-19.

Chen, R. (1983), Solution of minisum and minimax location-allocation problems with euclidean distances, *Naval Research Logistics Quaterly*, Vol. 30, pp. 449-459.

Chen, R. (1988), Conditional minisum and minimax location-allocation problems in Euclidean space, *Transportation Science*, Vol. 22, pp. 157-160.

Chen, P., Hansen, P., Jaumard, B. and Tuy, H. (1992), Weber's problem with attraction and repulsion, *Journal of Regional Science*, Vol. 32, pp. 467-409.

Chen, P., Hansen, P., Jaumard, B. and Tuy, H. (1994), Solution of the multi-source Weber and conditional Weber problems by D.C. Programming, Cahier du GERAD, G-92-35, Ecole Polytechnique, Montreal.

Drezner, Z. and Wesolowsky, G. (1990), The Weber problem on the plane with some negative weights, *INFOR*, Vol. 29, pp. 87-99.

Groch, A., Vidigal, L. and Director, S. (1985), A new global optimization method for electronic circuit design, *IEEE Trans. Circuits and Systems*, Vol. 32, pp. 160-170.

Hansen, P., Jaumard, B. and Tuy, H. (1995), Global optimization in location, in *Facility Location: a Survey of Applications and Methods*, edited by Dresner, Z., Springer, pp. 43-67.

Hansen, P., Peeters, D. and Thisse, J.F. (1982), An algorithm for a constrained Weber problem, *Management Science*, Vol. 28, pp. 1285-1295.

Hansen, P., Peeters, D., Richard, D. and Thisse, J.F. (1985), The minisum and minimax location problems revisited, *Operations Research*, Vol. 33, pp. 1251-1265.

Horst, R. and Tuy, H. (1996), *Global Optimization*, Springer, 3rd edition.

Konno, H., Thach, P.T. and Tuy, H. (1997), *Optimization on low rank nonconvex structures*, Kluwer.

Idrissi, H., Loridan, P. and Michelot, C. (1988), Approximation of Solutions for Location Problems, *Journal on Optimization Theory and Applications*, Vol. 56, pp. 127-143.

Love, R.F., Morris, J.G. and Wesolowsky, G.O. (1988), *Facilities Location: Models and Methods*, North-Holland.

Maranas, C.D. and Floudas, C.A. (1993), A global optimization method for Weber's problem with attraction and repulsion, in *Large Scale Optimization: State of the Art*, edited by Hager, W.W., Heran, D.W. and Pardalos, P.M., Kluwer, pp. 1-12.

Maranas, C.D. and Floudas, C.A. (1994), Global minimum potential energy conformations of small molecules, *Journal of Global Optimization*, Vol. 4, pp. 135-171.

Meggido, N. and Supowit, K.J. (1984), On the complexity of some common geometric location problems, *SIAM Journal on Computing*, Vol. 13, pp. 182-196.

Nguyen, V.H. and Strodiot, J.J. (1992), Computing a global optimal solution to a design centering problem, *Mathematical Programming*, Vol. 53, pp. 111-123.

Pardalos, P.M. and Xue, G.L. (eds.), (1997), *Journal of Global Optimization*, Special issue on 'Computer simulations in molecular and protein conformations', Vol. 11.

Plastria, F. (1992), The generalized big square small square method for planar single facility location, *European Journal of Operations Research*, Vol. 62, pp. 163-174.

Plastria, F. (1993), Continuous location anno 1992, A progress report, *Studies in Location Analysis*, Vol. 5, pp. 85-127.

Rockafellar, R.T. (1970), *Convex Analysis*, Princeton University Press.

Rosing, K.E. (1992), An optimal method for solving the generalized multi-Weber problem, *European Journal of Operations Research*, Vol. 58, pp. 414-426.

Saff, E.B. and Kuijlaars, A.B.J. (1997), Distributing many points on a sphere, *Mathematical Intelligencer*, Vol. 10, pp. 5-11.

Thach, P.T. (1993), D.C. sets, D.C. functions and nonlinear equations, *Mathematical Programming*, Vol. 58, pp. 415-428.

Thach, P.T. (1988), The design centering problem as a d.c. programming problem, *Mathematical Programming*, Vol. 41, pp. 229-248.

Tuy, H. (1986), A general deterministic approach to global optimization via d.c. programming, in *Fermat Days 1985: Mathematics for Optimization*, edited by Hiriart-Urruty, J.-B., North-Holland, pp. 137-162.

Tuy, H. (1987), Global minimization of a difference of convex functions, *Mathematical Programming Study*, Vol. 30, pp. 150-182.

Tuy, H. (1990), On a polyhedral annexation method for concave minimization, in *Functional Analysis, Optimization and Mathematical Economics*, edited by Leifman, L.J. and Rosen, J.B., Oxford University Press, pp. 248-260.

Tuy, H. (1991), Polyhedral annexation, dualization and dimension reduction technique in global optimization, *Journal of Global Optimization*, Vol. 1, pp. 229-244.

Tuy, H. (1992), The complementary convex structure in global optimization, *Journal of Global Optimization*, Vol. 2, pp. 21-40.

Tuy, H. (1992), On nonconvex optimization problems with separated nonconvex variables, *Journal of Global Optimization*, Vol. 2, pp. 133-144.

Tuy, H. (1995), D.C. optimization: theory, methods and algorithms, in *Handbook on Global Optimization*, edited by Horst, R. and Pardalos, P., Kluwer, pp. 149-216.

Tuy, H. (1996), A general D.C. approach to location problems, in *State of the Art in Global Optimization: Computational Methods and Applications*, edited by Floudas, C.A. and Pardalos, P.M., Kluwer, pp. 413-432.

Tuy, H. (1998), *Convex Analysis and Global Optimization*, Kluwer.

Tuy, H. and Al-Khayyal, Faiz A. (1992), Global optimization of a nonconvex single facility problem by sequential unconstrained convex minimization, *Journal of Global Optimization*, Vol. 2, pp. 61-71.

Tuy, H., Alkhayyal, Faiz A. and Zhou, F. (1995), D.C. optimization method for single facility location problem, *Journal of Global Optimization*, Vol. 7, pp. 209-227.

Tuy, H., Ghannadan, S., Migdalas, A. and Värbrand, P. (1993), Strongly polynomial algorithm for a production-transportation problem with concave production cost, *Optimization*, Vol. 27, pp. 205-227.

Tuy, H., Ghannadan, S., Migdalas, A. and Värbrand, P. (1995), Strongly polynomial algorithm for a production-transportation problem with a fixed number of nonlinear variables, *Mathematical Programming*, Vol. 72, pp. 229-258.

Tuy, H. and Thuong, N.V. (1988), On the global minimization of a convex function under general nonconvex constraints, *Applied Mathematics and Optimization*, Vol. 18, pp. 119-142.

Vidigal, L. and Director, S. (1982), A design centering problem algorithm for nonconvex regions of acceptability, *IEEE Transactions on Computer-aided Design of Integrated Circuits and Systems*, Vol. 14, pp. 13-24.

Xue, G.L. (1997), Minimum inter-particle distance at global minimizers of Lennard-Jones clusters, *Journal of Global Optimization*, Vol. 11, pp. 83-90.

2 BRANCH AND CUT METHODS FOR MIXED INTEGER LINEAR PROGRAMMING PROBLEMS

Louis Caccetta

School of Mathematics and Statistics,
Curtin University of Technology,
GPO Box U 1987,
Perth, W.A, 6845,
Australia.

Abstract: The discrete optimization problems arising in industry are typically very large and computationally difficult to solve. Over the past decade, the method of Branch and Cut has emerged as a powerful technique for solving large mixed integer linear programming problems. Advances in computer technology together with advances in computational algorithms now makes it possible to obtain provably good solutions for many industrial optimization problems. This paper focuses on the application of Branch and Cut. An example from the mining industry demonstrates the value of the method.

Key words: branch and cut, cutting planes, combinatorial optimization, mixed integer programming, scheduling.

X. Yang et al. (eds.), Progress in Optimization, 21–44.

1 INTRODUCTION

The theory and application of linear and integer programming has, over the past 50 years or so, undergone considerable development and advancement. This is clearly evident from the vast literature that has accumulated over this period and the current trend of increasing activity in this area. The growth, particularly in integer programming, has been greatly accelerated by advances in computer technology. These advances have facilitated the development of sophisticated mathematical techniques for solving many complex problems arising in modern business and industry. Further, the continued need for business and industry to efficiently utilize limited resources to survive in the present and future highly competitive environment will ensure that mathematical programming, in particular integer programming, remains active in the foreseeable future.

In many applications, we are typically dealing with problems in operational efficiency concerned with the utilization of limited resources. Very often, the problem that arises is one of optimizing a function (representing profit, output or cost) subject to a specified set of constraints (representing the limited resources or the operational requirements of the system). When all the mathematical expressions are linear in the variables, we have a Linear Program (LP). When, in addition, some of the variables are restricted to integer values, we have a Mixed Integer Linear Program ($MILP$). When all variables are integer restricted, we have a pure Integer (Linear) Program (IP).

Mathematically, the general $MILP$ can be formulated as:

$$\text{Minimize} \quad f = c^T x \tag{1.1}$$

subject to

$$Ax \leq b \tag{1.2}$$

$$x \geq 0 \tag{1.3}$$

$$\text{and} \quad x_j \text{ integer} \quad \text{for } j \in J \subseteq \{1, 2, \ldots, n\} \tag{1.4}$$

The input data consists of the matrices c $(1 \times n)$, A $(m \times n)$ and b $(m \times 1)$, and the n - vector x of variables whose values are to be determined. Note we have an LP when $J = \emptyset$ and a pure IP when $J = \{1, 2, \ldots, n\}$.

These linear optimization problems arise in many applications including: airline crew scheduling; data association; network design and analysis; network routing; production planning; resource allocation; rural land use planning; financial management and planning; facility layout design; design of automated production systems; human resource planning; location of service facilities; design of electoral systems; and many more.

Efficient commercial LP software packages, for example $CPLEX$, $LINDO$, $GAUSS$, $LAMPS$ and $MINOS$ (Moré and Wright (1993)), have been readily available for many years. Nowadays, many Operations Research texts (for example Bertsekas (1991); Martello and Toth (1990)) provide computer codes. The development of good $MILP$ and IP software has only recently progressed.

For example, in the early 1980's, mainframe computers could solve problems with up to 100 integer variables. Now days, powerful PC's and workstations can solve problems to optimality with thousands of integer variables. Furthermore, provably good solutions can be obtained for much larger problems. These advances have been made possible by advances in technology (computer hardware and software) and algorithmic development.

In this paper we focus on some recent advances in computational algorithms. Exact methods for the *MILP* (1.1) - (1.4) include: Lagrangian Relaxation; Branch and Bound; and Branch and Cut (and the recently introduced Branch and Price). The branching methods involve solving relaxed *LP's*. A discussion of these methods is given in Section 2.

The use of Branch and Cut methods have, over the past decade, emerged as extremely powerful techniques for solving large *MILPs*. Our objective in this paper is to review some of the recent work on the application of Branch and Cut methods to solve combinatorial optimization problems. We focus on three specific problems: the classical traveling salesman problem; the vehicle routing problem; and a mine scheduling problem. We devote a section to each of these. Our choice of applications is, of course, influenced by our own interests and the level of treatment is intended for a wide audience including non-specialists.

As noted earlier there is a vast literature in the area of linear and integer programming problems which includes many excellent books (Bertsekas (1991); Grötschel et al (1988); Ibaraki (1987); Ibaraki (1987); Johnson (1980); Lawler (1976); Lovász and Plummer (1986); Martello and Toth (1990); Moré and Wright (1993); Nemhauser and Wolsey (1988); Papadimitriou and Steiglitz (1982); Schrijver (1986)), collected works (Aarts and Lenstra (1997); Ball et al (1995); Dell'Amico et al (1997); Du and Pardalos (1998); Golden and Assard (1988); Graham et al (1995); Martello et al (1987); Nemhauser et al (1988)) and an uncountable number of research papers.

2 SOLUTION METHODS

In this section we briefly detail some basic approaches for solving *MILPs*. We focus on the exact methods of Lagrangian Relaxation, Branch and Bound, and Branch and Cut. As *MILPs* are for the most part *NP* - hard, many heuristic procedures have been proposed. In some cases, these search procedures produce good approximate solutions, particularly when the structure of the problem can be exploited. Fast heuristics capable of producing good approximations are important in the success of exact methods. For an excellent account of search methods in optimization we refer to the book Aarts and Lenstra (1997).

2.1 Lagrangian Relaxation

Many *MILPs* can be viewed as easy problems complicated by a relatively small set of difficult constraints. The Lagrangian Relaxation method dualizes these difficult constraints and then attempts to solve the resulting relaxed problem. For example, consider the *MILP* (1.1) - (1.4) written as:

$$\begin{array}{lll} \text{Minimize} & f = c^T x & (2.1) \\ \text{subject to} & & \\ & A^1 x \leq b^1 \quad \text{(difficult constraints)} & (2.2) \\ & A^2 x \leq b^2 \quad \text{(easy constraints)} & (2.3) \\ & x \geq 0 & (2.4) \\ \text{and} & x_j \text{ integer} \quad \text{for } j \in J \subseteq \{1, 2, \ldots, n\} & (2.5) \end{array}$$

where $A = \begin{pmatrix} A^1 \\ A^2 \end{pmatrix}$ and $b = \begin{pmatrix} b^1 \\ b^2 \end{pmatrix}$. The Lagrangian relaxation is:

$$\begin{array}{rcll} g(\lambda) & = & \min\{c^T x + \lambda(A^1 x - b^1)\} & (2.6) \\ \text{subject} & to & (2.3) - (2.1) & (2.7) \end{array}$$

where λ is a non-negative vector of Lagrangian multipliers.

For a given λ the relaxed problem is easy to solve. Observe that if x^* is an optimal solution for (2.1) - (2.1), then

$$g(\lambda) \leq c^T x^* + \lambda(A^1 x^* - b^1) \leq f(x^*)$$

Thus any solution to the relaxed problem provides a lower bound in the objective function value of the original *MILP*. The best choice for λ is that which yields the greatest lower bound, or equivalently any λ which is optimal in the dual problem. The optimality conditions are:

Theorem 2.1 *For a given $\overline{\lambda}$ suppose $\overline{x}$ satisfies:*

(i) $\overline{x}$ *is an optimal solution to (2.6), (2.3) - (2.1)*
(ii) $A^1\overline{x} \leq b^1$
(iii) $\overline{\lambda}(A^1\overline{x} - b^1) = 0$

Then $\overline{x}$ is an optimal solution to the original MILP (2.1) - (2.1).

The solution to the dual problem is difficult because of its discrete nature. In practice duality gaps may arise. The subgradient optimization method (Held and Karp (1970); Held and Karp (1971); Held et al (1974)) provides a numerical technique for reducing the duality gaps. Theoretically, it is guaranteed to determine the optimal λ. However, in practice convergence may be slow and the optimal λ may not be achievable. This is the major disadvantage of the method. An advantage of the method is that with a heuristic that generates a feasible solution one has both lower and upper bounds and thus a measure for evaluating the quality of the best available solution. Usually, once a feasible solution that is within a specific tolerance of the optimum is obtained the computation ceases and the current solution is accepted.

An alternative strategy for obtaining a solution is to implement a Branch and Bound method when a duality gap is encountered (branch on a fractional variable).

Numerous articles have reported on the successful application of subgradient optimization (Fisher (1981); Geoffrion (1974); Held and Karp (1970); Held and Karp (1971); Held et al (1974); Shapiro (1979)).

2.2 *Branch and Bound*

The method of Branch and Bound has been effectively used to tackle a number of computationally difficult problems. Basically, the idea is to subdivide (branch) the feasible solution set into successively smaller subsets, placing bounds on the objective function value over each subset, and using these bounds to discard subsets from further consideration and to select the next subset to further subdivide. The branching process is carried out by fixing the value of the branching variable. In the case of a feasible problem, the process stops when we have a solution to the original problem which has an objective function value, in the case of a minimization problem, less than or equal to all lower bounds of the generated subsets.

Consider the $MILP$ (2.1)-(2.1). A relaxed problem can be obtained by dropping the integer restrictions (2.1) and the difficult constraints (2.2). So the relaxed problem is the LP (2.1), (2.3) and (2.4). Note that in the context of LP, probably the constraints (2.2) would be kept. The optimal solution to the relaxed LP provides an initial lower bound for the objective function value of the original $MILP$. This solution is of course optimal if (2.2) and (2.1) are satisfied. Observe that any feasible solution of the $MILP$ provides a valid upper bound.

In many applications upper bounds are obtained by the application of a fast heuristic. Lower bounds can be used in a tree search technique to specify additional restrictions. Note that nodes in the tree correspond to subsets generated, the root node to the initial relaxed feasible solution set. If at any node in the search tree one has a lower bound for a subset that is greater than or equal to the current upper bound, then we do not need to consider this subset any further (*i.e.* no further branching from this node is done).

The Branch and Bound method can implemented in many ways. The success of the implementation is highly dependent on the:

(i) branching nodes
(ii) the search strategy and
(iii) the quality of the lower and upper bounds generated.

The usual search strategy is depth-first search, where a subset chosen from the list is explored until either it violates the lower bound criteria or an improved feasible solution is obtained. Branching occurs from the parent node in the branch that has not yet been fully explored. An alternative is breadth-first search where the branching is done from the subset with the lowest lower bound. Lower bounds can be obtained at each node by either solving the corresponding relaxed problem optimally or by the application of an approximation method (for example, application of subgradient optimization to a Lagrangian relaxation problem). There are many successful applications of Branch and Bound (Fisher (1981); Held et al (1974); Shapiro (1979)).

2.3 *Branch and Cut*

In recent years the method of Branch and Cut has emerged as a powerful technique for solving large $MILP$ problems (Achuthan et al (1995); Applegate et al (1995); Applegate et al (1998); Caccetta and Hill (1999a); Caccetta and Hill (1999b); Caccetta and Hill (1999c); Crowder et al (1983); Hoffman et al (1993)). Basically, this method attempts to strengthen the lower bounds by the addition of constraints (cuts) at each node within a Branch and Bound procedure. As our main focus is on the Branch and Cut method we provide a little more detail below. Our treatment follows that used in Caccetta and Hill (1999a).

Consider the $MILP$ (2.1) - (2.1); denote this problem by P. Let $\mathcal{K}$ be a set of valid inequalities for this problem. Now consider the relaxed LP problem P':

$$\text{Minimize} \qquad f = c^T x \tag{2.8}$$

$$\text{subject to}$$

$$Bx \leq d \tag{2.9}$$

$$x \geq 0 \tag{2.10}$$

Here (2.9) is a subset of the original constraints (2.2) and (2.3). Also $\{Ax \leq b\} \setminus \{Bx \leq d\} \subseteq \mathcal{K}$. A lower bound for (2.1) - (2.1) can be generated from following procedure adapted from Padberg et al (1991).

Cutting Plane Procedure:

Step 1. Set $\mathcal{L} = \emptyset$

Step 2. Solve P' with the additional constraints in $\mathcal{L}$, and let $\overline{x}$ be the optimal solution

Step 3. Find one or more inequalities in $\mathcal{K}$ that are violated by $\overline{x}$

Step 4. If none is found, stop. Otherwise add the violating inequalities to $\mathcal{L}$ and go to Step 2

The success of the above procedure is highly dependent upon efficiently finding sets of 'strong' inequalities of $\mathcal{K}$. The problem of finding a violating inequality of $\mathcal{K}$ or proving that no such inequality exists for solving $\overline{x}$ is commonly referred to as the 'separation problem'. Ideally an efficient exact method to solve the separation problem (Step 3) is required. Unfortunately such methods are usually unavailable or are computationally expensive and heuristic procedures are employed.

To assist in the discussion of 'strong' inequalities we introduce some basic polyhedral theory terminology (Nemhauser and Wolsey (1988)). For $S \subseteq \mathbb{R}^n$ we let conv(S) denote the convex hull of S. An important result is that for finite S, conv(S) can be described by a finite set of linear inequalities. Further

$$\min\{c^T x : x \in S\} = \min\{c^T x : x \in conv(S)\}$$

Thus any $MILP$ can be represented as an LP, provided we know a set of linear inequalities that represent the solution space. Note that such a system of inequalities is usually incredibly large in number and generally unknown. To overcome these problems, the approach is to use a subset of the constraints defining conv(S) and/or constraints which are redundant in a minimal representation. For a $MILP$, a linear constraint that does not exclude any integer feasible points is called a cutting plane. If $\pi x \leq \pi_0$ is a valid set of inequalities for P = $\{x \in \mathbb{R}^n : Ax \leq b\}$, and F = $\{x \in P : \pi x = \pi_0\}$, then F is called a face of P. A face of P is facet of P if dim(F) = dim(P) - 1. This leads to the result that for each facet F of P, one of the inequalities representing F is necessary in the description of P. Thus the use of facets in the description of the solution space yields a minimal system of inequalities. Also, if P defines the convex hull of integer solutions of a discrete optimization problem, then the use of facet defining inequalities is most likely to give the tightest lower bounds in the cutting plane procedure.

The above cutting plane procedure terminates when no further violation can be found or an optimal solution has been found. If an optimal solution is not obtained, then it is necessary to subdivide the solution space. To do this the cutting plane procedure is embedded in a search tree technique. In this case, a check that the lower bound generated at Step 2 is less than the best upper bound must be included. If the lower bound is at least equal to the best known upper bound then the subproblem is fathomed. Step 4 is modified so that an integral solution which satisfies all inequalities of $\mathcal{K}$ becomes the best known solution and the upper bound is set accordingly. Note that if the above procedure is terminated when further violations can be found (which in some cases may be useful) then the objective function value is still a valid lower bound.

So that the LP basis does not become too large, constraints of the LP are typically removed when they are found to no longer be effective, *i.e.* the slack variable is non-basic. This is either done as soon as the slack variable is found to be non-basic or after the slack variable has been non-basic for a prescribed number of iterations. Note this operation does not alter the feasibility nor the optimality of the LP. However, the removed constraints may need to be regenerated later in the search tree.

In a standard Branch and Bound method using an LP relaxation, the constraints present in the LP and the basis need to be stored at each child node. If the child node later needs to be explored then the LP tableau is recreated. This process is expensive both in terms of time and memory requirements. The method of Branch and Cut overcomes this difficulty by using constraints that are valid throughout the search tree, so the LP need not be solved from scratch.

For large problems, the number of variables in the LP relaxations may become large and thus reduce the effectiveness of the Branch and Cut procedure. This problem can be resolved by reducing the number of integer variables through preprocessing and probing (Savelsbergh (1994)). In some applications sparse graph techniques are useful (Crowder et al (1983)).

An important advantanage of the Branch and Cut method is that it benefits from further advances in the description of a problem's solution polytope or in the methods for detecting violations.

We conclude this section by noting that recently some good work has been done on the closely related method of Branch and Price (Barnhart et al (1998)).

This method focuses on column generation (variable introduction) rather than row generation (constraint introduction). In brief, columns are left out of the LP relaxations. To check for optimality of an LP solution, a subproblem (the pricing problem), which is a separation problem for the dual LP, is solved in an attempt to identify which columns are to enter the basis. If such columns are found, the LP is reoptimized. Otherwise branching occurs. This method is clearly suitable for problems with a very large number of variables. Implementation of Branch and Price is far from straight forward.

3 THE TRAVELING SALESMAN PROBLEM

The Traveling Salesman Problem (TSP) has been the classical testing problem for most integer programming techniques. This problem arises in many applications including: production scheduling; network design; vehicle routing; crystallography, etc. In this problem, a salesman needs to visit, starting at the home town T_1, n-1 neighboring towns $T_2, T_3, \ldots, T_n$ exactly once and then return home. Given the distance (cost) matrix $C = [c_{ij}]$, with c_{ij} being the distance between T_i and T_j, the objective is to find a tour that minimizes the total distance traveled. Let $G = (V, E)$ be the graph representing the network with $V = \{1, 2, \ldots, n\}$ and $E = \{(i, j) : i, j \in V, i < j\}$. This problem can be formulated as follows:

$$\text{Minimize } f = \sum_{i=1}^{n} \sum_{i<j} c_{ij} x_{ij} \tag{3.1}$$

subject to

$$\sum_{j<i} x_{ji} + \sum_{j>i} x_{ij} = 2, \quad 1 \leq i \leq n \tag{3.2}$$

$$\sum_{i,j \in S} x_{ij} \leq |S| - 1, \quad S \subset V, \quad 3 \leq |S| \leq n-2 \tag{3.3}$$

and

$$x_{ij} = 0,\ 1 \quad \text{for all } i, j \tag{3.4}$$

In the above, constraints (3.2) ensure that every town is visited exactly once. The subtour elimination constraints (3.3) ensure that the salesman's route is specified by exactly one tour; note that there are an exponential number of these constraints.

The early work of Dantzig et al (1954, 1959) motivated research into the cutting plane procedures. In the early 1970's a number of authors (Balas and Toth (1985)) tackled the TSP by considering a Lagrangian Relaxation procedure which relaxes the degree constraints (3.2). The minimum weight spanning tree problem is the relaxed subproblem and subgradient optimizations are used to determine multipliers. This approach does not yield the most powerful method and has been superseded by cutting plane approaches.

Early results of Edmonds (1965) on the symmetric TSP polytope provided a complete linear description of the 2-matching (perfect 2-matching) polytope; a 2-matching is a set of edges such that every vertex of the graph is incident to at most two edges (the set is perfect if every vertex is incident to exactly two edges. This work was further developed in Caccetta and Giannini (1990); Grötschel et al (1988); Grötschel and Pulleybank (1986). Constraints that are currently used in Branch and Cut algorithms were available by 1986. Though additional constraints such as : Regular Star; Chain; and Ladder inequalities, are facet defining they have not yet been successfully incorporated into current computational methods (see Chvátal (1973); Grötschel and Padberg (1985); Laporte (1992a); Grötschel and Pulleybank (1986)).

In the following we discuss those constraints that are used in computational procedures. For the TSP (3.1)-(3.4) the initial LP relaxation is to drop (3.3) and replace (3.4) with $0 \leq x_{ij} \leq 1$ for all i, j. This provides the initial lower bound on the optimal objective function value. This bound is improved by adding violating constraints (Step 3 of Cutting Plane Procedure) of the following types.

Subtour Elimination Constraints (Dantzig et al (1954)):

$$\sum_{i,j \in S} x_{ij} \leq |S| - 1, \quad S \subset V, \quad 3 \leq |S| \leq n-2 \tag{3.5}$$

Comb Inequalities (Grötschel and Padberg (1985)):
Let $H, T_1, T_2, ..., T_k \subset V, k \geq 1$, be vertex sets satisfying :

(i) $|H \cap T_i| \geq 1, \; 1 \leq i \leq k$

(ii) $|T_i \backslash H| \geq 1, \; 1 \leq i \leq k$

(iii) $T_i \cap T_j = \phi$, for $i \neq j$

(iv) $k \geq 3$ and odd

The comb inequalities are given by :

$$\sum_{i,j \in H} x_{ij} + \sum_{p=1} \sum_{i,j \in T_p} x_{ij} \leq |H| + \sum_{i=1}^{k} |T_i| - \frac{1}{2}(3k+1) \tag{3.6}$$

Note that for the special case $|T_i| = 2$ for each i we have the 2-matching inequalities.

Clique Tree Inequalities (Grötschel and Pulleybank (1986)) :
Let $H_1, H_2, ..., H_r, T_1, T_2, ..., T_s, s$ odd, be vertex sets satisfying

(i) $|T_i \cap T_j| = 0$, for $i \neq j$

(ii) $|H_i \cap H_j| = 0$, for $i \neq j$

(iii) $2 \leq |T_i| \leq n-2$ and $|T_i \backslash \bigcup_{j=1}^{r} H_j| \geq 1$, for each i

(iv) Each H_i intersects an odd number (≥ 3) of T_i's

(v) The intersection graph of the sets H_i's and T_j's, forms a tree

Let t_i denote the number of H_j's that T_i intersects. Then the clique tree inequalities are given by :

$$\sum_{p=1}^{r} \sum_{i,j \in H_p} x_{ij} + \sum_{q=1}^{s} \sum_{i,j \in T_q} x_{ij} \leq \sum_{p=1}^{r} |H_p| + \sum_{q=1}^{s} (|T_q| - t_q) - \frac{1}{2}(s+1) \qquad (3.7)$$

Note that the set of subtour elimination constraints and comb inequalities are subsets of the clique tree inequalities.

Exact methods for finding subtour elimination constraints are available (see Padberg and Rinaldi (1990a)). Further, exact methods for finding comb inequalities are known for the 2-matching constraints case (Padberg and Rao (1982)). For other cases heuristics are used; often, for computational reasons, heuristics are used prior to or in place of exact methods. The usefulness of good heuristics for these separation problems is best demonstrated in Applegate et al (1995); Applegate et al (1998).

Over the past two decades there has been considerable improvement in the size of TSP's solved exactly; from a 120-city problem (Grötschel (1980)) in 1977 to a 13,509-city problem (Applegate et al (1998)) in 1998. The latter involved a parallel implementation of an algorithm using 48 $UNIX$ workstations; a 'pool' of constraints as well as a 'gluing' operation was used to find 'clique tree like' inequalities which are a large subset of the clique tree inequalities. Test problems are available (Reinelt (1991)).

4 VEHICLE ROUTING PROBLEM

Vehicle Routing Problems (VRP's) are concerned with the delivery of some commodities from one or more depots to a number of customer locations with known demand. Such problems arise in many physical systems dealing with distribution. The specific problem which arises is dependent upon the type of constraints and management objective. The constraints of the problem arise from: vehicle capacity; distance/time restriction; number of customers serviced by a vehicle; and other practical requirements. The management objectives usually relate to the minimization of cost/distance or fleet size. Much of the literature on vehicle routing has been concerned with problems having the following features :

(i) a single commodity is to be distributed from a single depot to customers with known demand
(ii) each customer's demand is serviced by one vehicle
(iii) each vehicle has the same capacity and makes one trip
(iv) the total distance traveled by each vehicle cannot exceed a specified limit
(v) each customer must be serviced within a specified time window
(vi) the objective is to minimize the total distance traveled by all vehicles.

Relaxing restrictions (iv) and (v) gives rise to the so called Capacitated Vehicle Routing Problem ($CVRP$), the most studied VRP. Relaxing (v) gives rise to the Distance Capacity-Restricted Vehicle Routing Problem ($DCVRP$). These NP-hard problems have attracted considerable attention over the past two decades resulting in many exact and heuristic algorithms (Achuthan et al (1995); Achuthan et al (1999a); Achuthan et al (1998); Achuthan et al (1999b); Araque et al (1994); Augerat et al (1995); Ball et al (1995); Desrochers et al (1992); Laporte (1992b); Laporte and Bourjolly (1984); Laporte et al (1986); Laporte et al (1992); Laporte et al (1984); Laporte and Nobert (1984); Laporte et al (1980)). A good deal of the recent work has been motivated by the success of the Branch and Cut Methods in solving large TSP's. Unfortunately the success for VRP's has not been so dramatic. For $CVRP$ the largest literature problem solved to date is a 134-city problem (Achuthan et al (1998); Augerat et al (1995)), accomplished by using Branch and Cut. We now give a $MILP$ formulation of the $CVRP$.

We denote the depot by 1 and the set of customer locations by $\mathcal{C} = \{2, 3, ..., n\}$. Thus the graph $G = (V, E)$ representing the vehicle network has $V = \{1, 2, ..., n\}$ and $E = \{(i, j) : i, j \in V, i < j\}$. We adopt the following notation :

q_j : Demand of customer $j, j \in \mathcal{C}$
c_{ij} : Distance between locations i and j
m : Number of delivery vehicles
Q : Common vehicle capacity

We assume throughout that $Q \geq max\{q_j\}$, and the distance matrix $C = (c_{ij})$ is symmetric and its elements satisfy the triangle inequality. For $S \subseteq \mathcal{C}$ we let $\ell(S)$ be a lower bound on the number of vehicles required to visit all locations of S in an optimal solution. Note that $\ell(S) \geq 1$. We write $\bar{S}$ for the complement of $\mathcal{C}$ of S. For $i, j \in V$ our decision variables are defined as

$$x_{ij} = \begin{cases} 1, & \text{if a vehicle travels on a single trip between } i \text{ and } j \\ 2, & \text{if } i = 1 \text{ and } (1, j, 1) \text{ is a route} \\ 0, & \text{otherwise} \end{cases}$$

Note that x_{ij} only needs to be defined for $i < j$, and $q_i + q_j \leq Q$. The following formulation is from Laporte et al (1985):

$$\text{Minimize} \qquad f = \sum_{i<j} c_{ij} x_{ij} \tag{4.1}$$

subject to

$$\sum_{j \in \mathcal{C}} x_{1j} = 2m \tag{4.2}$$

$$\sum_{j<k} x_{ik} + \sum_{j>k} x_{kj} = 2, \text{ for each } k \in \mathcal{C} \tag{4.3}$$

$$\sum_{ij \in S} x_{ij} \leq |S| - \ell(S), \text{ for all } S \subseteq \mathcal{C} \text{ with } 3 \leq |S| \leq n-2 \tag{4.4}$$

$$x_{ij} = \begin{cases} 0, 1, \text{ or } 2 & \text{for } i = 1 \\ 0, 1 & \text{otherwise} \end{cases} \tag{4.5}$$

$$m \text{ is a positive integer.} \tag{4.6}$$

In the above formulation, m can be fixed or variable. Constraints (4.2) and (4.3) are degree constraints on the depot and the customer locations, respectively. Constraints (4.4) eliminate subtours, that is they prohibit subtours that are free from the depot, or connected to the depot but violate the capacity or distance restrictions. In an actual implementation of the above formulation $\ell(S)$ is only calculated when required and typically $\left\lceil \frac{1}{Q} \sum_{j \in S} q_j \right\rceil$ is used. Violations of (4.3) using $\ell(S) = \frac{1}{Q} \sum_{j \in S} q_j$ were found exactly in Augerat et al (1995). Another approach for finding violations of (4.3) is to search for violations of equivalent constraints. Two computationally useful constraints (Achuthan et al (1995); Achuthan et al (1999a)) are :

$$\sum_{i,j \in S} x_{ij} + \sum_{i \in S} x_{1j} - m \leq |S| - l(\bar{S}) \qquad \text{where} \quad \bar{S} = C \setminus S \tag{4.7}$$

$$\sum_{i \in P} x_{ij} + \sum_{i,j \in P} x_{ij} + \sum_{i,j \in S} x_{ij} + \sum_{E[P,\bar{S}]} x_{ij} \leq |S| + |P| - l(\bar{P}) \tag{4.8}$$

where $\qquad P \subseteq S \subseteq C, \bar{P} = S \setminus P, \bar{S} = C \setminus S \qquad$ and

$$E[S,T] = \{(i,j) : i \in S, j \in T\}$$

Recently many cutting planes have been developed for the $CVRP$. These include

Comb Inequalities(Cornuéjols and Harche (1993)) :

Let $W_0, W_1, ..., W_s \in C$ satisfy :
1. $|W_i \setminus W_0| \geq 1, i = 1, \ldots, s$
2. $|W_i \cap W_0| \geq 1, i = 1, \ldots, s$
3. $|W_i \cap W_j| = 0, 1 \leq i \leq j \leq s$ and
4. $s \geq 3$ and odd

The comb inequality is given by

$$\sum_{p=0}^{s} \sum_{i,j \in W_p} x_{ij} \leq \sum_{p=0}^{s} |W_p| - \frac{3s+1}{2} + \alpha(m-1) \tag{4.9}$$

where

$$\alpha = \begin{cases} 0, & \text{if } 1 \notin \bigcup_{i=0}^{s} W_i \\ 1, & \text{if } 1 \in W_0 \setminus \bigcup_{i=1}^{s} W_i \text{ or } 1 \in W_j \setminus W_0 \text{ for some } j = 1, \ldots, s \\ 2, & \text{if } 1 \in W_j \cap W_0 \text{ for some } j = 1, \ldots, s \end{cases}$$

If in addition to conditions 1 - 4, the comb also satisfies $1 \in W_1 \setminus W_0$, then (4.9) can be strengthened (Cornuéjols and Harche (1993)) to :

$$\sum_{p=0}^{s} \sum_{i,j \in W_p} x_{ij} \leq \sum_{p=0}^{s} |W_p| - \frac{3s+1}{2} + m - \ell(C \setminus W_1) \tag{4.10}$$

Computational results using (4.4) and (4.9) with $\alpha = 0$ were reported in Cornuéjols and Harche (1993).

Bin Packing Inequalities (Achuthan et al (1998)):

The Bin Packing Problem (BPP) has an objective of packing k items having positive weights $w_1, w_2, ..., w_k$ into the least number, M, of bins $B_1, B_2, ..., B_M$ such that the total weight in each bin is at most Q. We denote this problem by $BPP\{ w_1, w_2, ..., w_{ki}; Q\}$. Consider a set of customers $S \subseteq C$ and a partition $P_S = (P_1, ..., P_u)$ of S such that :

$$\sum_{k \in P_i} q_k \leq Q, \quad 1 \leq i \leq u$$

Let $\hat{M}(P_S)$ be a lower bound to the minimum number of bins M^*, required for BPP $\{\hat{w}_1, \hat{w}_2, \ldots, \hat{w}_k; Q\}$, where $\hat{w}_i = \sum_{k \in P_i} q_k$. Then

$$\sum_{i,j \in S} x_{ij} + \sum_{k=1}^{u} \sum_{i,j \in P_k} x_{ij} \leq 2|S| - u - \hat{M}(P_S) \tag{4.11}$$

is a valid inequality for the $CVRP$.

Generalized Connectivity Constraints (Achuthan et al (1998)):

Let $S, T_1, T_2, ..., T_k \subseteq C$ be such that :

1. $k \geq v > 0$ and $\sum_{i \in S} q_i + \sum_{j=1}^{v} \sum_{i \in T_{p_j}} q_i > \ell(S)Q$, with $1 \leq p_i \neq p_j \leq k$
2. $T_o \cap T_j = \phi, i \neq j$
3. $S \cap T_i = \phi, q \leq i \leq k$ and
4. $T = \bigcup_{i=1}^{k} T_i$

Then any feasible solution (x_{ij}) of the $CVRP$ satisfies :

$$\begin{aligned} & A(S,v,k)\sum_{i,j\in S} x_{ij} + \sum_{E[S,T]} x_{ij} + \sum_{p=1}^{k}\sum_{i,j\in T_p} x_{ij} \\ \leq\ & A(S,v,k)(|S| - \ell(S)) + |T| - k + v - 1 \end{aligned} \tag{4.12}$$

where

$$A(S,v,k) = \begin{cases} \min\{k - v + 1, 2\ell(S) - v + 3\}, & \text{if } 2\ell(S) - v + 3 > 1 \\ k - v + 1 & \text{otherwise} \end{cases}$$

In addition, for $R \subseteq S$ any feasible solution of the $CVRP$ satisfies

$$\begin{aligned} & (2\ell(R) + v - 1)\sum_{i,j\in S} x_{ij} + 2\sum_{i,j\in R} x_{ij} + \sum_{E[R,T]} x_{ij} + \sum_{p=1}^{k}\sum_{i,j\in T_p} x_{ij} \\ & \leq (2\ell(R) + v - 1)(|S| - \ell(S)) + 2(|R| - \ell(R)) + |T| - k + v - 1 \end{aligned} \tag{4.13}$$

All the above cutting planes are based on feasibility requirements. Recently, cutting planes based on a specified structure of an optimal solution were established. These cutting planes are given below :

Optimality Based Constraints (Achuthan et al (1999a)) :

1. There exists an optimal solution $X = (x_{ij})$ of the $CVRP$ (4.1) to (4.6) with variable m satisfying :

$$\sum_{i,j\in S} x_{ij} + \sum_{j\in S} x_{1j} \leq |S| + 1 \tag{4.14}$$

   for all $S \in C$ with $2 \leq |S| \leq |C|$ and $\sum_{i\in S} q_i \leq Q$

2. There exists an optimal solution $X = (x_{ij})$ of the $CVRP$ (4.1) to (4.6) with variable m satisfying (4.14) and :

$$\sum_{i,j\in S} x_{ij} + \sum_{i\in S} x_{1i} \leq |S| + \left\lfloor 2(\sum_{i\in S} q_i + \delta)/(Q + 1 + \delta) \right\rfloor \tag{4.15}$$

for all $S \subseteq C$ with $2 \leq |S| \leq |C|$ and $\sum_{i\in S} q_i > Q$,
$\delta = 0, 1$ according as Q is odd or even.

Note that any feasible solution which does not satisfy (4.14) or (4.15) need not be considered whilst searching for an optimal solution. A further important point is that inequalities (4.14) and (4.15) yield the only available non trivial upper bound for m. Computational results using a Branch and Cut algorithm incorporating (4.1) - (4.8) and (4.11) - (4.15) were given in Achuthan et al (1995); Achuthan et al (1999a); Achuthan et al (1998).

The $DCVRP$ was studied in Achuthan et al (1999b). A number of other variations of VRP's have been studied including : time windows (Boyd (1996); Desrochers et al (1992)); stochastic demand (Golden and Assard (1988)); and many more (Golden and Assard (1988)).

5 A MINE SCHEDULING PROBLEM

The operation and management of a large open pit mine is an enormous and complex task, particularly for mines having a life of many years. Optimization techniques can be successfully applied to resolve a number of important problems that arise in the planning and management of a mine. These applications include: ore-body modeling and ore reserve estimation; the design of optimum pits; the determination of optimal production schedules; the determination of optimal operating layouts; the determination of optimal blends; the determination of equipment maintenance and replacement policies; and many more (Caccetta and Giannini (1986); Caccetta and Giannini (1990); Caccetta et al (1994); Caccetta et al (1998)).

Optimization techniques are applied to the 'block model' of an ore body. This model is based on the ore body being divided into fixed-size blocks. The block dimensions are dependent on the physical characteristics of the mine, such as pit slopes, dip of deposit and grade variability as well as the equipment used. The center of each block is assigned a grade representation of the whole block, based on drill hole data and a numerical technique. Using the financial and metallurgical data the net profit of each block is determined.

A fundamental problem in mine planning is that of determining the optimum ultimate pit limit of a mine. The optimum ultimate pit of a mine is defined to be that contour which is the result of extracting the volume of material which provides the total maximum profit whilst satisfying the operational requirement of safe wall slopes. The ultimate pit limit gives the shape of the mine at the end of its life. Usually this contour is smoothed to produce the final pit outline. This problem has been efficiently solved using the Lerchs-Grossmann graph theoretic algorithm (Lerchs and Grossmann (1965)) or the network flow method of Picard (Picard (1976)).

The open pit mine production scheduling problem can be defined as specifying the sequence in which blocks should be removed from the mine in order to maximize the total discounted profit from the mine subject to a variety of constraints. The constraints may involve the following : mill throughput (mill feed and mill capacity); volume of material extracted per period; blending constraints; stockpile related constraints; logistic constraints.

We now present a simple mixed integer linear programming formulation that incorporates the mill throughput and volume of material extracted constraints;

this formulation was detailed in Caccetta and Hill (1999c). We begin with some notation:

T is the number of periods over which the mine is being scheduled.
N is the total number of blocks in the ore body.
c_i^t is the profit (in NPV sense) resulting from the mining of block i in period t.
O is the set of ore blocks.
W is the set of waste blocks.
t_i is the tonnage of block i.
m^t is the tonnage of ore milled in period t.
S_i the set of blocks that must be removed prior to the mining of block i.
$x_i^t = \begin{cases} 1, & \text{if block } i \text{ is mined in periods 1 to } t \\ 0, & \text{otherwise} \end{cases}$
ℓ_0^t lower bound on the amount of ore that is milled in period t.
u_0^t upper bound on the amount of ore that is milled in period t.
u_w^t upper bound on the amount of waste that is mined in period t.

Then the $MILP$ formulation is :

$$\text{Maximize } Z = \sum_{t=2}^{T}\sum_{i=1}^{N}(c_i^{t-1} - c_i^t)x_i^{t-1} + \sum_{i-1}^{N} c_i^T x_i^T \tag{5.1}$$

subject to

$$\sum_{i\in O} t_i x_i^1 = m^1 \tag{5.2}$$

$$\sum_{i\in O} t_i(x_i^t - x_i^{t-1}) = m^t, \quad t = 2,3,\ldots,T \tag{5.3}$$

$$\sum_{i\in W} t_i x_i^1 \le u_w^1 \tag{5.4}$$

$$\sum_{i\in W} t_i(x_i^t - x_i^{t-1}) \le u_w^t \tag{5.5}$$

$$x_i^{t-1} \le x_i^t, \qquad t = 2,3,\ldots,T \tag{5.6}$$

$$x_i^t \le x_j^t, \quad t = 1,2,\ldots,T, \ j \in S_i, \quad i = 1,2,\ldots,T \tag{5.7}$$

$$\ell_0^t \le m^t \le u_0^t, \qquad t = 1,2,\ldots,T \tag{5.8}$$

$$x_i^t = 0,1, \qquad \text{for all } i,\ t. \tag{5.9}$$

Constraints (5.2), (5.3) and (5.8) ensure that the milling capacities hold. Constraints (5.4) and (5.5) ensure that the tonnage of waste removed does not exceed the prescribed upper bounds. Constraints (5.6) ensures that a block is removed in one period only. Constraints (5.7) are the wall slope restrictions.

The above formulation has NT 0-1 variables, and $(N+2)T+N(d-1)$ linear constraints, where d is the average number of elements in a cone. Typically T is around 10, N is 100,000 for a small pit and over 1,000,000 for a larger pit. Consequently the $MILP$'s that arise are much too large for direct application of commercial packages. However, as demonstrated in Caccetta and Hill (1999c), the structure of the problem can be exploited to develop computational strategies that produce provably good solutions. Prior to this, schedules have been generated manually or via some crude heuristic (Caccetta and Hill (1999c)).

The $MILP$ was tackled using the Lagrangian Relaxation technique in Caccetta et al (1998). At each step a problem similar to the ultimate pit limit problem, with the additional constraints dualized, is solved using the Lerchs-Grossmann algorithm. Subgradient optimization is used to reduce the duality gaps. The method was tested on a real ore body with 20,979 blocks and 6 time periods. The schedules obtained are within 5% of the theoretical optimum. The main problem with the method is resolving the duality gaps. However, the subproblems are useful in producing solutions using a heuristic. In fact, the heuristic solution obtained for the real ore body is within 2% of the theoretical optimum.

Recently (Caccetta and Hill (1999c)) a Branch and Cut procedure for solving the $MILP$ (5.1) - (5.9) was developed and implemented. The important features of the method are summarized below.

Key Features of the Branch and Cut Procedure

1. The block model is reduced to only include blocks inside the final pit design developed from the ultimate pit. Further reductions are made through consideration of (5.2) - (5.5) and (5.7).

2. The $MILP$ has strong branching variables due to the dependencies between variables (5.6) and (5.7). Note that setting a variable to 0 or 1 will fix a potentially large number of other variables. Consequently the subsequent LP relaxations are significantly smaller in size. This motivates more branching compared to typical Branch and Cut methods

3. Cutting planes involving Knapsack constraints are identified using the capacity upper bounds (5.2)- (5.5) and the block removal dependencies (5.7). Also cuts are identified through material removal dependencies between benches.

4. The search strategy involves a combination of best first search and depth first search. The motivation for this is to achieve a 'good spread' of possible pit schedules (best first search) whilst benefiting from using depth first search where successive LP's are closely related from one child node to the next. For large problems this often results in provably good solutions being found earlier than a search method geared to establishing an optimal solution.

5. Good lower bounds are generated through the use of an LP-Heuristic. The method works by considering each period in turn and fixing in and out sets of free variables. Cutting planes are then generated for the period, further LP's are solved and further fixing occurs. Throughout the fixing

of variables feasibility checks are used. If the heuristic succeeds, or fails due to an inferior lower bound being found, then periods are considered in the same direction, otherwise the direction is reversed. The heuristic is called for the first five levels of the search tree and every eighth node created thereafter.

6. Standard fixing of non-basic variables using reduced costs is carried out. Because of the block dependencies this may lead to the *LP* solution losing its optimality. In this case the *LP* solver is called and the cutting plane generation phase is re-entered without branching.

7. Many branching rules were tested and the following proved to be the best. The free variables are considered and a subset of these is chosen on the basis of closeness to the value of 0.5. For each variable in the subset we calculate the sum change in the fractional values of all variables dependent on the inclusion and exclusion of the branching variable. Choose the one with the highest minimal sum change in both directions of branching. Strong branching is used if the gap between the lower and upper bound is sufficiently small.

8. If the *LP* subproblem is not solved within a prescribed maximum time (2 minutes), then the *LP* optimization is terminated and branching is performed using the rules in (7) above. An attempt is then made to solve the resulting *LP*'s with in the specified time. This process is repeated as long as necessary.

9. The cutting plane phase is terminated early if tailing-off is detected or if the *LP* subproblem is solved optimally in more than a prescribed time (1 minute). Note that adding further cutting planes, even with purging of ineffective constraints, tends to increase the solution times for successive calls to the *LP* solver.

10. When branching we probe a random subset of variables having the same time index as the branching one. Bounds on variables may also be updated during this process.

11. All The *LP* subproblems are solved using *CPLEX* Version 6.0.

The Branch and Cut algorithm was implemented in *C++* on an *SGI* Origin 2000 dual processor computer. The dual processor was used only to solve the relaxed *LP*'s.

Extensive testing was carried on industry generated data as well as data from producing mines. The models ranged from 26,208 to 209,664 blocks. In all cases $T = 10$. For the smaller models solutions guaranteed to be within 0.4% of the optimum were obtained within 12 minutes. For the largest model, solutions guaranteed to be within 2.5% of the optimum were obtained within 4 hours. For these larger models the computations were continued for a further 16 hours with negligible change in the gap. The method generates tight bounds.

However, establishing optimality (except, of course, for small problems) is difficult because once a near optimal solution is achieved there are no available cutting planes to remove fractional variables occurring in the same bench level. Note that (5.6) and (5.7) give dependencies between variables corresponding to block removal in time and the vertical dimension, but not horizontally.

On the producing mine the software generated 7 good schedules (validated by mining engineers) within a total time of 10 minutes. The best schedule was within 0.27% of the optimum. The application of the Branch and Cut algorithm resulted in an improvement in net present value (NPV) of at least 15% over the schedule used by the company and designed by a commercially available package.

We remark that the model used incorporated a number of other requirements such as: processing different ore types; maximum vertical depth; minimum pit bottom width; and stockpiles (see Caccetta and Hill (1999c) for details).

6 CONCLUSIONS

In this paper we considered the computational aspects of solving large mixed integer programming problems. The technique that emerges most favorably is the method of Branch and Cut. This method draws on many areas in combinatorial optimization. Thus any advances in a range of fields and for different problems may result in significant reductions in solution times for an unrelated problem. In particular we can expect to see reductions in solution times and larger problems being solved through improvements in such areas as LP solution times; heuristic techniques; preprocessing and reformulation of the problems; methods for reducing the size of the search tree; development of exact and heuristic methods for solving separation problems; parallelization and the discovery of new cutting planes. Since this is the case, we can be assured that Branch and Cut will remain the method of choice for many combinatorial optimization problems and an active field of research for many years to come.

References

Aardal, K, and Van Hoesel, S., (1996), *Polyhedral techniques in combinatorial optimization I : Theory,* Statistica Neerlandica, Vol. 50, pp. 4-26.

Aardal, K, and Van Hoesel, S., *Polyhedral techniques in combinatorial optimization II : Computations,* Statistica Neerlandica (to appear).

Aarts, E. and Lenstra J.K., (Eds.), (1997), *Local search in combinatorial optimization,* J Wiley and Sons, Chichester.

Achuthan, N.R., Caccetta, L. and Hill, S.P., (1995), A new subtour elimination constraint for the vehicle routing problem, *European Journal of Operations Research*, Vol. 91, pp. 573-586.

Achuthan, N.R., Caccetta, L. and Hill, S.P., (1999a), An improved branch and cut algorithm for the capacitated vehicle routing problem, (submitted for publication).

Achuthan, N.R., Caccetta, L. and Hill, S.P., (1998), The capacitated vehicle routing problem : Some new cutting planes, *Asia Pacific Journal of Operational Research*, Vol. 15, pp. 109-123.

Achuthan, N.R., Caccetta, L. and Hill, S.P., (1999b), The vehicle routing problem with capacity and distance restrictions (submitted for publication).

Applegate, D., Bixby, R., Chvatal, V. and Cook, W. (1995), Finding cuts in the TSP (A preliminary report). *DIMACS Technical Report*

Applegate, D., Bixby, R., Chvatal, V. and Cook, W. (1998), On the solution of traveling salesman problems, *Documenta Mathematica (Journal der Deutschen Mathematiker-Vereinigung), International Congress of Mathematicians,* pp. 645-656.

Araque, J.R., Kudva, G, Morin, T.L., and Pekny, J.F., (1994), A branch-and-cut algorithm for vehicle routing problems, *Annals of Operations Research*, Vol. 50, pp. 37-59.

Augerat, P., Belengeur, J.M., Benavent, E., Corberan, A., Naddef N., and Rinaldi, G., (1995), *Computational results with a branch and cut code for the capacitated vehicle routing problem,* Research Report 949-M, Universite Joseph Fourier, Grenoble, France.

Barnhart, C. , Johnson, E.L., Nemhauser, G.L., Savelsberg, M.W.P., and Vance, P.H., (1998), Branch and Price : Column generation for solving huge integer programs, *Oper. Res.*, Vol. 46, pp. 316-329.

Balas, E. and Toth, P. (1985), *Branch and bound methods, in The Traveling Salesman Problem (Lawler, E.L., Lenstra, J.K., Rinnoy Kan A.G.H. and Shmoys, D.B., Editors)* John Wiley and Sons, pp. 361 - 401.

Balas, E., Ceria, S. and Cornuéjols, G., (1993), A lift-and-project cutting plane algorithm for mixed 0-1 programs, *Mathematical Programming*, Vol. 58, pp. 295 - 324.

Balas, E., Ceria, S. and Cornuéjols, G., (1996), Mixed 0-1 programming by lift-and-project in a branch-and-cut framework, *Management Science*, Vol. 42, pp. 1229 - 1246.

Ball, M.O. Magnanti, T.L., Monma, C.L., and Nemhauser, G. (Eds)., (1995), *Network routing, Vol 8 of Handbooks in Operations Research and Management Science,* North Holland, Amsterdam.

Bertsekas, D.P., (1991), *Linear network optimization : algorithms and codes,* MIT Press, Cambridge, MA.

Boyd, E. D., (1996) On the complexity of a cutting plane algorithm for solving combinatorial linear programs, *SIAM Journal on Discrete Mathematics*, pp. 365-376.

Boyd, S.T. and Cunningham, W., (1991), Small traveling salesman polytopes, *Math. Ops. Res.*, Vol. 16, pp. 259 - 271.

Caccetta, L. and Giannini, L.M., (1986), Optimization techniques for the open pit limit problem, *Proc. Australas. Inst. Min. Metall.*, Vol. 29, pp. 57-63.

Caccetta, L. and Giannini, L.M., (1990), *Application of operations research techniques in open pit mining, in Asian-Pacific Operations Research : APORS'88 (Byong-Hun Ahn Ed.),* Elseview Science Publishers BV, pp. 707-724.

Caccetta, L. and Giannini, L.M.,and Kelsey, P., (1994), On the implementation of exact optimization techniques for open pit design, *Asia-Pacific Journal of Operations Research*, Vol. 11, pp. 155-170.

Caccetta, L. and Hill, S.P., (1999a) *Branch and cut methods for network optimization,* Mathematical and Computer Modeling (in press).

Caccetta, L. and Hill, S.P., (1999b) *A branch and cut method for the degree constrained minimum spanning tree problem,* Networks, (to appear).

Caccetta, L. and Hill, S.P., (1999c) *An application of branch and cut to open pit mine scheduling* (submitted for publication).

Caccetta, L., Kelsey, P. and Giannini, L.M., (1998), *Open pit mine production scheduling, in Computing Applications in the Minerals Industries International Symposium (3rd Regional APCOM) (A.J. Basu, N. Stockton and D. Spottiswood, Eds.),* Austral. Inst. Min. Metall. Publication Series, Vol. 5 , pp. 65-72.

Caprara, A. and Fischetti, M., (1997), *Branch-and-cut algorithms, Annotated Bibliographies in Combinatorial Optimization, (M. Dell'Amico, F. Maffioli and S. Martello, Editors),* J. Wiley and Sons, Chichester.

Chvátal, V., (1973), Edmonds polytopes and weakly hamiltonian graphs, *Math. Programming*, Vol. 5, pp. 29 - 40.

Cook, W., Cunningham, W.H. , Pulleybank, W.R. and Schrijver, A. (1998), *Combinatorial Optimization,* Wiley, New York.

Cornuéjols, G., Fonlupt J., and Naddef, D., (1985), The Traveling salesman problem on a graph and some related integer polyhedra, *Mathematical Programming*, Vol. 33, pp. 1 - 27.

Cornuéjols, G., and Harche, F., (1993), Polyhedral study of the capacitated cehicle routing problem, *Mathematical Programming*, Vol. 60, pp. 21 - 52.

Crowder, H., Johnson, E.L. and Padberg, M. (1983), Solving large-scale zero one linear programming problems, *Oper. Res.*, Vol. 33, pp. 803 834.

Crowder, H. and Padberg, M.W., (1980), Solving large-scale symmetric traveling salesman problems to optimality, *Management Science*, Vol. 26, pp. 495 - 509.

Dantzig, G.B.,. Fulkerson, D.R and Johnson, S.M., (1954), Solution of a large scale traveling salesman problem, *Operations Research*, Vol. 2, pp. 393 - 410.

Dantzig, G.B., Fulkerson, D.R and Johnson, S.M., (1959), On a linear-programming combinatorial approach to the traveling-salesman problem, *Operations Research*, Vol. 7, pp. 58-66.

Dell'Amico, M., Maffioli, F. and Martello, S. ,(Eds.), (1997), *Annotated biobliographies in combinatorial optimization,* John Wiley and Sons, Chichester.

Desrochers, M. , Desrosiers, J. and Solomon, M. (1992), A new optimization algorithm for the vehicle routing problem with time windows, *Operations Research*, Vol. 40, pp. 342-354.

Du, D.Z. and Pardalos, P.M. ,(Eds.), (1998), *Handbook on combinatorial optimization (3 volumes),* Kulwer Academic Publishers, Boston.

Edmonds, J., (1965), Maximum matching and a polyhedron with 0,1 vertices, *Journal of Research of the National Bureau of Standards*, Vol. 69B, pp. 67 - 72.

Fisher, M. L., (1981), The Lagrangian relaxation method for solving integer programming problems, *Management Science*, Vol. 27, pp. 1 - 17.

Fisher, M. L., (1994), Optimal solution of vehicle routing problems using minimum K-trees, *Operations Research*, Vol. 42, pp. 626 - 642.

Fisher, M. L., Northup, W.D. and Shapiro, J.F., (1975), Using duality to solve discrete optimization problems : theory and computational experience, *Math. Programming Study*, Vol. 3, pp. 56-94.

Fleishmann, B., (1988), A new class of cutting planes for the symmetric traveling salesman problem, *Mathematical Programming*, Vol. 40, pp. 225 - 246.

Garvin, W.M. , Crandall, H.W. John, J.B. and Spellman, R.A., (1957), Applications of linear programming in the oil industry, *Management Science*, Vol. 3, pp. 407 - 430.

Geoffrion, A.M., (1974), Lagrangean relaxation for integer programming, *Math. Programming Study*, Vol. 2, 82-114.

Golden, B.L. and. Assad, A.A.,(Eds.), (1988), *Vehicle routing : methods and studies,* North-Holland, Amsterdam.

Graham, R., Grötschel, M. and Lovász, L (Eds.), (1995), *Handbook of combinatorics,* North-Holland, Amsterdam.

Grötschel, M. (1980), On the symmetric travelling salesman problem : solution of a 120-city problem, *Mathematical Programming Studies*, Vol. 12, pp. 61-77.

Grötschel, M. and Holland, O. (1991), Solution of large-scale symmetric travelling salesman problems, *Mathematical Programming*, Vol. 51, pp. 141 - 202.

Grötschel, M. and Holland, O. (1987), A cutting plane algorithm for minimum perfect two matching, *Computing*, Vol. 39, pp. 327-344.

Grötschel, M. , Lovász, L. and Schriver, A., (1988), *Geometric algorithms and combinatorial optimization,* Springer-Verlag, Berlin.

Grötschel, M. and Padberg, M.W. (1985), *Polyhedral theory, in The travelling salesman problem (E.L. Lawler, J.K. Lenstra, A.G.H. Rinnooy Kan and D.B. Shmoys, Eds.),* John Wiley and Sons, pp. 251-305.

Grötschel, M. and Pulleybank, W.R. (1986), Clique tree inequalities and the symmetric travelling salesman problem, *Mathematics of Operations Research*, Vol. 11, pp. 537 - 569.

Held, M., and Karp, R.M. (1970), The travelling salesman problem and minimum spanning trees, *Oper. Res.*, Vol. 18, pp. 1138-1162.

Held, M and Karp, R.M., (1971), The travelling salesman problem and minimum spanning trees : part II, *Math. Programming*, Vol. 1, pp. 6-25.

Held, M. , Wolfe, P.and Crowder, H.D., (1974), Validation of subgradient optimization, *Math. Programming*, Vol. 6, pp. 62-88.

Hoffman, K. and Padberg, M., (1991), Improving LP-representations of zero-one linear programs for branch and cut, *ORSA Journal on Computing*, Vol. 3, pp. 121-134.

Hoffman, K. and Padberg, M., (1993), Solving airline crew scheduling problems by branch-and-cut, *Management Science* 39, pp. 657-682.

Ibaraki, T., (1987), *Enumerative approaches to combinatorial optimization - Part I.,* Annals of Operations Research 10, Baltzer Basel.

Ibaraki, T., *Enumerative approaches to combinatorial optimization - Part II.*, Annals of Operations Research 11, Baltzer, Basel.

Johnson, E.L., (1980), Integer programming, *SIAM CBMS-NSF* Series No. 32.

Laporte, G., (1992a), The travelling salesman problem : an overview of exact and approximate algorithms, *European Journal of Operational Research,* Vol. 59, pp. 231-247.

Laporte, G., (1992b), The vehicle routing problem : an overview of exact and approximate algorithms, *European Journal of Operations Research* , Vol. 59, pp. 345-358.

Laporte, G. and Bourjolly, J.M., (1984), *Some further results on k-star constraints and comb inequalities,* Cahiers Du GERAD, G-82-10, Ecole des Hautes Etudes Commerciales de Montral.

Laporte, G. , Mercure, H. , and Nobert, Y., (1986), An exact algorithm for the asymmetrical capacitated vehicle routing problem, *Networks,* Vol. 16, pp. 33 - 46.

Laporte, G. , Mercure, H. , and Nobert, Y., (1992), A branch and bound algorithm for a class of asymmetrical vehicle routing problems, *Journal of the Operational Research Society,* Vol. 43, pp. 469 - 481.

Laporte, G. , Desrochers M. , and Nobert, Y., (1984), Two exact algorithms for the distance constrained vehicle routing problem, *Networks,* Vol. 14, pp. 161 - 172.

Laporte, G. and Nobert, Y., (1984), Comb inequalities for the vehicle routing problem, *Methods of Operational Research,* Vol. 51, pp. 271 - 276.

Laporte, G. and Nobert, Y., (1980), A cutting plane algorithm for the *m*-salesman problem, *Journal of the Operational Research Society,* Vol. 31, pp. 1017 - 1023.

Laporte, G., Nobert, Y. and Desrochers, M., (1985), Optimal routing under capacity and distance restrictions, *Oper. Res.*, Vol. 33, pp. 1050 - 1073.

Lawler, E.L., (1976), *Combinatorial optimization : networks and matroids,* Holt Rinehart and Winston, New York.

Lawler, E.L. , Lenstra, J.K. , Rinnooy Kan A.H.G. ,and Shmoys, D.B., (1985), *The travelling salesman problem,* John Wiley & Sons.

Lerchs, H., and Grossmann, L.F., (1965), *Optimum design of open pit mines,* Canad. Inst. Mining Bull., Vol. 58, pp. 47-54.

Lovász, L. and Plummer, M., (1986), *Matching theory,* North-Holland, Amsterdam.

Martello, S., Laporte, G., Minoux M. and Ribeiro C. ,(Eds.), (1987), *Surveys in combinatorial optimization,* North-Holland, Amsterdam.

Martello, S. and Toth, P. (1990), *Knapsack problems : algorithms and computer implementations,* Wiley, New York.

Moré J.J. and Wright, S.J. (1993), Optimization software guide, *SIAM Frontiers in Applied Mathematics* 14.

Nemhauser G.L., Rinnooy Kan A.G.H. and Todd M.J. (Eds.), (1988), *Optimization* **Vol** 1 of *Handbooks in Operations Research and Management Science,* North-Holland, Amsterdam.

Nemhauser G.L. and Wolsey, L.A. (1988), *Integer and combinatorial optimization,* John Wiley and Sons.

Padberg M. and Hong, S. (1980), On the symmetric travelling salesman problem : a computational study, *Mathematical Programming Studies*, Vol. 12, pp. 78-107.

Padberg M.W. and Rao, M.R. (1982), Odd minimum cut sets and b-matchings, *mathematics of operations research*, Vol. 7, pp. 67 - 80.

Padberg M. and Rinaldi, G. (1987), Optimization of a 532-symmetric travelling salesman problem, *Operations Research Letters*, Vol. 6, pp. 1 - 7.

Padberg M. and Rinaldi, G. (1990a), An efficient algorithm for the minimum capacity cut problem, *Mathematical Programming*, Vol. 47, pp. 19 - 36.

Padberg M. and Rinaldi, G. (1991), A branch and cut algorithm for the resolution of large scale travelling salesman problems, *SIAM Review*, Vol. 33 No. 1, pp. 60 - 100.

Padberg M. and Rinaldi, G. (1990b), Facet identification for the symmetric travelling salesman polytope, *Mathematical Programming*, Vol. 47, pp. 219 - 257.

Papadimitriou C.H. and Steiglitz, K. (1982), *Combinatorial optimization : algorithms and complexity,* Prentice-Hall Inc., Englewood Cliffs, NJ.

Picard, J.C. (1976), Maximum closure of a graph and applications to combinatorial problems, *Management Sc.*, Vol. 22, pp. 1268 - 1272.

Reinelt, G. (1991), A travelling salesman problem library, *ORSA Journal of Computing*, Vol. 3, pp. 376 - 384.

Savelsbergh, M.W.P. (1994), Preprocessing and probing techniques for mixed integer programming problems, *ORSA J. on Computing*, Vol. 6, pp. 445-454.

Schrijver, A. (1986), *Theory of Linear and Integer Programming,* Wiley, New York.

Shapiro, J.F., (1979), A survey of lagrangean techniques for discrete optimization, *Annals of Discrete Math*, Vol. 5, pp. 113-138.

3 SEPARABILITY OF STAR-SHAPED SETS WITH RESPECT TO INFINITY

A. M. Rubinov

School of Information Technology and Mathematical Sciences,
University of Ballarat, Victoria 3353, Australia

and A. P. Shveidel

Department of Mathematics,
Karaganda State University, Karaganda, 470074 Kazakhstan

Abstract: We study the separability of star-shaped sets with respect to infinity by finite collections of linear functions.

Key words: star-shaped sets with respect to infinity, Minkowski gauge, cone-separability

X. Yang et al. (eds.), Progress in Optimization, 45–63.

1 INTRODUCTION

Many applications of convexity are based on the separation property: if the intersection of two convex sets is either empty or coincides with the intersection of their boundaries, then these sets can be separated by a linear function (geometrically, by a hyperplane) under fairly mild assumptions. In particular, the separation property plays a crucial role in the study of convex optimization problems. The separation property for two convex sets easily follows from a simpler fact: if a point does not belong to the closed convex set then this point can be separated from this set. Generalizations of this assertion are studied in the framework of *abstract convexity* where not necessarily linear functions are used for separation (see Fan (1963)). Note that in contrast with the classical case, the nonlinear separation of a point from a set does not imply the separation property for two sets.

One more generalization of the separation property is based on the separation by a finite collection of linear functions. This approach has been considered in Shveidel (1997) for special classes of star-shaped sets, which are called *strongly star-shaped* in this paper. Properties of these sets were discussed in Rubinov and Yagubov (1986). Actually this kind of separation can be studied in the framework of abstract convexity, by assuming that finite minima of linear functions are used as nonlinear separation functions (see Rubinov and Vladimirov (1998) for details). As it turned out (Shveidel (1997)) the separation of a point from a strongly star-shaped set by finite collections of linear functions can be extended to the separation of two strongly star-shaped sets, hence this kind of separation can be applied in the study of some nonconvex problems of global optimization (see Shveidel (1997), where possible applications are outlined).

In this paper we study the separation by finite collections of linear functions of the so-called star-shaped sets with respect to (w.r.t.) infinity (for definitions and simple properties of these sets see for example Rubinov and Glover (1995)). First we introduce *strongly* star-shaped sets w.r.t. infinity and establish their connections with strongly star-shaped sets. Then we consider the so-called cone separation of a point from both star-shaped and strongly star-shaped sets w.r.t. infinity. We point out necessary and also sufficient conditions for such a separation. Then we demonstrate that for star-shaped sets w.r.t. infinity, the cone-separation of a point from a set is equivalent to the separation by a finite collection of linear functions.

In order to extend the results obtained for the separation of two sets, we need to establish the properties of the sum of these sets. We show that the sum possesses good properties if it is closed and does not contain zero. The separation of two strongly star-shaped sets w.r.t infinity can be proved under different sets of conditions. We give only two of them: for the separation of nonintersecting sets and for sets such that their intersection coincides with the intersection of their boundaries.

2 PRELIMINARIES

Let $x \in \mathbb{R}^n, x \neq 0$. We will denote by R_x the ray $\{\lambda x : \lambda \geq 0\}$. Recall the well-known definition of star-shaped sets. Note that we consider only closed star-shaped sets.

Definition 2.1 *1) Let A be a nonempty subset of $\mathbb{R}^n$. The set kern$A$ of all $a \in A$ such that $(x \in A, 0 \leq \lambda \leq 1) \Longrightarrow a + \lambda(x - a) \in A$ is called the kernel of A;*
2) A closed nonempty set A is called a star-shaped set if kern$A \neq \emptyset$;

The totality of all star-shaped sets will be denoted by $\mathcal{A}$. A set $A \in \mathcal{A}$ is called *star-shaped* with respect to (w.r.t.) a point $a \in A$ if $a \in$ kern$A$. In particular a set $U \in \mathcal{A}$ is star-shaped w.r.t. zero if $\lambda x \in U$ for all $x \in U$ and $\lambda \in [0, 1]$. The totality of all star-shaped sets w.r.t. zero will be denoted by $\mathcal{U}$. Let $U \in \mathcal{U}$. The function

$$\mu_U(x) = \inf \{\lambda > 0 : x \in \lambda U\} \quad (x \in \mathbb{R}^n)$$

is called the Minkowski gauge of the set U. (It is assumed that the infimum over the empty set is equal to $+\infty$.)

Let $\mathbb{R}_{+\infty} = \mathbb{R} \cup \{+\infty\}$. The function $\mu_U : \mathbb{R}^n \to \mathbb{R}_{+\infty}$ is nonnegative. Clearly $\mu_U(0) = 0$. For $x \neq 0$ we have $\mu_U(x) = 0$ if and only if $R_x \subset U$ and $\mu_U(x) = +\infty$ if and only if $R_x \cap U = \{0\}$. The function μ_U is positively homogeneous (of the first degree), that is $\mu_U(\lambda x) = \lambda\mu_U(x)$ for all $x \in \mathbb{R}^n$ and $\lambda > 0$. It follows from the definition that $U = \{x : \mu_U(x) \leq 1\}$. Since level sets $S_c(x) = \{x : \mu_U(x) \leq c\} = cU$ are closed for $c \geq 0$ (by the definition $0 \cdot U = \cap_{c>0}(c \cdot U)$), it follows that μ_U is a lower semicontinuous (l.s.c) function.

Proposition 2.1 *(see, for example Rubinov and Yagubov (1986)) Let $p : \mathbb{R}^n \to \mathbb{R}_{+\infty}$ and $U = \{x : p(x) \leq 1\}$. Then the following are equivalent:*
1) p is positively homogeneous, nonnegative, l.s.c and $p(0) = 0$;
2) U is a nonempty star-shaped w.r.t. zero set and $p = \mu_U$.

Definition 2.2 *1) Let B be a nonempty subset of $\mathbb{R}^n$. The set kern$_\infty B$ of all $u \in \mathbb{R}^n$ such that $u \notin B$ and $(x \in B, \lambda \geq 1) \Longrightarrow u + \lambda(x - u) \in B$ is called the ∞- kernel of the set B.*

2) A closed nonempty set $B \in \mathbb{R}^n$ is called star-shaped w.r.t. infinity if kern$_\infty B \neq \emptyset$.

(Star-shaped w.r.t. infinity sets are known also as *co-star-shaped* Penot(1997)). We will denote the totality of all star-shaped w.r.t. infinity sets by $\mathcal{B}$ and the totality of all $V \in \mathcal{B}$ such that $0 \in$ kern$_\infty V$ by $\mathcal{V}$. Thus a closed nonempty $V \in \mathcal{V}$ if and only if $0 \notin V$ and $(x \in V, \lambda \geq 1) \Longrightarrow \lambda x \in V$.

Definition 2.3 *(see for example Barbara and Crouzeix (1994); Rubinov and Glover (1995)). Let $V \in \mathcal{V}$. The function $\nu : \mathbb{R}^n \to \mathbb{R}$ defined by*

$$\nu_V(x) = \sup\{\lambda > 0 : x \in \lambda V\}$$

is called ∞*-gauge of* V*. (It is assumed that the supremum over the empty set is equal to zero.)*

Let us describe some properties of the ∞-gauge (see Rubinov and Glover (1995) for details).

Proposition 2.2 *Let* $V \in \mathcal{V}$*, that is* V *is a star-shaped set w.r.t infinity and* $0 \in kern_{\infty} V$*. Then* ν_V *is a real-valued, upper semicontinuous (u.s.c), nonnegative, positively homogeneous function with* $\nu_V(0) = 0$*. We also have:*

$$(x \neq 0, \quad \nu_V(x) = 0) \Longleftrightarrow R_x \cap V = \emptyset.$$

Proposition 2.3 *(Rubinov and Glover (1995)) Let* $p : \mathbb{R}^n \to \mathbb{R}$ *and* $V = \{x : p(x) \geq 1\}$*. Then the following are equivalent:*
a) $p \neq 0$ *is u.s.c , nonnegative, positively homogeneous,* $p(0) = 0$*;*
b) V *is a star-shaped set w.r.t. infinity,* $0 \in kern_{\infty} V$ *and* $p = \nu_V$*.*

3 STRONGLY STAR-SHAPED SETS

Definition 3.1 *1) Let* A *be a nonempty subset of* $\mathbb{R}^n$*. The set* $kern_* A$ *of all* $a \in intA$ *such that for each* $x \in \mathbb{R}^n$ *the ray* $a + R_x$ *does not intersect the boundary* bdA *more than once is called the lower kernel of* A*.*

2) A closed set $A \subset \mathbb{R}^n$ *is called strongly star-shaped if this set is* proper, *that is* $A \neq \emptyset A \neq \mathbb{R}^n$*, and its lower kernel* $kern_* A$ *is nonempty.*

3) A closed proper set $U \subset \mathbb{R}^n$ *is called strongly star-shaped w.r.t zero if* $0 \in kern_* A$

Strongly star-shaped sets w.r.t. zero and the space of such sets have been studied in Rubinov and Yagubov (1986). We will denote the totality of all strongly star-shaped subsets by $\mathcal{A}_s$ and the totality of all strongly star-shaped w.r.t zero sets by $\mathcal{U}_s$. If $A \in \mathcal{A}_s$ then $u \in \text{kern}_* A$ if and only if $A - u \in \mathcal{U}_s$. Clearly each strongly star-shaped set is star-shaped.

Proposition 3.1 *(see, for example Rubinov and Yagubov (1986)) Let* $p : \mathbb{R}^n \to \mathbb{R}$ *and* $U = \{x : p(x) \leq 1\}$*. Then the following are equivalent:*
1) $p \neq 0$ *is positively homogeneous nonnegative and continuous;*
2) U *is strongly star-shaped w.r.t. zero and* $p = \mu_U$*.*

Definition 3.2 *1) The upper kernel* $kern^* B$ *of a nonempty set* $B \subset \mathbb{R}^n$ *is the set of all* $u \in \mathbb{R}^n$ *such that* $u \notin B$ *and the following holds:*
if the ray R starting from u intersects B, then the intersection $R \cap B$ contains at least two points and the intersection $\text{bd}B \cap R$ consists of exactly one point.
2) A closed nonempty set $B \subset \mathbb{R}^n$ *is called strongly star-shaped w.r.t. infinity if* $kern^* B \neq \emptyset$*.*

We present an important example of a strongly star-shaped set w.r.t. infinity.

Example 3.1 *Let* U *be an open proper convex set. Then its complement* $B = \mathbb{R}^n \setminus U$ *is strongly star-shaped w.r.t.* ∞*. This statement follows from Corollary 3.2 below.*

The totality of all strongly star-shaped sets w.r.t. infinity will be denoted by $\mathcal{B}_s$. The totality of all $V \in \mathcal{B}_s$ such that $0 \in \text{kern}^* V$ will be denoted by $\mathcal{V}_s$. Clearly $B \in \mathcal{B}_s$ if and only if B is a shift of a set $V \in \mathcal{V}_s$.

Proposition 3.2 *Let $V \in \mathcal{V}_s$. Then $V \in \mathcal{V}$ and $(x \in V, \lambda > 1) \Longrightarrow \lambda x \in \text{int}V$.*

Proof: 1) Let $x \in V$ and $R = R_x = \{\lambda x : \lambda \geq 0\}$. Since $R \cap V \neq \emptyset$ it follows that $R \cap \text{bd}V$ contains exactly one point $\tilde{\lambda} x$. Assume without loss of generality that $\tilde{\lambda} = 1$. Let $\Lambda = \{\lambda > 1 : \lambda x \notin V\}$. We have to show that Λ is empty. On the contrary assume, that $\Lambda \neq \emptyset$. Let $\lambda_o = \inf \Lambda$. Since the set V is closed it follows that Λ is open so $\lambda_o \notin \Lambda$, that is $\lambda_o x \in V$. Clearly $\lambda_o x \in \text{bd}V$. Since $\text{bd}V \cap R$ only contains the point x, it follows that $\lambda_o = 1$. Assume there exists a segment $(\lambda_1, \lambda_2) \subset \Lambda$ such that $\lambda_1 > 1$ and $\lambda_1 \notin \Lambda$. Then $\lambda_1 x$ is a boundary point of V, which is a contradiction. Hence Λ consists of the unique segment $(1, \lambda')$. If $\lambda' \neq +\infty$ we have a contradiction as well, so $\Lambda = (1, +\infty)$. Thus $\lambda x \notin V$ for $\lambda > 1$. Let $\bar{\lambda} = \inf\{\lambda > 0 : \lambda x \in V\}$. Since $0 \notin V$ it follows that $\bar{\lambda} > 0$. Since $\lambda x \notin V$ for $\lambda < \bar{\lambda}$, it follows that $\bar{\lambda} x \in \text{bd}V$ and so $\bar{\lambda} = 1$. This $\lambda x \notin V$ with $\lambda \neq 1$. We arrive at a contradiction since $R \cap V$ should contain at least two points. Hence $\Lambda = \emptyset$ and $\lambda x \in V$ for all $\lambda > 1$. We have proved that V is a star-shaped w.r.t. infinity set and $0 \in \text{kern}_\infty V$.

2) We now show that $\lambda x \in \text{int}V$ if $x \in V$ and $\lambda > 1$. It follows from the first part of the proposition that $\inf\{\lambda > 0 : \lambda x \in V\} := \bar{\lambda} \leq 1$ and $\lambda x \in V$ for all $\lambda \geq \bar{\lambda}$. Since $\bar{\lambda} x \in \text{bd}V$ and $R_x \cap V$ consist of a unique point, it follows that $\lambda x \in \text{int}V$ for all $\lambda > \bar{\lambda}$, in particular for $\lambda > 1$. □

Corollary 3.1 *If $B \in \mathcal{B}_s$ then B is star-shaped w.r.t. infinity and $\text{kern}^* B \subset \text{kern}_\infty B$.*

We now describe some properties of ∞-gauges for $V \in \mathcal{V}_s$.

Proposition 3.3 *Let $V \in \mathcal{V}_s$. Then ∞-gauge ν_V is a real-valued, nonzero and continuous function with the following properties:*

$$\nu_V(x) > 0 \Longrightarrow \nu_V(x) = \max\{\lambda > 0 : x \in \lambda V\}, (x \neq 0, \nu_V(x) = 0) \Longrightarrow R_x \cap V = \emptyset; \tag{3.1}$$

$$V = \{x \in \mathbb{R}^n : \nu_V(x) \geq 1\}; \tag{3.2}$$

$$\text{int}V = \{x \in \mathbb{R}^n : \nu_V(x) > 1\}, \quad \text{bd}V = \{x \in \mathbb{R}^n : \nu_V(x) = 1\}. \tag{3.3}$$

Proof: The inequality $\nu_V(x) < +\infty$ follows from the definition of ν_V, the first implication in (3.1) is evident, the second implication follows from Proposition 2.2. The equality (3.2) follows from Proposition 2.3. Let us prove (3.3). From (3.1) we have: $\{x : \nu_V(x) = 1\} \subset \text{bd}V$. Assume now that $x \in V$ and $\nu_V(x) \neq 1$. Then $\alpha = \nu_V(x) > 1$. Let $x' = \alpha^{-1} x$. Since ν_V is positively homogeneous, it follows that $\nu_V(x') = 1$, hence $x' \in V$. By applying Proposition 3.2 we can conclude that $x = \alpha x' \in \text{int}V$. Thus (3.3) has been proved. Since the level set $\{x : \nu_V(x) \geq 1\} = V$ of the positively homogeneous function ν_V is closed and the level set $\{x : \nu_V(x) > 1\} = \text{int}V$ is open, it follows that ν_V is continuous. □

Proposition 3.4 *Let $p \neq 0$ be a nonnegative positively homogeneous real-valued continuous function. Then $p = \nu_V$ where $V = \{x : p(x) \geq 1\}$ is a strongly star-shaped w.r.t. infinity set with $0 \in \text{kern}^* V$.*

Proof: Following from Proposition 2.3, we know that $p = \nu_V$ and V is star-shaped w.r.t. infinity with $0 \in \text{kern}_\infty V$. We now show that V is strongly star-shaped w.r.t. infinity. Let $R = R_x$ be a ray such that $R \cap V \neq \emptyset$. Since p is real-valued we can assume without loss of generality that $\nu_V(x) = 1$. Since V is star-shaped w.r.t. infinity with $0 \in \text{kern}_\infty V$, it follows that $x \in \text{bd}V$ and $\lambda x \in V$ for all $\lambda > 1$. The set $\{x : p(x) > 1\}$ is open due to the continuity of p, hence $\lambda x \in \text{int}V$ for $\lambda > 1$. Thus the intersection $R \cap V$ contains more than one point and the intersection $R \cap \text{bd}V$ contains the unique point x. □

Let $X \subset \mathbb{R}^n$. The complement $\mathbb{R}^n \setminus X$ of the set X will be denoted by $\mathcal{C}X$.

A closed set $Z \subset \mathbb{R}^n$ is called regular if this set is proper and $Z = \text{clint}Z$. It easily follows from Proposition 3.2 that a set $B \in \mathcal{B}_s$ is regular. It is easy to show that a set $A \in \mathcal{A}_s$ is regular as well. (Recall that $\mathcal{A}_s$ is the totality of all strongly star-shaped sets, $\mathcal{B}_s$ is the totality of all strongly star-shaped sets w.r.t. infinity, $\mathcal{U}_s = \{U \in \mathcal{A}_s : 0 \in \text{kern}_* U\}$, $\mathcal{V}_s = \{V \in \mathcal{B}_s : 0 \in \text{kern}^* V\}$.) If Z is regular then the set

$$\mathcal{C}_r Z = cl(\mathcal{C}Z) \tag{3.4}$$

is also regular. This set is called the regular complement of the set Z. It is well known that $\mathcal{C}_r(\mathcal{C}_r Z) = Z$ for a regular set Z.

We now present the main result of this section.

Theorem 3.1 *1) $V \in \mathcal{V}_s$ if and only if there exists a nonnegative, positively homogeneous, real-valued, continuous function $p \neq 0$ defined on $\mathbb{R}^n$ such that $V = \{x \in \mathbb{R}^n : p(x) \geq 1\}$;*
2) $U \in \mathcal{U}_s$ if and only if $V = \mathcal{C}_r U \in \mathcal{V}_s$; the equality $\mu_U = \nu_V$ holds;
3) $V \in \mathcal{V}_s$ if and only if $U = \mathcal{C}_r V \in \mathcal{U}_s$; the equality $\nu_V = \mu_U$ holds.

Proof: The first statement follows directly from Propositions 3.3 and 3.4. Let us prove the second statement. Let $U \in \mathcal{U}_s$. Then (see Proposition 3.1) the Minkowski gauge μ_U is a nonnegative, positively homogeneous, real-valued, continuous function, $\mu_U \neq 0$. Hence the set $V = \{x \in \mathbb{R}^n : \mu_U(x) \geq 1\}$ belongs to $\mathcal{V}_s$. Since $\mathcal{C}U = \{x : \mu_U(x) > 1\}$ and μ_U is continuous and positively homogenous it follows that $V = \text{cl}(\mathcal{C}U)$. We have also $\mathcal{C}V = \text{int}U$ so

$$\begin{aligned}
\nu_V(x) &= \sup\{\lambda > 0 : \frac{x}{\lambda} \in V\} = \inf\{\lambda > 0 : \frac{x}{\lambda} \notin V\} \\
&= \inf\{\alpha > 0 : x \in \alpha \text{int}U\} \\
&= \inf\{\alpha > 0 : x \in \alpha U\} \\
&= \mu_U(x).
\end{aligned}$$

Thus the second statement is proved. The third statement can be proved in the same manner. □

Lemma 3.1 *1) If $A = a+U$ with $U \in \mathcal{U}_s$ then $\mathcal{C}_r A = a + \mathcal{C}_r U$. 2) If $B = b+V$ where $V \in \mathcal{V}_s$ then $\mathcal{C}_r B = b + \mathcal{C}_r V$.*

Proof: We only prove the first part of the lemma. We have

$$\mathcal{C}_r A = \mathrm{cl}(\mathcal{C}(a+U)) = \mathrm{cl}(a + \mathcal{C}U) = a + \mathrm{cl}(\mathcal{C}U) = a + \mathcal{C}_r U. \qquad \square$$

Corollary 3.2 *1) $A \in \mathcal{A}_s$ if and only if $\mathcal{C}_r A \in \mathcal{B}_s$; 2) $B \in \mathcal{B}_s$ if and only if $\mathcal{C}_r B \in \mathcal{A}_s$.*

We now compare the lower kernel of a strongly star-shaped w.r.t. zero set and the upper kernel of its complement (a strongly star-shaped with respect to ∞ set).

Proposition 3.5 *1) If $B \in \mathcal{B}_s$ then $\mathrm{kern}^* B = \mathrm{kern}_* \mathcal{C}_r B$. 2) If $A \in \mathcal{A}_s$ then $\mathrm{kern}_* A = \mathrm{kern}^* \mathcal{C}_r A$.*

Proof: We only prove the first part of the proposition. Let $b \in \mathrm{kern}_* B$ then $B = b + V$ where $V \in \mathcal{V}_s$. We have $\mathcal{C}_r B = b + \mathcal{C}_r V$. Since $\mathcal{C}_r V \in \mathcal{U}_s$ it follows that $b \in \mathrm{kern}_* \mathcal{C}_r B$. Thus $\mathrm{kern}_* B \subset \mathrm{kern}_* \mathcal{C}_r B$. In the same manner we can prove the reverse inclusion. □

The following result holds (see the paper Shveidel (1997) and the discussion therein).

Theorem 3.2 *If A is a strongly star-shaped set then $\mathrm{kern}_* A$ is convex.*

Corollary 3.3 *If $B \in \mathcal{B}_s$ then the upper kernel $\mathrm{kern}^* B$ is convex.*

Let $X \subset \mathbb{R}^n$. We denote the conic hull $\mathrm{cone} X = \bigcup_{\lambda>0} \lambda X$ of the set X by K'_X. We also denote the cone $K'_X \cup \{0\}$ by K_X.

Proposition 3.6 *Let $V \in \mathcal{V}$. Then $z \in \mathrm{cl} K_V$ if and only if there exist sequences $v_k \in V$ and $\lambda_k \downarrow 0$ such that $\lambda_k v_k \to z$.*

Proof: Let $z \in \mathrm{cl} K_V$. Then $z = \lim \lambda_k v_k$ where $v_k \in V$ and $\lambda_k > 0$ for all $k = 1, \ldots$. Clearly $\|z\| = \lim \lambda_k \|v_k\|$. If the sequence v_k is unbounded then we can assume without loss of generality that $\lambda_k \downarrow 0$. Let v_k be bounded. Assume that $v_k \to v \in V$. Since $V \in \mathcal{V}$ it follows that $v \neq 0$. Clearly $\lambda_k \to \lambda = \|z\| \|v\|^{-1}$ so $z = \lambda v$. Assume without loss of generality that $\lambda > 0$. Then $z = \lambda v = \lim_k \frac{1}{k}(k\lambda v)$. Since $k\lambda > 1$ for large k it follows that $k\lambda v \in V$. Thus the desired result follows. □

It follows from this proposition that for a convex set $V \in \mathcal{V}$ the cone $\mathrm{cl} K_V$ coincides with the recession cone $\mathrm{rc} V$ of the set V.

Definition 3.3 *Let B be a star-shaped with respect to ∞ set. The cone $rcB = \{x = \lim \lambda_i b_i : \lambda_i \downarrow 0, b_i \in B\}$ is called the recession cone of the set B.*

Clearly rc$B$ is a cone, rc$B$ = rc(αB) for $\alpha > 0$ and rc$(x + B)$ = rc$B$ for each x. If $V \in \mathcal{V}$ then (see Proposition 3.6) rc$V$ = clK_V. Thus the following assertion holds:

Proposition 3.7 *If $B \in \mathcal{B}$ and $b \in kern_\infty B$ then $rcB = clK_{B-b}$.*

4 SEPARATION

In this section we shall study the separation by cone.

Definition 4.1 *Let $X \subset \mathbb{R}^n$ and $x \in \mathbb{R}^n$. Let K be a closed solid convex cone.*
1) The set X and the point x are said to be separated by the cone K if $(x + intK) \cap X = \emptyset$.
2) The set X and the point x are said to be cone-separated if there exists a closed convex solid cone K, which separates X and x.

Let $x \neq 0$. Denote the open ray $\{\lambda x : \lambda > 0\}$ by $\tilde{R}_x$.

Proposition 4.1 *Let X be a closed set and $x \notin X$. Let there exist $y \notin clK_X$ such that $(x + \tilde{R}_y) \cap X = \emptyset$. Then X and x can be separated by a cone K such that $y \in intK$.*

Proof: Since $x \notin X$ it follows that $(x + R_y) \cap X$ is empty. Let K_i be a sequence of closed, solid, convex cones such that $\cap_i$int$K_i = R_y$ and int$K_i \supset K_{i+1}$. Assume that the intersection $(x + \text{int}K_i) \cap X$ nonempty for all i. Then there exists a sequence $u_i \in X$ such that $u_i - x \in \text{int}K_i$, $i = 1, 2, \ldots$. Since $\cap_i$int$K_i = R_y$ it follows that there exists a sequence $y_i \to y$ and a sequence of positive numbers λ_i such that $u_i = x + \lambda_i y_i$. If λ_i is bounded, we can assume without loss of generality that u_i tends to a point on the ray $x + R_y$. Since $u_i \in X$, it follows that $\lim_i u_i \in X$, which is a contradiction. If λ_i is unbounded, we have

$$\frac{u_i}{\lambda_i} = \frac{x}{\lambda_i} + y_i \to y.$$

Since $u_i \in X$, it follows that $y \in$ clK_X, which is again a contradiction. Thus there exists i such that $(x + K_i) \cap X = \emptyset$. It follows from the definition of the cone K_i that $y \in$ intK_i. □

The following example shows that there exists a closed set V, a point $x \notin V$ and a point y such that $(x + \tilde{R}_y) \cap V = \emptyset$ and $y \in$ clK_V. Moreover, there exists a strongly star-shaped w.r.t. infinity set V such that for some x, y this property is valid.

Example 4.1 *Let $n = 2, V = \{(x_1, x_2) \in \mathbb{R}^2 : x_2 \geq a - \exp(-|x_1|)\}$ where $a > 1$. It is easy to check that $V \in \mathcal{V}_s$ and $0 \in intkern^*V$. If $x = (0, (a-1)/2)$ and $y = (1, 0)$ then $(x + R_y) \cap V$ is empty and $y \in clK_V$.*

Corollary 4.1 *Let X be a closed set and $x \notin X, x \notin clK_X$. Then X and x can be separated by a cone K such that $x \in intK$. Indeed we have $x + \tilde{R}x = \{\lambda x : \lambda > 1\}$, so if $(x + \tilde{R}x) \cap X \neq \emptyset$ then there exists $\lambda > 1$ such that $\lambda x \in X$. In such a case $x \in clK_X$ and we have a contradiction. Thus $(x + \tilde{R}x) \cap X = \emptyset$ and we can apply Proposition 4.1.*

Proposition 4.2 *Let $V \in \mathcal{V}$, $x \notin V$ and $-x \notin clK_V$. Then V and x can be cone-separated by the cone K such that $-x \in intK$.*

Proof: In order to apply Proposition 4.1 we need to show that $(x+\tilde{R}_{-x}) \cap V = \emptyset$. It is easy to check that $x + \tilde{R}_{-x} = \{\lambda x : \lambda < 1\}$. If $(x + \tilde{R}_{-x}) \cap V \neq \emptyset$ then there exists $\lambda < 1$ such that $\lambda x \in V$. Since $V \in \mathcal{V}$ it follows that $0 \notin V$ so either $0 < \lambda < 1$ or $\lambda < 0$. Assume $0 < \lambda < 1$ and let $x' = \lambda x$. Since $x' \in V, x = \lambda^{-1}x'$ and $V \in \mathcal{V}$ it follows that $x \in V$, which is a contradiction. If $\lambda < 0$ then $-x \in |\lambda|^{-1}V \subset \text{cl}K_V$, which is again a contradiction. □

The following assertion demonstrates, in particular, that for $V \in \mathcal{V}_s$ and $x \notin V$ conditions in Proposition 4.1 are not only sufficient but also necessary for the separability by a cone.

Proposition 4.3 *Let $V \in \mathcal{V}_s$ and $x \in \mathbb{R}^n$. If V and x can be cone-separated then there exists $y \notin clK_V$ such that $(x + \tilde{R}_y) \cap V = \emptyset$.*

Proof: Let K be a closed, solid, convex cone such that $(x+\text{int}K) \cap V$ is empty. Let $y \in \text{int}K$. Then $x + \tilde{R}_y \subset \text{cl}(\mathcal{C}V) = \mathcal{C}_r V$. Theorem 3.1 demonstrates that $\mathcal{C}_r V$ is a strongly star-shaped w.r.t. zero set so $\tilde{R}_y \subset \mathcal{C}_r V$. Thus $\tilde{R}_y \cap \text{int}V = \emptyset$. Since $V \in \mathcal{V}_s$ it follows that $\tilde{R}_y \cap V$ is empty. We have proved that $\text{int}K$ does not intersect V. Since $\text{int}K$ is an open cone we have $\text{int}K \cap \text{cl}K_V = \emptyset$. Hence, if $y \in \text{int}K$ then $y \notin \text{cl}K_V$ and $(x + \tilde{R}_y) \cap V = \emptyset$. □

Let $V \in \mathcal{V}$ and $x \notin V$. Corollary 4.1 and Proposition 4.2 show that V and x can be cone-separated if either x or $-x$ do not belong to $\text{cl}K_V$. Assume that $V = \mathcal{C}U$ where U is an open, bounded, convex set with $0 \in U$. Then $K_V = \mathbb{R}^n$ so $x, -x \in \text{cl}K_V$ for all $x \in \mathbb{R}^n$. Clearly V and each point $x \notin V$ can not be cone-separated. We now give an example of the set V such that $\text{cl}K_V$ is a proper cone and a point $x \notin V$ exists such that $x \in \text{cl}K_v, -x \in \text{cl}K_V$ and V and x cannot be separated.

Example 4.2 *Let φ be a positive, decreasing, convex function defined on $\mathbb{R}_+ = \{x \in \mathbb{R} : x \geq 0\}$ and such that $\varphi(0) = 2$, $\varphi(2) < \frac{1}{3}$, $\varphi(x) \to 0$ as $x \to +\infty$. Consider the following sets:*

$$Y' = \{(x, y) \in \mathbb{R}^2 : x \geq 0, 0 \leq y \leq \varphi(x)\},$$

$$Y'' = \{(x, y) \in \mathbb{R}^2 : x \geq 0, -\frac{1}{3}x - \varphi(x) \leq y \leq 0\}.$$

Let $C_2 = \{(x,y) \in \mathbb{R}^2 : x^2 + y^2 \leq 4\}$. It is easy to show that the set $U = Y' \cup Y'' \cup C_2$ is strongly star-shaped w.r.t. zero so $V = C_r U$ belongs to $\mathcal{V}_s$. It is easy to check that $clK_V = CK$ where

$$K = \{(x,y) \in \mathbb{R}^2 : x > 0, -\frac{1}{3}x < y < 0\}.$$

Consider the point $x = (0,1)$. We have $x \notin V, x \in clK_V, -x \in clK_V$. It is easily seen that $(x + \tilde{R}_y) \cap V \neq \emptyset$ for each $y \in K$. It follows from Proposition 4.3 that x and V cannot be separated.

Let K be a convex cone. Denote the conjugate cone to K by K^*. By definition, $K^* = \{l \in \mathbb{R}^n : [l,x] \geq 0\}$ for all $x \in K$.

Proposition 4.4 *Let $V \in \mathcal{V}$ be a set such that clK_V is a proper, convex cone. Then V and each point $x \notin V$ can be cone-separated.*

Proof: Following from Proposition 4.2 and Corollary 4.1, we can consider only elements x such that $x \in \text{cl}K_V, -x \in \text{cl}K_V$. First assume that $x \neq 0$. Since $\text{cl}K_V$ is a proper cone, it easily follows that $x \in \text{bdcl}K_V$, so there exists $l \in (clK_V)^*, l \neq 0$ such that $[l,x] = 0$. We have $-l \notin \text{cl}K_V$ and $[x - \lambda l, l] < 0$ for all $\lambda > 0$ that is $(x + \tilde{R}_{-l}) \cap V = \emptyset$. Applying Proposition 4.1, we can conclude that V and x can be cone-separated. Let $x = 0$. Consider a nonzero vector $l \in (\text{cl}K_V)^*$. Since $-l \notin \text{cl}K_V$ we have $(0 + \tilde{R}_{-l}) \cap V = \emptyset$ so 0 and V can be cone-separated by Proposition 4.1. □

We now study the cone-separation of the set $V \in \mathcal{V}$ and a point $x \in V$. It is clear that it is impossible to separate by cone the set and an interior point of this set. So we assume that x is a boundary point of V.

Let $x \in X$ where $X \subset \mathbb{R}^n$. An element $u \in \mathbb{R}^n$ is called a feasible direction at the point x with respect to the set X if there exist sequences $\lambda_k > 0$, $\lambda_k \to 0$ and $u_k \to u$ such that $x + \lambda_k u_k \in X$ for all $k = 1, 2, \ldots$

Proposition 4.5 *Let $V \in \mathcal{V}$. Let $x \in V$ be a vector such that $\nu_V(x) = 1$ and $-x \notin clK_V$. Then the following assertions are equivalent:*
1) V and x can be cone-separated by the cone K such that $-x \in intK$;
2) $-x$ is not a feasible direction at the point x.

Proof: 1)$\Longrightarrow$ 2) Let K be a cone such that

$$(x + \text{int}K) \cap V = \emptyset \tag{4.1}$$

and $-x \in \text{int}K$. If $-x$ is a feasible direction at the point x w.r.t. the set V then there exist sequences $u_k \to -x$ and $\lambda_k \to 0$, $\lambda_k > 0$ such that $x + \lambda_k u_k \in V$. It follows from (4.1) that $x + \lambda_k u_k \notin (x + \text{int}K)$, hence $u_k \notin \text{int}K$. Thus $\lim u_k = -x \notin \text{int}K$ and we have a contradiction.

2)$\Longrightarrow$ 1). Assume that $-x$ is not a feasible direction at the point x. Let K_i be a sequence of closed, solid, convex cones such that $\text{int}K_i \supset K_{i+1} \setminus \{0\}$

and $\cap_i \text{int} K_i = \tilde{R} - x$. Assume that $(x + \text{int} K_i) \cap V \neq \emptyset$ for all i. Then there exists a sequence $y_i \in V$ such that $y_i - x \in \text{int} K_i, i = 1, 2, \ldots$. Since $\cap_i \text{int} K_i = R_{-x}$ it follows that there exist sequences $z_i \to -x$ and $\lambda_i > 0$ such that $y_i = x + \lambda_i z_i$. If λ_i is bounded, we can assume without loss of generality that there exists $\lambda \geq 0$ such that $y_i \to x + \lambda(-x) = (1 - \lambda)x$. Since $-x$ is not a feasible direction at x it follows that $\lambda > 0$. Assume that $0 < \lambda < 1$. Then $y = \lim y_i = (1 - \lambda)x \in V$. We have $\nu_V(y) = (1 - \lambda)\nu_V(x) < 1$ which is impossible since $V = \{x : \nu_V(x) \geq 1\}$. Let $\lambda \geq 1$. Since $-(\lambda - 1)x \in V$ it follows that $-x \in \text{cl} K_V$, which is again a contradiction. Thus λ_i is unbounded. Assume without loss of generality that $\lambda_i \to +\infty$. Then

$$\frac{y_i}{\lambda_i} = \frac{x}{\lambda_i} + z_i \to -x.$$

Since $y_i \in V$ it follows that $-x \in \text{cl} K_V$. We have got a contradiction, which shows that there exists i such that the intersection $(x + K_i) \cap V$ is empty. □

Corollary 4.2 *Let $V \in \mathcal{V}_s$. Then the condition $\nu_V(x) = 1$ can be expressed in the form $x \in bdV$. This follows from Proposition 3.3.*

Proposition 4.6 *Let $V \in \mathcal{V}_s$ and $p = \nu_V$ be the ∞-gauge of the set V. Let $x \in bdV$. Then $-x$ is a feasible direction for the set V at the point x if and only if the upper Hadamard directional derivative $p'_H(x, -x)$ at the point x in the direction $-x$ is non-negative.*

The proof is similar to that of Lemma 3.1 in Rubinov and Vladimirov (1998) and we omit it.

Corollary 4.3 *Let $V \in \mathcal{V}_s$ and ν_V be a locally Lipschitz function. Then $-x$ is not a feasible direction for all $x \in bdV$. Indeed since $p := \nu_V$ is positively homogeneous and locally Lipschitz, it follows that $p'_H(x, -x) = -1$ (Compare with the proof of Remark 3.2 in Rubinov and Vladimirov (1998)).*

Theorem 4.1 *Let $V \in \mathcal{V}_s$ and ∞-gauge ν_V of V be locally Lipschitz. Then the set V and each boundary point x of this set can be cone-separated by a cone K such that $x \in intK$.*

Proof: This follows from Proposition 4.5 and Corollary 4.3. □

We now consider another set of conditions ensuring cone-separability of a boundary point.

Theorem 4.2 *Let $V \in \mathcal{V}$ and $0 \in intkern_{\infty} V$, let $x \in bdV$ and assume that there exists $y \notin clK_V$ such that $(x + \tilde{R}_y) \cap V = \emptyset$. Then V and x can be cone-separated.*

Proof: Since $y \notin \text{cl} K_V$, it follows that there exists a closed, solid, convex cone $\tilde{K}$ such that $y \in \text{int}\tilde{K}$ and the intersection $\tilde{K} \cap V$ is empty. Assume

$(x + \tilde{R}_y) \cap \mathrm{int}\tilde{K} = \emptyset$. Applying the separation theorem, we can find a vector $l \neq 0$ such that

$$\sup\{[l, z] : z \in \tilde{K}\} \leq 0 \leq \inf\{[l, x + \lambda y] : \lambda > 0\}. \tag{4.2}$$

It easily follows from (4.2) that $[l, y] = 0$. Since $l \in -\tilde{K}^*$ and $y \in \mathrm{int}\tilde{K}$ we can conclude that $l = 0$. Thus we get a contradiction which shows that there exists $\lambda' > 0$ such that $x + \lambda' y \in \mathrm{int}\tilde{K}$. On the other hand the inclusion $y \in \mathrm{int}\tilde{K}$ implies $x + \lambda' y \in x + \mathrm{int}\tilde{K}$. Clearly there exists positive $\alpha < 1$ such that

$$z := \alpha(x + \lambda' y) \in (\mathrm{int}\tilde{K}) \cap (x + \mathrm{int}\tilde{K}). \tag{4.3}$$

Let $\tilde{G}$ be an open convex subset of $\mathrm{kern}_\infty V$ with $0 \in \tilde{G}$. Define

$$D = \bigcup_{\lambda>0, 0<\beta<1} ((1-\beta)(x + \lambda y) + \beta\tilde{G}).$$

It is easy to check that D is an open convex set. We now check that the intersection $D \cap V$ is empty. On the contrary, assume that there exists $v \in V$ such that $v = (1-\beta)(x + \lambda y) + \beta g$ with $0 < \beta < 1, \lambda > 0$ and $g \in \tilde{G}$. Since $g \in \mathrm{kern}_\infty V$ it follows that $g + \gamma(v - g) \in V$ for all $\gamma > 1$. In particular,

$$x + \lambda y = \frac{v - \beta g}{1 - \beta} = g + \frac{1}{1-\beta}(v - g) \in V.$$

We arrive at a contradiction, since $(x + \tilde{R}_y) \cap V = \emptyset$. Hence $D \cap V = \emptyset$. It is clear that the point z defined by (4.3) belongs to D. It follows from (4.3) that

$$z \in \mathrm{int}\tilde{K} \cap \mathrm{int}(x + \tilde{K}) \cap D. \tag{4.4}$$

There exists a positive $r < \|x - z\|$ such that the intersection in (4.4) contains the closed ball $G = z + N_r$, where $N_r = \{x' : \|x'\| \leq r\}$. Consider the solid convex cone

$$K = K_{G-x} = \bigcup_{\lambda \geq 0} \lambda(z - x + N_r) \tag{4.5}$$

It is easily seen that

$$x + K = \bigcup_{\lambda \geq 0} ((1-\lambda)x + \lambda G)$$

We now check that $(x + \mathrm{int}K) \cap V = \emptyset$. Let $x' = (1-\lambda)x + \lambda g \in x + \mathrm{int}K$. Here $\lambda > 0$ and $g \in G$. First assume that $\lambda > 1$. We have $x' = (1-\lambda)x + \lambda g = g + (\lambda - 1)(g - x)$. Since $g \in G \subset x + \tilde{K}$ and $\tilde{K}$ is a cone, it follows that $(\lambda - 1)(g - x) \in \tilde{K}$. Since $g \in \tilde{K}$ we have $x' \in \tilde{K}$, hence $x' \notin V$. Assume now that $0 < \lambda \leq 1$. An easy computation shows that $x' \in D$ in such a case, hence $x' \notin V$ again. □

Proposition 4.7 *Let $V \in \mathcal{B}$ and $\mathrm{intkern}_\infty V \neq \emptyset$, let $x \in \mathrm{bd}V$. Assume that there exists $y \notin \mathrm{rc}V$ such that $(x + \tilde{R}_y) \cap V = \emptyset$. Then V and x can be cone-separated.*

Proof: Let $a \in \text{intkern}_\infty V$. Then $-a + V \in \mathcal{V}$ and

$$-a + x \in \text{bd}(-a + V), \quad ,0 \in \text{intkern}_\infty(V - a), \quad y \notin \text{rc}(-a + V) = \text{cl}K_{-a+V}.$$

It follows from Theorem 4.2 that there exists a closed solid convex cone K which separates $-a + x$ and $-a + V$. Clearly K separates x and V as well. □

Proposition 4.8 *Let $V \in \mathcal{V}$ and $\text{intkern}_\infty V \neq \emptyset$. Let $\nu_V(x) = 1$ be a point such that $-x \notin \text{cl}K_V$. Then V and x can be cone-separated. If $0 \in \text{intkern}_\infty V$ then V and x can be separated by the cone K such that $-x \in \text{int}K$.*

Proof: Let $y = -x$. First we show that $(x + \tilde{R}_y) \cap V = \emptyset$. We have $x + \tilde{R}_y = \{\lambda x : \lambda < 1\}$. Since $V \in \mathcal{V}$ it follows that $\lambda x \notin V$ for $\lambda \in [0, 1)$. Clearly $V \cap \tilde{R}_y = \emptyset$, otherwise $-x \in \text{rc}K_V$. Since $V \in \mathcal{V}$ we can conclude that $-x \in \text{cl}K_V$. Proposition 4.7 shows that V and x can be cone-separated.

Let $0 \in \text{intkern}_\infty V$. Applying proof of Theorem 4.2 (see (4.5) and (4.3)) we can conclude that V and x can be separated by the convex cone K where

$$K = \bigcup_{\lambda > 0} \lambda(z - x + N_r) \quad \text{with} \quad N_r = \{x : \|x\| < r\}$$

and

$$z = \alpha(x + \lambda' y), 0 < \alpha < 1, \lambda' > 0.$$

Since $y = -x$ it follows that $z - x = (\alpha - \alpha\lambda' - 1)x$. Since $\alpha - 1 < 0$ and $-\alpha\lambda' < 0$ and $0 \in N_r$ it follows that $-x \in \text{int}K$. □

Corollary 4.4 *Let $V \in \mathcal{V}$ and $0 \in \text{intkern}_\infty V$. Let $\nu_V(x) = 1$ be a point such that $-x \notin \text{cl}K_V$. Then $-x$ is not a feasible direction at the point x. This follows from Proposition 4.5 and Proposition 4.8*

Theorem 4.3 *Let $V \in \mathcal{V}$ with $\text{intkern}_\infty V \neq \emptyset$. Let the cone $\text{cl}K_V$ be proper and convex. Then each point x with $\nu_V(x) = 1$ can be cone - separated from V.*

Proof: If $-x \notin \text{cl}K_V$ then we can apply Proposition 4.8. Assume $-x \in \text{cl}K_V$. Let $g \in (\text{cl}K_V)^*, g \neq 0$. Since both $x, -x \in \text{cl}K_V$, it follows that $[x, g] = 0$. Thus we have $[x - \lambda g, g] = -\lambda\|g\|^2 < 0$ for all $\lambda > 0$. On the other hand $[g, v] \geq 0$ for all $v \in V$ so $(x + R_{-g}) \cap V = \emptyset$. Since $-g \notin \text{cl}K_V = \text{rc}K_V$, we can apply Proposition 4.7. The proof is complete. □

Consider a collection of m elements $l = (l_1, \dots, l_m)$ and let

$$T_l = \{x : [l_i, x] < 0 (i = 1, \dots, m)\}, \qquad T^l = \{x : [l_i, x] > 0 (i = 1, \dots, m)\}. \tag{4.6}$$

Clearly both T_l and T^l are open convex cones. If $m \leq n$ and elements l_i are linearly independent then both T_l and T^l are not empty.

Proposition 4.9 *Let K be a solid cone and $x \in intK$. Then there exists a collection $l = (l_1, \dots, l_n)$ of n linearly independent elements such that $T_l \subset intK$ and $[l_i, x] = -1$ for all $i = 1, \dots, n$.*

Proof: Denote by H the hyperplane $\{y : [x, y] = 0\}$. Since $x \in \text{int}K$, it follows that there exists a simplex $S \subset H$ such that zero is an interior point of S (with respect to H) and $x + S \subset \text{int}K$. Let $y_1, \dots, y_n$ be vertices of S, then points $x + y_1, \dots, x + y_n$ are vertices of $x + S$. It is easy to check that elements $x, x + y_1, \dots, x + y_{k-1}, x + y_{k+1}, \dots, x + y_n$ are linearly independent for each $k = 1, \dots, n$. Then the system of linear equations

$$[l, x] = -1, [l, x + y_i] = 0 (i \neq k)$$

has the unique solution l_k. It is easy to check that the collection $l = (l_1, \dots, l_n)$ consists of linearly independent elements. Since $[l_k, y_i] = 1$ with $i \neq k$ it follows that the simplex S can be represented in the following form: $S = \{y : [l_k, y] \leq 1, k = 1, \dots, n, [x, y] = 0\}$. Since also $[l_k, x] = -1$ we have

$$\begin{aligned} x + S &= \{z : z = y + x; [l_k, y] \leq 1, k = 1, \dots, n; [x, y] = 0\} \\ &= \{z : [l_k, z] \leq 0, k = 1, \dots, n; [x, z] = \|x\|^2\} \end{aligned}$$

and $K_{x+S} = \{z : [l_k, z] \leq 0\}$. Since $x + S \subset \text{int}K$ it follows that $T_l = \{z : [l_1, z] < 0, \dots, [l_n, z] < 0\} \subset \text{int}K$. □

Remark 4.1 Let l be the collection as in Proposition 4.9 and $l' = \{l'_1, \dots, l'_n\}$ with $l'_i = -l_i$ for all i. Then $[l', x] = 1$ and $T^{l'} \subset \text{int}K$.

Denote by L_m the totality of all collections $l = (l_i)_{i=1}^m$ of m linearly independent elements. If $l \in L_m$ then the cone T^l is nonempty. We shall use the following notation in the sequel: if $l = (l_1, \dots, l_m) \in (\mathbb{R}^n)^m$ and $x \in \mathbb{R}^n$ then

$$\langle l, x \rangle = \min_{i=1,\dots,m} [l_i, x]; \qquad \langle l, x \rangle^+ = \max_{1=1,\dots,m} [l_i, x]. \tag{4.7}$$

Proposition 4.10 *Let V be a closed set. Then the following assertions are equivalent:*
1) $V \in \mathcal{V}$ and $C(V) \cap (-clK_V) = \emptyset$;
2) For each nonzero $x \notin V$ there exists a collection $l = (l_1, \dots, l_n) \in L_n$ such that

$$0 < [l_1, x] = \dots = [l_n, x] < 1 \leq inf_{v \in V} \langle l, v \rangle^+ \tag{4.8}$$

Proof: 1)$\Longrightarrow$ 2) Let $x \notin V$. There exists $\varepsilon > 0$ such that $(1 + \varepsilon)x \notin V$ as well. Clearly $-(1 + \varepsilon)x \notin \text{cl}K_V$. In light of Proposition 4.2, there exists a closed solid cone K such that $((1 + \varepsilon)x + \text{int}K) \cap V = \emptyset$ and $-x \in \text{int}K$. It follows from Proposition 4.9 that there exists a collection $\tilde{l} = (\tilde{l}_i)_{i=1}^n$ such that $[\tilde{l}_i, -x] = -1$ for all i and $T_{\tilde{l}} \subset \text{int}K$. It is easy to check that

$$(1+\varepsilon)x + T_{\tilde{l}} = \{y : [\tilde{l}_i, y] < 1+\varepsilon, (i = 1, \dots, n)\} = \{y : \langle \tilde{l}, y\rangle^+ < 1+\varepsilon\}.$$

Since $(1+\varepsilon)x + T_l \subset (1+\varepsilon)x + \text{int}K$, it follows that $\langle \tilde{l}, y\rangle^+ \geq 1+\varepsilon$ for all $v \in V$. Let $l = (1/(1+\varepsilon))\tilde{l}$. Then $0 < l_1(x) = \dots = l_n(x) < 1$ and $\inf_{v\in V}\langle l, v\rangle^+ \geq 1$.
2)$\Longrightarrow$ 1) First we check that

$$(v \in V, \lambda > 1) \Longrightarrow \lambda v \in V. \tag{4.9}$$

Indeed, if $\lambda v \notin V$ then we can find a collection $l = (l_i)_{i=1}^n$ such that for all i:

$$\lambda\langle l, v\rangle^+ = \lambda l_i(v) = l_i(\lambda v) < 1 \leq \inf_{v'\in V}\langle l, v'\rangle^+ \leq \langle l, v\rangle^+.$$

We have a contradiction which shows that the implication (4.9) is valid. If $0 \in V$ then $\inf_{v\in V}\langle l, v\rangle^+ \leq 0$ for each collection l which is impossible. Thus V is a star-shaped set w.r.t infinity and $0 \in \text{kern}_\infty V$.

We now check that $C(V) \cap (-\text{cl}K_V) = \emptyset$. Indeed let $x \notin V$. Then there exists a collection $l = (l_i)_{i=1}^n$ such that $\langle l, v\rangle^+ \geq 1$ for each $v \in V$ and $[l_i, x] = \lambda > 0$ for all i. We have $\langle l, y\rangle^+ \geq 0$ for each $y \in \text{cl}K_V$. On the other hand $\langle l, x\rangle^+ = \max_i [l_i, -x] = -\lambda < 0$. Thus $-x \notin \text{cl}K_V$. □

Theorem 4.4 *Let $V \in \mathcal{V}$ and $clK_V \cap (-clK_V) = \{0\}$. Then for each nonzero $x \notin V$ there exists a collection $l = (l_i)_{i=1}^n \in L_n$ such that*

$$[l_1, x] = \dots [l_n, x] \leq \inf_{v\in V}\langle l, v\rangle^+$$

Proof: Let $x \in \text{cl}K_V$. Then $-x \notin \text{cl}K_V$. In such a case the proof is similar to the proof of 1) $\Longrightarrow$ 2) in Proposition 4.10. Assume now that $x \notin \text{cl}K_V$. Then we can conclude, by applying Corollary 4.1, that V and x can be separated by a closed convex solid cone K such that $x \in \text{int}K$. It follows from Proposition 4.9 that there exists a collection l of linearly independent vectors $l_1, \dots, l_n$ such that

$$[l_1, x] = \dots = [l_n, x] = -1 \quad \text{and} \quad T_l := \{x' : [l_i, x'] < 0, i = 1, \dots, n\} \subset \text{int}K.$$

Clearly $x + T_l = \{y : [l_i, y] < -1, i = 1, \dots, n\}$. Since $(x + \text{int}K) \cap V = \emptyset$ it follows that $(x + T_l) \cap V = \emptyset$, hence for each $v \in V$ there exists i such that $[l_i, v] \geq -1$. Thus $\langle l, v\rangle^+ \geq -1$ for each $v \in V$. □

Proposition 4.11 *Let a closed set X and a point x be separated by $l = (l_i)_{i=1}^n \in L_n$, that is $c := \langle l, x\rangle^+ \leq \inf_{y\in X}\langle l, y\rangle^+$. Then X and x can be cone-separated.*

Proof: Consider the polyhedron $P = \{y : \langle l, y\rangle^+ \leq c\} = \{y : [l_i, y] \leq c, i = 1, \dots, n\}$. Clearly $x \in P$. It is easy to check that the recession cone rc$P$ of this polyhedron has a form rc$P = \{y : [l_i, y] \leq 0, i = 1, \dots, n\}$. Since intrc$P = T_l \neq \emptyset$ it follows that rc$P$ is a closed solid convex cone. We have

$$x + T_l \subset \{y : [l_i, y] < c, i = 1, \dots, n\} = \{y : \langle l, y\rangle^+ < c\},$$

hence $(x + T_l) \cap X = \emptyset$. □

5 SUM OF STRONGLY STAR-SHAPED SETS W.R.T. INFINITY

It it easy to check that the sum $A_1 + A_2$ of star-shaped sets A_1 and A_2 is again a star-shaped set if and only if this sum is closed; we have also $\text{kern}(A_1 + A_2) \supset \text{kern}A_1 + \text{kern}A_2$. Similarly the sum $B_1 + B_2$ of star-shaped sets w.r.t. infinity is star-shaped w.r.t. infinity if and only if this sum is closed; the inclusion $\text{kern}_\infty(B_1 + B_2) \supset \text{kern}_\infty B_1 + \text{kern}_\infty B_2$ holds.

The following simple assertion demonstrates in particular that the sum of strongly star-shaped sets is also strongly star-shaped if this sum is closed.

Proposition 5.1 *Let $A, X \in \mathcal{A}$. Then $A + X \in \mathcal{A}$ if and only if $A + X$ is closed. If $A \in \mathcal{A}_s, X \in \mathcal{A}$ and $A + X$ is closed, then $A + X \in \mathcal{A}_s$.*

Proof: We prove only the second part of Proposition. Consider the sets $U = A - a$ and $Y = X - x$ where $a \in \text{kern}_* A$ and $x \in X$. Clearly $0 \in \text{int}(U+Y)$. Assume there exists a point $z \neq 0$ such that the intersection $R_z \cap \text{bd}(U + Y)$ contains at least two points: z and λz with $0 < \lambda < 1$. Since $z \in \text{bd}(U + Y)$ it follows that there exist $u \in \text{bd}U$ and $y \in \text{bd}Y$ such that $z = u + y$. Since $U \in \mathcal{U}_s$ it follows that $\lambda u \in \text{int}U$, since $Y \in \mathcal{U}$ it follows that $\lambda y \in Y$. So $\lambda z = \lambda u + \lambda y \in \text{int}(U+Y)$. Thus we have a contradiction and $U+Y$ is strongly star-shaped with respect to zero. Clearly $A + X$ is strongly star-shaped with respect to $a + x$. □

We will now study in detail the sum of strongly star-shaped sets w.r.t infinity.

Proposition 5.2 *Let Z_1 and Z_2 be regular sets and $Z_1 \cap Z_2 \neq \emptyset$. Then either the intersection of these sets is equal to the intersection of their boundaries or $\text{int}Z_1 \cap \text{int}Z_2$ is not empty.*

Proof: It follows directly from the definition of regular sets. □

Lemma 5.1 *Let $V_1, V_2 \in \mathcal{V}_s$ and b is a vector such that the intersection of sets $b + V_1$ and V_2 is either empty or equal to the intersection of their boundaries. Then $(\lambda b + V_1) \cap V_2 = \emptyset$ for each $\lambda \in (0, 1)$.*

Proof: Assume that there exists $\lambda < 1$ and $v_1 \in V_1$ such that $c = \lambda b + v_1 \in V_2$. Since $\lambda^{-1} > 1$ it follows that $\lambda^{-1} v_1 \in \text{int}(b + V_1)$ so $\lambda^{-1} c = b + \lambda^{-1} v \in \text{int}(b + V_1)$. On the other hand $\lambda^{-1} c \in \text{int}V_2$. Thus the intersection $\text{int}(b + V_1) \cap \text{int}V_2$ is not empty. □

Proposition 5.3 *Let $V_1, V_2 \in \mathcal{V}_s, B_1 = b_1 + V_1, B_2 = b_2 + V_2$ and $B_1 \cap B_2 \neq \emptyset$. Then the following are equivalent:*

1) $B_1 \cap B_2 = (\text{bd}B_1) \cap (\text{bd}B_2)$; 2) $b_1 - b_2 \in \text{bd}(V_2 - V_1)$.

Proof: 1) $\Longrightarrow$ 2). Since $B_1 \cap B_2 \neq \emptyset$ it easily follows that $b_1 - b_2 \in V_2 - V_1$. We have also that $V_2 \cap ((b_1 - b_2) + V_1) = \text{bd}V_2 \cap \text{bd}((b_1 - b_2) + V_1)$. It follows from Lemma 5.1 that $(\lambda(b_1 - b_2) + V_1) \cap V_2 = \emptyset$ for all $\lambda \in (0, 1)$ therefore $\lambda(b_1 - b_2) \notin V_2 - V_1$ for all such λ.

2) $\Longrightarrow$ 1) If $B_1 \cap B_2 \neq \mathrm{bd}B_1 \cap \mathrm{bd}B_2$ then (see Proposition 5.2) there exists $c \in \mathrm{int}B_1 \cap \mathrm{int}B_2$. We have $c - b_1 \in \mathrm{int}V_1, c - b_2 \in \mathrm{int}V_2$ so $b_1 - b_2 \in \mathrm{int}V_2 - \mathrm{int}V_1 \subset \mathrm{int}(V_2 - V_1)$ which contradicts 2). □

Proposition 5.4 *Let $V_1, V_2 \in \mathcal{V}$. Then $V_1 + V_2 \in \mathcal{V}$ if and only if $V_1 + V_2$ is closed and $0 \notin V_1 + V_2$. If $V_1 \in \mathcal{V}_s, V_2 \in \mathcal{V}, 0 \notin V_1 + V_2$ and $V_1 + V_2$ is closed, then $V_1 + V_2 \in \mathcal{V}_s$.*

Proof: We prove only the second part of Proposition. Let $V_1 + V_2$ be closed and $0 \notin V_1 + V_2$. Let $R = R_x$ be a ray such that $R \cap (V_1 + V_2) \neq \emptyset$. It easily follows from properties of $V_1 + V_2$ that there exists λ such that $\lambda x \in \mathrm{bd}(V_1 + V_2)$. If $\lambda' > \lambda$ then we can easily conclude, by applying Proposition 3.2, that $\lambda' x \in \mathrm{int}V_1 + \mathrm{int}V_2 \subset \mathrm{int}(V_1 + V_2)$. Thus $V_1 + V_2 \in \mathcal{V}_s$. □

The further study of the sum of two strongly star-shaped sets w.r.t. infinity is based on the properties of the conic hull of these sets.

Proposition 5.5 *Let $V_1 \in \mathcal{V}_s, V_2 \in \mathcal{V}$. The following assertions are eqivalent:*

1) $0 \in V_1 + V_2$ 2) $V_1 + V_2 = \mathbb{R}^n$ 3) $K_{V_1} \cap (-K_{V_2}) \neq \{0\}$.

Proof: 1) $\Longrightarrow$ 2). Let $0 \in V_1 + V_2$. Then there exists an element v such that $v \in V_1, -v \in (-V_2)$. Then $2v \in \mathrm{int}V_1, -2v \in \mathrm{int}V_2$ so $0 \in \mathrm{int}(V_1 + V_2)$. Thus for each $z \in \mathbb{R}^n$ there exists $\varepsilon \in (0,1)$ such that $\varepsilon z \in V_1 + V_2$. Hence $z = \frac{1}{\varepsilon}\varepsilon z \in V_1 + V_2$.

2) $\Longrightarrow$ 3) Since $0 \notin V_1, 0 \notin V_2$ it follows from 2) that there exists $v \neq 0$ such that $v \in V_1, -v \in V_2$. Clearly $v \in K_{V_1} \cap (-K_{V_2})$.

3) $\Longrightarrow$ 1) Let $v \neq 0, v \in K_{V_1}, -v \in K_{V_2}$ There exist $\lambda_i > 0$ such that $v \in \lambda_1 V_1$, $-v \in \lambda_2 V_2$. If $\lambda_1 \geq \lambda_2$ then $v \in \lambda_2 V_1$ and $v \subset \lambda_2 V_2$, hence $0 \in V_1 + V_2$. □

Remark 5.1 If $V_1, V_2 \in \mathcal{V}$, then only 1) and 3) are equivalent.

Proposition 5.6 *Let $V_1, V_2 \in \mathcal{V}$ and*

$$clK_{V_1} \cap (-clK_{V_2}) = \{0\}. \tag{5.1}$$

Then $V_1 + V_2 \in \mathcal{V}$. If at least one of the sets V_1, V_2 belongs to $\mathcal{V}_s$, then $V_1 + V_2 \in \mathcal{V}_s$.

Proof: Due to Proposition 5.4 we have only to check the following: 1) $0 \notin V_1 + V_2$ and 2) the set $V_1 + V_2$ is closed. It follows from (5.1) that $K_{V_1} \cap (-K_{V_2}) = \{0\}$, so (see Proposition 5.5) $0 \notin V_1 + V_2$. Let $v_k = v_k^1 + v_k^2 \to v$ with $v_k^i \in V_i, i = 1,2$. It easily follows from (5.1) that sequences $(v_k^i)i = 1,2$ are bounded, so we can assume without loss of generality that there exists $\lim_k v_k^i \in V_i, i = 1,2$. Thus the set $V_1 + V_2$ is closed. □

Remark 5.2 The equality (5.1) is a sufficient but not necessary condition for $V_1 + V_2 \in \mathcal{V}_s$. The following example confirms this statement. Consider subsets V_1 and V_2 of the space $\mathbb{R}^2$: $V_1 = (1,1) + K_1, V_2 = (1,-1) + K_2$ where $K_1 = \{(x,y) : x \geq 0\}, K_2 = \{(x,y) : x \geq 0, y \leq 0\}$. Clearly $V_1 + V_2 = (2,0) + \{(x,y) : x \geq 0\} \in \mathcal{V}_s$, but $\text{cl}K_{V_1} \cap \text{cl}K_{V_2} \neq \{0\}$.

Proposition 5.7 *If $B_1 \in \mathcal{B}_s$, $B_2 \in \mathcal{B}$ and $rcB_1 \cap rc(-B_2) = \{0\}$ then $B_1 + B_2 \in \mathcal{B}_s$ and $rc(B_1 + B_2) = rcB_1 + rcB_2$.*

Proof: Let $b_i \in \text{kern}^* B_i, V_i = B_i - b_i, i = 1, 2$. Since $\text{rc}V_i = \text{rc}B_i$ it follows that

$$\text{rc}V_1 \cap (-\text{rc}V_2) = \{0\}. \tag{5.2}$$

Applying Proposition 5.6 we can conclude that $V_1 + V_2 \in \mathcal{V}_s$ so $B_1 + B_2 \in \mathcal{B}_s$. It easily follows from (5.2) that $\text{rc}(V_1 + V_2) = \text{rc}V_1 + \text{rc}V_2$. Thus

$$\text{rc}(B_1 + B_2) = \text{rc}(V_1 + V_2) = \text{rc}V_1 + \text{rc}V_2 = \text{rc}B_1 + \text{rc}B_2.$$ □

6 SEPARATION OF STRONGLY STAR-SHAPED SETS W.R.T. INFINITY

In this section we shall study the separation by a collection of linear functions. We begin with a definition.

Definition 6.1 *Star-shaped sets w.r.t.∞ B_1 and B_2 are said to be* separated *by a collection $l = (l_1, \dots, l_m) \in L_m$ if for each $b_1 \in B_1, b_2 \in B_2$ there exists i such that $[l_i, b_1] \leq [l_i, b_2]$.*

Collection l in this definition is called separating for sets B_1 and B_2.

Applying results from previous sections we can obtained different sufficient conditions for separation of two star-shaped w.r.t. ∞ sets. We now present some of these conditions.

The following assertion is obvious.

Proposition 6.1 *If $B_1, B_2 \in \mathcal{B}$ and $intB_1 \cap intB_2 \neq \emptyset$ then these sets can not be separated.*

Proposition 6.2 *Let $B_1, B_2 \in \mathcal{B}, B_1 \cap B_2 = \emptyset$ and*

$$(rcB_1) \cap (rcB_2) = \{0\}, \qquad (rcB_1 - rcB_2) \cap (rcB_2 - rcB_1) = \{0\} \tag{6.1}$$

Then B_1 and B_2 are separated by a collection $l \in L_n$.

Proof: Let $b_i \in \text{kern}_\infty B_i$ and $V_i = B_i - b_i, i = 1, 2$. Then $V_i \in \mathcal{V}_s$ and $B_2 - B_1 = b_2 - b_1 + (V_2 - V_1)$. It follows from Propositions 3.7 and 5.6 that $V := V_2 - V_1 \in \mathcal{V}$ so V is closed and $0 \notin V$. By applying Propositions 3.7 and 5.7, we have also,

$$\text{cl}K_V = \text{rc}(B_2 - B_1) = \text{rc}B_2 - \text{rc}B_1$$

Since B_1 and B_2 are disjoint sets it follows that $b_1 - b_2 \notin V$. Assume that $b_2 \neq b_1$. Applying Theorem 4.4 we can conclude that there exists a collection

$l = (l_i)_{i=1}^n \in L_n$ such that $[l_1, b_1 - b_2] = \ldots = [l_n, b_1 - b_2] \leq \inf_{v \in V} \langle l, v \rangle^+$. It is easy to check that the B_1 and B_2 are separated by the collection l. If $b_2 = b_1$ then we can obtain the required collection l by applying Theorem 4.4 for arbitrary $x \notin V$, $x \neq 0$. □

Remark 6.1 Conditions in Proposition 6.2 are satisfied for example if there exists a convex closed cone K such that $K \cap (-K) = \{0\}$ and $\mathrm{rc}B_1 = \mathrm{rc}(-B_2) = K$

Proposition 6.3 *Let $B_1, B_2 \in \mathcal{B}_s$, $B_1 \cap B_2 = (bdB_1) \cap (bdB_2)$ and at least one of the sets intkern$^*B_i (i = 1, 2)$ is nonempty. Let $rcB_1 \cap rcB_2 = \{0\}$ and $rcB_2 - rcB_1$ be a proper convex cone. Then B_1 and B_2 are separated.*

Proof: Assume that $\mathrm{intkern}^* B_2 \neq \emptyset$. Let $b_1 \in \mathrm{kern}^* B_1, b_2 \in \mathrm{intkern}^* B_2$ and $V_i = B_i - b_i$. It follows from Propositions 3.7 and 5.6 that $V := V_2 - V_1 \in \mathcal{V}_s$. In light of Proposition 5.3, we have $b_1 - b_2 \in \mathrm{bd}V$. Clearly $0 \in (\mathrm{kern}^* V_1) \cap (\mathrm{intkern}^* V_2)$, hence $0 \in \mathrm{intkern}^* V_2 - \mathrm{kern}^* V_1$. Since $\mathrm{kern}^* V_2 - \mathrm{kern}^* V_1 \subset \mathrm{kern}^*(V_2 - V_1)$ we can conclude that $0 \in \mathrm{intkern}^* V$. It follows from Propositions 5.7 and 3.7 that $\mathrm{cl}K_V = \mathrm{rc}B_2 - \mathrm{rc}B_1$ is a proper convex cone. It follows from Theorem 4.3 that the point $b_1 - b_2$ can be cone-separated from the set $V_2 - V_1$. Due to Proposition 4.9 we can find a collection $l = (l_i)_{i=1}^n \in L_n$ such that for each $v = v_2 - v_1 \in V$ there exists i for which $[l_i, b_1 - b_2] \leq [l_i, v_2 - v_1]$. Clearly l is a separating collection for sets B_1 and B_2. □

References

Barbara, A. and Crouzeix, J.-P. (1994), Concave gauge functions and applications, *Zeitschrift fur Operations Research*, Vol. 40, pp. 43 -74.

Fan, K. (1963), On the Krein-Milman theorem, in *Convexity, Proceed. of Sympos. in Pure Math.* Vol. 7, American Math. Soc., Providence, pp. 211-220.

Penot, J.- P. (1997), Duality for radiant and shady programs, *Acta Mathematica Vietnamica* Vol. 22, pp. 541-566.

Rockafellar, R. T. (1970), *Convex Analysis*, Princeton University Press, Princeton, NJ.

Rubinov, A. M. and Glover, B. M. (1995), Characterization of optimality for homogeneous programming problems with applications, in *Recent Advances in Nonsmooth Optimization*, edited by Du, D.-Z., Qi, L. and Womersley, R.S., World Scientific Publishing.

Rubinov, A. M. and Vladimirov, A. A. (1998), Convex-along-rays functions and star-shaped sets, *Numerical Functional Anlysis and Optimization*, Vol. 19, pp. 593-614.

Rubinov, A. M. and Yagubov, A. A. (1986), The space of sar-shaped sets and its application in nonsmooth optimization, *Mathematical Programming Study*, Vol. 29, pp. 176-202.

Shveidel, A. (1997), Separability of star-shaped sets and its applicatiion to an optimization problem, *Optimization*, Vol. 40, pp. 207-227.

4 NONLINEAR UNCONSTRAINED OPTIMIZATION METHODS: A REVIEW

A. M. Rubinov,
School of Information Technology and Mathematical Sciences,
University of Ballarat, Victoria 3353, Australia

X. Q. Yang
Department of Applied Mathematics,
The Hong Kong Polytechnic University,
Kowloon, Hong Kong

and B. M. Glover
Department of Mathematics,
Curtin University of Technology,
Perth, WA 6845, Australia

Abstract: In this review paper, we report recent results in the study of nonlinear unconstrained optimization methods, such as nonlinear penalty function method and nonlinear Lagrangian method. One important feature of this line of research is that unconstrained optimization problems may be nonlinear in the cost function of the original constraint problem. Results such as zero duality gaps, exact penalization, and the existence of nonlinear Lagrange multipliers are reviewed.

Key words: modified penalty function, modified Lagrangian function, zero duality gap.

X. Yang et al. (eds.), Progress in Optimization, 65–77.

1 INTRODUCTION

Consider the following inequality constrained optimization problem:

$$(P) \quad f_0(x) \to \inf \quad \text{subject to } x \in X,\ f_i(x) \leq 0, i \in I \tag{1.1}$$

where $X \subseteq \mathbb{R}^n$, $I = \{1, \ldots, m\}$, and $f_i : X \to \mathbb{R}, i \in \{0\} \cup I$ are real-valued functions.

The set of all feasible solutions for (P) is defined by

$$X_o = \{x \in X : f_i(x) \leq 0, i \in I\}.$$

Throughout this paper, we assume that X is a connected set and that

$$\inf_{x \in X} f_0(x) > 0.$$

We shall study nonlinear analogues of Lagrange and penalty functions for the problem (1.1). The construction of various types of modified Lagrange functions is based on a convolution of the objective function f_0 and constraints f_i, while the construction of penalty functions based on a convolution of the objective function and positive parts f_i^+ of constraints. (Here $f_i^+(x) = \max(f_i(x), 0)$.) Unfortunately the classical sum convolution does not guarantee the zero duality gap property unless e.g. the problem is convex and a constraint qualification holds. Thus the class of convolution functions should be extended in order to obtain zero duality gap or, more useful, exact Lagrange or penalty functions.

The nonlinear Lagrangian function with maximum structures was initially studied in Goh and Yang (1997) where a zero duality gap property was obtained. This was motivated by noting that there is always a supporting cone to the epigraph of the nonincreasing perturbation function. More general schemes were proposed in Rubinov, et al (1997a) for a unified approach of nonlinear Lagrangian and penalty functions by noting that the dual function introduced in Goh and Yang (1997) is an increasing positively homogenous (IPH) function. So the study has been advanced with the theory of IPH functions established in Rubinov and Glover (1997) and the references therein. It was shown in Rubinov, et al (1997a); Rubinov, et al (1997b) that increasing (isotonic) convolution functions with some additional properties can be used to construct Lagrange and penalty functions with useful properties. We now give corresponding definitions.

Let $\mathbb{R}^s$ be a Euclidean space, $\mathbb{R}^s_+$ be the cone of all vectors from $\mathbb{R}^s$ with nonnegative coordinates and $\mathbb{R}^s_{++}$ be the cone of all vectors from $\mathbb{R}^s$ with positive coordinates. We shall use the usual partial ordering in the space $\mathbb{R}^s$:

$$x^1 \geq x^2 \Longleftrightarrow x^1 - x^2 \in \mathbb{R}^s_+ \text{ for } x^1, x^2 \in \mathbb{R}^s.$$

Let $\mathbb{R}^{1+m} := \mathbb{R} \times \mathbb{R}^m$ and $Y \subseteq \mathbb{R}^{1+m}$. A function $p : Y \to \mathbb{R}_{+\infty} \equiv \mathbb{R} \cup \{+\infty\}$ is called *increasing* if $y^1 \geq y^2$ implies $p(y^1) \geq p(y^2)$.

We shall consider mainly the so-called IPH functions. A function p defined on either the cone $\mathbb{R}^I_{++}$ or the cone $\mathbb{R}^I_+$ and mapping into $\mathbb{R}_{+\infty} := \mathbb{R}_+ \cup \{+\infty\}$ is called an IPH function if p is increasing, positively homogeneous ($p(\lambda x) = \lambda p(x)$ for $\lambda > 0$) and there is a point $y \in \mathbb{R}^I_{++}$ such that $p(y) < +\infty$.

A theory of IPH functions based on abstract convexity was developed in Rubinov and Glover (1997); Rubinov, et al (1997b). We only indicate that the set of all IPH functions is a conditionally complete lattice. More precisely, if $\{p_\alpha\}$ is an arbitrary family of IPH functions then $p_*(x) = \inf_\alpha p_\alpha(x)$ is an IPH function as well; if $p^*(x) = \sup_\alpha p_\alpha(x)$ is not equal identicall to $+\infty$ then p^* is also an IPH function. Note that the infimum of a family of convex functions is not necessarily convex and supremum of a family of concave functions is not necessarily concave.

Moreover, these studies have been extended to reduce the optimal value of (P) to the (least) root of an increasing continuous function in Goh and Yang (1997); Yang and Li (1999). This continuous function is of a single variable and is evaluated by solving an unconstrained optimization problem.

The closure of a set X is denoted by $\overline{X}$.

2 NONLINEAR PENALTY FUNCTIONS

In this section we shall consider continuous increasing functions p that are defined on the space $\mathbb{R}^{1+m}_+$ which enjoy the following properties:

1) There exist positive numbers $a_o, a_1, \dots a_m$ with $a_o = 1$ such that for $\bar{y} = (y_o, y_1, \dots, y_m) \in \mathbb{R}^{1+m}_+$ we have:

$$p(\bar{y}) \geq \max_{0 \leq i \leq m} a_i y_i. \tag{2.1}$$

2) For $y_o \in \mathbb{R}_+$

$$p(y_o, 0, \dots, 0) = y_o. \tag{2.2}$$

The set X_o of feasible solutions for (P) can be represented in the following form:

$$X_o = \{x \in X : f_i^+(x) = 0, i \in I\}.$$

where $f_i^+(x) = \max\{f_i(x), 0\}$. Let

$$F^+(x,d) = (d_o f_0(x), d_1 f_1^+(x), \dots, d_m f_m^+(x)), \qquad x \in X, d \in \mathbb{R}^{1+m},$$

and let p be defined on $\mathbb{R}^{1+m}_+$ with properties (2.1) and (2.2). The nonlinear (isotonic) penalty function is given by

$$\mathcal{L}^+(x,d) = p(F^+(x,d)), \qquad x \in X, d \in \mathbb{R}^{1+m}_+, \tag{2.3}$$

and the associated dual cost function is defined by

$$q^+(d) = \inf_{x \in X} \mathcal{L}^+(x,d). \tag{2.4}$$

The dual nonlinear penalty problem is defined by

$$(D^+) \quad q^+(d) \longrightarrow \sup \text{ subject to } d = (d_o, d_1, \ldots, d_m) \in \mathbb{R}_+^{1+m}, d_o = 1.$$

Denote the optimal value of (P) by

$$M_P = \inf\{f_0(x) : x \in X_o\}.$$

The following convergence for the nonlinear penalty problem is established.

Theorem 2.1 *Rubinov, et al (1997a) Consider the problem (P) such that the functions $f_0, f_1, \ldots, f_m$ are continuous, the set X_o is compact and (if X is unbounded) the function f_0 enjoys the following property:*

$$\lim_{x \to \infty, x \in X} f_0(x) = +\infty. \tag{2.5}$$

Then

$$q^+(d) \to M_P \text{ as } d \to \infty. \tag{2.6}$$

Note that (2.6) is equivalent to

$$M_p = \sup_{d \in \mathbb{R}_+^{1+m}} q^+(d).$$

The following corollaris are obtained from Theorem 2.1 where p is chosen as k-th power and maximum functions, respectively.

Corollary 2.1 *Rubinov, et al (1997a) Let $0 < k < +\infty$. Define the k-th power nonlinear penalty problem as*

$$q_k^+(d) = \inf_{x \in X} \left(\sum_{i=0}^{m} d_i [f_i^+(x)]^k \right)^k, \quad d = (d_o, d_1, \ldots, d_m) \in \mathbb{R}_+^{1+m}, d_o = 1.$$

Assume that the conditions in Theorem 2.1 hold. Then,

$$q_k^+(d) \to M_P \text{ as } d \to \infty.$$

Corollary 2.2 *Rubinov, et al (1997a) Define the maximum penalty problem as*

$$q_\infty^+(d) = \inf_{x \in X} \max_{0 \le i \le m} d_i f_i^+(x), \quad d = (d_o, d_1, \ldots, d_m) \in \mathbb{R}_+^{1+m}, d_o = 1.$$

Assume that the conditions in Theorem 2.1 hold. Then,

$$q_\infty^+(d) \to M_P \text{ as } d \to \infty.$$

3 PENALTY FUNCTIONS FOR PROBLEMS WITH ONE CONSTRAINT

We now assume that there is only one constraint so the problem (1.1) has the following form:

$$(P) \ : \quad f_0(x) \longrightarrow \inf \text{ subject to } x \in X, \ f_1(x) \leq 0, \tag{3.1}$$

where $X \subseteq \mathbb{R}^n$ and $f_i : X \to \mathbb{R}, i = 0, 1$, are real-valued functions.

We shall denote the problem (3.1) by (f_0, f_1). The optimal value of (f_0, f_1) is denoted by $M_P(f_0, f_1)$.

Remark 3.1 A more general problem

$$f_0(x) \longrightarrow \inf \text{ subject to } \ x \in X, \ g_i(x) \leq 0, i \in I \tag{3.2}$$

can be represented in the form of (3.1) if all constraints g_i are convoluted to a single constraint f_1. Various type of convolutions are studied in Yevtushenko and Zhadan (1990). The simplest way is to use operation of taking of supremum: $f_1(x) = \sup_{i \in I} g_i(x)$.

We make the following assumption that shows that the constraint f_1 is essential.

(A) There exists a sequence $x_k \in X$ such that $f_1(x_k) > 0$, $f_1(x_k) \to 0$ and $f_0(x_k) \to M_P(f_0, f_1)$.

Let

$$\beta(\varepsilon) = \inf\{f_0(x) : x \in X, f_1(x) \leq \varepsilon\} \qquad (\varepsilon \geq 0). \tag{3.3}$$

The function β defined on $\mathbb{R}_+$ by (3.3) is called the perturbation function of the problem (3.1). Clearly $\beta(0)$ coincides with the value $M_P(f_0, f_1)$ of the problem (3.1).

We denote by $\mathcal{P}_+$ the set of all continuous IPH functions defined on $\mathbb{R}^2_+$ such that

$$p(1,0) = 1, \qquad \lim_{u \to +\infty} p(1, u) = +\infty.$$

In this section we denote the function defined by (2.4) with the help of (2.3) by $q^+_{(p)}(d)$. Note that d is a positive number here.

Theorem 3.1 *Rubinov, et al (1997b) Consider a problem (f_0, f_1) such that Assumption (A) holds. Let $p \in \mathcal{P}_+$. The following*

$$q^+_{(p)}(d) \longrightarrow M_P(f_0, f_1) \text{ as } d \to \infty \tag{3.4}$$

or

$$M_p(f_0, f_1) = \sup_{d \in \mathbb{R}^{1+m}_+} q^+_{(p)}(d).$$

holds if and only if the perturbation function β is l.s.c at zero.

It is not very hard to give simple sufficient conditions for lower semicontinuity of the function β (see, for example Rubinov, et al (1997b)). It follows from Theorem 3.1 that we should restrict ourselves by considering the class C_X of all problems (f_0, f_1) with f_0, f_1 defined on the set X, such that Assumption (A) holds and the perturbation function of the problem (f_0, f_1) is lower semicontinuous at the origin. The class C_X was introduced and studied in Rubinov, et al (1998). The following result shows that we also should restrict ourselves by consideration of convolution functions $p \in \mathcal{P}_+$.

Theorem 3.2 *Rubinov, et al (1998) Let p be a continuous IPH function defined on the cone $\mathbb{R}^2_+$ and such that the convergence (3.4) holds for all $(f_0, f_1) \in C_X$. Then $p \in \mathcal{P}_+$.*

We now study conditions which guarantee the exactness of a penalization; in other words we will study when the supremum $\sup_{d>0} q^+_{(p)}$ is attained.

Theorem 3.3 *Rubinov, et al (1998) Let X be a set with the following property: there exists a function defined on X and mapping onto $\mathbb{R}_+$. Then for each $p \in \mathcal{P}_+$ there exists a problem $(f_0, f_1) \in C_X$ such that the supremum $\sup_{d>0} q^+_{(p)}(d)$ is not attained, that is $q^+_{(p)}(d) < M_P$ for all $d > 0$.*

Theorem 3.4 *Rubinov, et al (1997b) For each $(f_0, f_1) \in C_X$ there exists a function $p \in \mathcal{P}_+$ and a number $\bar{d} > 0$ such that $M(f_0, f_1) = q^+_{(p)}(1, \bar{d})$.*

Theorem 3.5 *Rubinov, et al (1997b) Let $k > 0$ and the function p_k be defined by*

$$p_k(\delta, y) = (\delta^k + y^k)^{\frac{1}{k}}.$$

Then there exists a number $\bar{d} > 0$ such that $M_P = q^+_{p_k}(1, \bar{d})$ if and only if

$$\liminf_{\epsilon \to +0} \frac{\beta(\epsilon) - \beta(0)}{\epsilon^k} > -\infty$$

where $\beta(\epsilon)$ is the perturbation function of (P).

A special case ($k = 1$) of this theorem easily follows from well-known results related to the notion of calmness (see, for example Burke (1991); Rockafellar (1993) and reference therein).

4 NONLINEAR LAGRANGIAN FUNCTIONS

Consider continuous increasing functions p defined on the space $\mathbb{R}^{1+m}$ which enjoy the following properties:

1) There exist positive numbers $a_o, a_1, \dots a_m$ with $a_o = 1$ such that for $\bar{y} = (y_0, y_1, \dots, y_m) \in \mathbb{R}^{1+m}$ with $y_o \in \mathbb{R}_+$ we have:

$$p(\bar{y}) \geq \max_{0 \leq i \leq m} a_i y_i. \tag{4.1}$$

2) For $y_o \in \mathbb{R}_+$

$$p(y_o, 0, \ldots, 0) = y_o. \tag{4.2}$$

Let us consider the vector-valued function $F : X \times \mathbb{R}_+^{1+m} \longrightarrow \mathbb{R}^{1+m}$ defined as follows, for $x \in X, d \in \mathbb{R}_+^{1+m}$:

$$F(x, d) = (d_o f_0(x), d_1 f_1(x), \ldots, d_m f_m(x)).$$

Let us compare the set of conditions (4.1) and (4.2) with (2.1) and (2.2). Since the zero duality gap is not valid for the classical Lagrange function defined by means of a linear convolution it is necessary to have conditions which exclude linearity. This is the main difference between the penalty approaches given in sections 2 and 3 and the Lagrange approach to be given in this section. It is clear that (2.1) and (4.1) have the same form, but for linear positive functions (2.1) holds while (4.1) does not hold. Actually (2.1) is not restrictive for IPH functions. Indeed let $p(e_k) = a_k > 0$ where $e_k = (0, \cdots, 1, \cdots, 0)$ with only kth component being 1. Then for each $y \geq 0$ we have $y \geq y_k e_k$, so $p(y) \geq p(y_k e_k) = y_k p(e_k)$. Thus $p(y) \geq \max_k a_k y_k$. At the same time (4.1) is very restrictive.

Let p be defined on $\mathbb{R}^{1+m}$ with properties (4.1) and 4.2). The modified Lagrange function corresponding to p for the problem (P) is defined by,

$$\mathcal{L}(x, d) = p(F(x, d)), \qquad x \in X, \quad d \in \mathbb{R}_+^{1+m}.$$

The dual function for (P) corresponding to p is defined by,

$$q(d) = \inf_{x \in X} \mathcal{L}(x, d), \quad d \in \mathbb{R}_+^{1+m}.$$

The following problem:

$$(D) \qquad q(d) \longrightarrow \sup \quad \text{subject to } d \in \mathbb{R}_+^{1+m}, d_o = 1$$

is called the dual problem to the problem (P) corresponding to the function p.

Let

$$M_D = \sup\{q(d) : d \in \mathbb{R}_+^{1+m}, d_o = 1\}.$$

Lemma 4.1 *(Weak duality). Goh and Yang (1997); Rubinov, et al (1997a)* $M_P \geq M_D$.

Theorem 4.1 *(Zero duality gap). Rubinov, et al (1997a) Consider a problem (P) such that function $f_0, f_1, \ldots, f_m$ are continuous, the set X_o is compact and (if X is unbounded) the function f_0 enjoys the following property:*

$$\lim_{x \to \infty, x \in X} f_0(x) = +\infty. \tag{4.3}$$

Then $M_P = M_D$.

The following gives a nonlinear Lagrangian with a maximum structure.

Theorem 4.2 *Goh and Yang (1997) Let the maximum Lagrangian function* $q_{\infty}^{+}(d)$ *is defined as*

$$\mathcal{L}(x,d) = \max\{f_0(x), d_1 f_1(x), \cdots, d_m f_m(x)\}.$$

Assume that $f_1, \ldots, f_m$ *are continuous on* X *and that* f_0 *enjoys the property (2.5), if* X *is unbouded:*

$$\lim_{x \to \infty, x \in X} f_0(x) = +\infty.$$

Then, $M_P = M_D$.

It is possible to give a different set of conditions, which guarantees zero duality gap (see Andramonov (1997)).

Let the following assumptions hold:

1) $p(y)$ is an increasing function on $\mathbb{R}_+^{m+1}$.

2) $p(y_o, y_1, \ldots, y_m) \to +\infty$ whenever $y_o \geq 0$ and $\max_{i \in I} y_i \to +\infty$.

3) $p(y_o, y_1, \ldots, y_m) = y_o$ for all $y_o \geq 0$ and for all y such that $y_i \leq 0, \quad i = 1, \cdots, m$

Consider the following point-to-set-map:

$$A(\varepsilon) = \{x \in X : f_i(x) \leq \varepsilon \quad \forall i = 1, \cdots, m\}. \tag{4.4}$$

Note that $A(0)$ is the feasible set of the problem (P).

Theorem 4.3 *Andramonov (1997) Suppose that the set-valued mapping* $A(\varepsilon)$ *is upper semicontinuous at* 0 *and* f_0 *is continuous. Then for any* $\varepsilon > 0$ *there exists* $\overline{d} = (1, \overline{d}_1, \overline{d}_2, \ldots, \overline{d}_m) \in \mathbb{R}_+^{m+1}$ *such that*

$$q(d) \geq M_p - \varepsilon.$$

Analogous results can be obtained for problems with equality constraints. For details, see Andramonov (1997).

5 LAGRANGE FUNCTIONS FOR PROBLEMS WITH ONE CONSTRAINT

Consider the class C_X of problems with one constraint (see Section 3). For constructing the Lagrange function for problems belonging to C_X we need a convolution function p defined on the half-plane $\{(u,v) : u \geq 0\}$.

It has been shown (see Theorem 3.1) that for each problem $(f_0, f_1) \in C_X$ and each IPH function $p \in \mathcal{P}_+$ the zero duality gap property is valid. We now describe all IPH functions p defined on the half-plane $\{(u,v) : u \geq 0\}$ and such that the zero duality gap property for the corresponding Lagrange functions holds for each $(f_0, f_1) \in C_X$. Let $\mathcal{P}_*$ be the class of functions p defined on $\{(u,v) : u \geq 0\}$ with the following properties: the restriction p on $\mathbb{R}^2_+$ belongs to $\mathcal{P}_+$ and $p(u,v) = u$ if $v \leq 0$. It is easy to check that each function $p \in \mathcal{P}_*$ is IPH.

Theorem 5.1 *Rubinov, et al (1998) Let p be a continuous IPH function defined on $\{(u,v) : u \geq 0\}$. Then the zero duality gap property holds for all $(f_0, f_1) \in C_X$ if and only if $p \in \mathcal{P}_*$.*

Theorem 5.2 *Rubinov, et al (1998) Let $p \in \mathcal{P}_*$ and p_+ be the restriction of the function p on $\mathbb{R}^2_+$. Let q_p be the dual extended Lagrange function with respect to p be defined by*

$$q_p(1,d) = \inf_{x \in X} p(f_0(x), df_1(x))$$

and $q^+_{p_+}$ be the dual extended penalty function with respect to p_+ be defined by

$$q^+_{p_+}(1,d) = \inf_{x \in X} p_+(f_0(x), df_1(x)).$$

Then

$$q_p(1,d) = q^+_{p_+}(1,d) \quad \text{for all } d > 0.$$

Thus the extended dual Lagrange function provides the zero duality gap property for all problems from C_X if and only if this function coincides (in the sense of Theorem 5.2) with the dual penalty functions

6 PENALTY AND LAGRANGE

The scheme considered in Andramonov (1997) leads to combination of nonlinear penalty functions with classical Lagrange functions.

The modified Lagrange function is defined by the formula:

$$U(x,d,r) = G(L(x,d), r_1 f_1(x), \ldots, r_m f_m(x))$$

where $L(x,d) = f_0(x) + \sum_{i=1}^{m} d_i f_i(x)$ is a classical Lagrange function. Then the auxilary problem to be solved is:

$$\sup_{d \in \mathbb{R}^m_+} \inf_{x \in X} U(x,d,\bar{r}) \tag{6.1}$$

The zero duality gap property can be proved under some assumptions. In particular, it is assumed that the following hold:

1) G is an increasing function on $\mathbb{R}^{m+1}$.

2) All the constraints f_i are continuous and bounded from below on $\mathbb{R}^n$, the objective function f_0 is continuous.

3) $G(y_0, y_1, \ldots, y_m) \to +\infty$ whenever $\max_{i \in I} y_i \to +\infty$ and y_0 is bounded from below.

There are also some other necessary assumptions (see Andramonov (1997) for details). A similar scheme was considered in Yevtushenko and Zhadan (1990). However, the scheme from Andramonov (1997) does not require that a saddle point of the classical Lagrangian should exist. The following results can be found in Andramonov (1997).

Theorem 6.1 *Suppose that a saddle point (x^*, d^*) for $L(x, d)$ exists, $x^* \in X_0$ for all $i \in I$ and the following complementary slackness condition holds:*

$$\forall i \in I, \quad d_i^* f_i(x^*) = 0.$$

Then (x^*, d^*) is also a saddle point for the function $U(x, d, \overline{r})$ for any fixed vector $\overline{r}$ of penalty parameters.

Even if no saddle points exist, the optimal value of the auxiliary problem (6.1) tends to the optimal value M_P of (P) under mild assumptions.

Theorem 6.2 *Let the point-to-set map $A(\varepsilon)$ defined by (4.4) be upper semi-continuous. Then for any $\varepsilon > 0$ there exists $Q > 0$ such that whenever $\min_{i \in I} \overline{r}_i > Q$ we have $M_P - \varepsilon \leq \sup_{d \in \mathbb{R}_+^m} \inf_{x \in X} U(x, d, \overline{r}) \leq M_P$.*

There exists a relation between the optimal solutions of the initial constrained program and the saddle points of the auxiliary problem.

Theorem 6.3 *Let (x^*, d^*) be a saddle point for the problem (6.1). Then x^* is an optimal solution of (P).*

Note that the scheme has improved numerical efficiency for a number of non-smooth constrained optimization problems, see Andramonov (1997). Probably, this is due to the fact that ill-conditioning of the auxiliary function does not appear.

7 PARAMETER NONLINEAR UNCONSTRAINED OPTIMIZATION METHODS

In this section we discuss the parameter methods presented in Goh and Yang (1997); Yang and Li (1999). These approaches transform problem (P) into the problem of finding a solution of an equation where the function of a single variable in this equation is obtained by solving a nonlinear unconstrained optimization problem.

Let the function $\phi : \mathbb{R}^{1+m} \to \mathbb{R}$ satisfy the following properties, for $y = (y_0, y_1, \cdots, y_m)$:

(a) $\phi(y) \geq \max_{0 \leq i \leq m} y_i, \quad \forall y \in \mathbb{R}^{1+m}$;

(b) $\phi(y) = \max\{0, y_0\}, \quad \forall y_0 \in \mathbb{R}, (y_1, \cdots, y_m) \in \mathbb{R}_-^m$;

(c) ϕ is an increasing function with respect to its first component y_0.

It is clear from (b) that $\phi(0,0,\cdots,0) = \max\{0,0\} = 0$.

Example 7.1 *The following functions satisfy properties (a), (b) and (c):*

$$\phi_\infty(y) = \max\{0, y_0, y_1, \cdots, y_m\};$$

$$\phi_k(y) = \left(\sum_{i=0}^{m} [\max\{0, y_i\}]^k\right)^{\frac{1}{k}}, \quad 0 < k < +\infty;$$

$$\phi'_k(y) = \max\{0, y_0\} + \left(\sum_{i=1}^{m} [\max\{0, q_i(y_i)\}]^k\right)^{\frac{1}{k}}, \quad 0 < k < +\infty,$$

where $q_i : \mathbb{R} \to \mathbb{R}$ can be any continuous function that satisfies $q_i(z) \le 0$, if $z < 0$ and $q_i(z) \ge z$, if $z \ge 0$. For example, $q_i(z) = \exp(z) - 1$.

Let

$$H(x, \theta_0) = \left(\frac{f_0(x)}{\theta_0} - 1, f_1(x), \cdots, f_m(x)\right), \quad x \in X, \theta_0 > 0.$$

Let ϕ be a function that satisfies properties (a), (b) and (c) and $\theta_0 > 0$. Define an auxiliary function $\Phi : \mathbb{R}^n \times \mathbb{R} \to \mathbb{R}$ as

$$\Phi(x, \theta_0) = \phi(H(x, \theta_0)).$$

An auxiliary problem $P_\phi(\theta_0)$ is defined as follows

$$\inf_{x \in X} \Phi(x, \theta_0).$$

Define $\varphi(\theta_0)$ to be the global optimal value of $P_\phi(\theta_0)$, i.e.,

$$\varphi(\theta_0) = \inf_{x \in X} \phi(H(x, \theta_0)).$$

Theorem 7.1 *Yang and Li (1999) Consider the problem (P) such that $f_i, i = 0, 1, \cdots, m$ are continuous and X is closed (and f_0 satisfies the condition (2.5) if X is unbounded). Let x^* solve the problem (P). The following hold:*

(i) If $0 < \theta_0 < f_0(x^*)$, then $\varphi(\theta_0) > 0$;

(ii) If $\theta_0 \ge f_0(x^*)$, then $\varphi(\theta_0) = 0$;

(iii) $\varphi(\theta_0)$ is a continuous, nonincreasing and nonnegative function of θ_0;

Theorem 7.2 *Yang and Li (1999) Consider the problem (P) such that $f_i, i = 0, 1, \cdots, m$ are continuous and X is closed (and f_0 satisfies the condition (2.5) if X is unbounded). θ_0 is equal to the optimal value of (P) if and only if θ_0 is the solution of the following least root problem*

$$\theta_0 = \min\{\theta \mid \varphi(\theta) = 0\}. \qquad (7.1)$$

It is worth noting that the function with a maximum structure defined by

$$\phi_1(y) = \max\{y_0, y_1, \cdots, y_m\},$$

does not satisfy property (b) and that this nonsmooth function is used in Goh and Yang (1997) to formulate another type of the auxiliary parametric optimization problems. We shall describe this in the following.

Let $b = [b_1, \cdots, b_m]^\top$. Consider the following inequality constrained optimization problem (P(b)):

$$f_0(x) \longrightarrow \inf \ \text{subject to}\ x \in X,\ f_i(x) \leq b_i, i \in I$$

where $f_i : \mathbb{R}^n \rightarrow \mathbb{R}$, $i = 0, 1, 2, \cdots, m$ are real-valued functions, and X is a subset of $\mathbb{R}^n$. Without loss of generality, we assume that the parameter vector $b \in \text{int}\ \mathbb{R}^m_+$. Define the set of all feasible solutions for (P(b)) to be

$$X_0(b) = \{x \in X \mid f_i(x) \leq b_i, i = 1, 2, \cdots, m\}.$$

Let

$$F(x, \theta_0) = \left(\frac{f_0(x)}{\theta_0}, \frac{f_1(x)}{b_1}, \cdots, \frac{f_m(x)}{b_m}\right), \quad x \in X, \theta_0 > 0.$$

Define an auxiliary function $\Psi : \mathbb{R}^n \times \mathbb{R} \rightarrow \mathbb{R}$ as

$$\Psi(x, \theta_0) = \phi_1(F(x, \theta_0)).$$

Define $\psi(\theta_0)$ as follows

$$\psi(\theta_0) = \inf_{x \in X} \Psi(F(x, \theta_0)).$$

Theorem 7.3 *Goh and Yang (1997) Assume that $f_i, i \in \{0\} \cup I$ are continuous on the closed set X (and f_0 satisfies the condition (2.5) if X is unbounded). Let $\theta_0 > 0$. Assume that (P(θ)) admits a solution x^* with $f_0(x^*) > 0$ and that for any $\theta_0 > f_0(x^*)$, there exists $x \in X$ such that $f_i(x) < \theta_i$, $f_0(x^*) < f_0(x) < \theta_0$. Then*

(i) If $0 < \theta_0 < f_0(x^*)$, then $\psi(\theta_0) > 1$;

(ii) If $\theta_0 > f_0(x^*)$, then $\psi(\theta_0) < 1$;

(iii) $\psi(\theta_0)$ is a monotone non-increasing function of θ_0.

Theorem 7.4 *Goh and Yang (1997) Assume that the conditions of Theorem 7.3 hold. θ_0 is equal to the optimal value of (P) if and only if $\psi(\theta_0) = 1$.*

References

Andramonov, M. Yu. (1997), An approach to constructing generalized penalty functions, Research Report 21/97, University of Ballarat.

Burke, J. V. (1991), An exact penalization viewpoint of constrained optimization, *SIAM J. Control and Optimization*, Vol. 29, pp. 968-998.

Goh, C. J. and Yang, X. Q. (1997), A sufficient and necessary condition for nonconvex constrained optimization, *Applied Mathematics Letters*, Vol. 10, pp. 9-12.

Goh, C. J. and Yang, X. Q. (1996) A nonlinear Lagrangian theory for nonconvex optimization problems, JOTA (revised).

Rockafellar, R. T. (1993), Lagrange multipliers and optimality, *SIAM Review*, Vol. 35, pp. 183-238.

Rubinov, A. M. and Glover, B. M. (1997), Duality for increasing positively homogeneous functions and normal sets, *RAIRO Operations Research*, Vol. 32, pp. 105 -123.

Rubinov, A. M. Glover, B. M. and Yang, X.Q. (1997a), Modified Lagrange and penalty functions in continuous optimization, *Optimization* (to appear).

Rubinov, A. M. Glover, B. M. and Yang, X.Q. (1997b), Decreasing functions with applications to penalization, *SIAM Journal on Optimization* (to appear).

Rubinov, A. M., Yang, X.Q. and Glover B. M. (1998), Extended Lagrange and penalty functions in optimization, (submitted).

Yang, X.Q. and Li, D. (1999), Successive global optimization method via parametric monotone composition formulation, *Journal of Global Optimization*, (to appear).

Yevtushenko, Yu. G. and Zhadan, V. G. (1990), Exact auxiliary functions in optimization problems, *U.S.S.R Comput. Maths. Math. Phys.*, Vol. 30, pp. 31- 42.

5 NEW DUAL FORMULATIONS IN CONSTRAINED INTEGER PROGRAMMING

Xiaoling Sun

Department of Mathematics,
Shanghai University,
Jiading, Shanghai, China

and Duan Li

Department of Systems Engineering and Engineering Management,
The Chinese University of Hong Kong,
Shatin, N. T., Hong Kong

Abstract: While the Lagrangian relaxation method (Geoffrion 1974 and Fisher 1981) has been one of the most powerful solution schemes in solving constrained integer programming problems, this conventional dual search method often fails to identify an optimal solution of the primal constrained integer programming problem. Recent research results by Li (1997b), Yang and Goh (1997), and Sun and Li (1998) represent an extension from the traditional linear Lagrangian theory to nonlinear Lagrangian theory in an advancement to achieve a more satisfactory dual with a guarantee of the identification of an optimal solution of the primal problem via dual search. This paper summarizes recent progress of the new dual formulations for constrained integer programming with clear motivation and full geometric interpretation in order to better our understanding of the fundamental properties in constrained integer programming and in the newly developed nonlinear Lagrangian duality theory. Prominent features in both the theoretical achievements and computational implementation of the new dual formulations are addressed in this paper.

Key words: constrained integer programming, dual formulations, p-th power transformation, minimax-type Lagrangian function, log-exp dual formulation.

X. Yang et al. (eds.), Progress in Optimization, 79–91.

1 INTRODUCTION

Consider the following general constrained integer programming (P):

$$\min \ f(x) \tag{1.1}$$
$$\text{s. t. } x \in S = \{x \in X \mid g_i(x) \leq 0, \ i = 1, \ldots, m\}, \tag{1.2}$$

where $f : \mathbb{R}^n \mapsto \mathbb{R}$ and $g_i : \mathbb{R}^n \mapsto \mathbb{R}$, $i = 1, \ldots, m$, are continuous functions, and X is a finite integer set. Without loss of generality, we assume that $f(x) > 0$ for all $x \in X$.

Various methods for solving problem (P) have been proposed during the last two decades. The branch and bound method (Gupta and Ravindran (1985) and Körner (1988)) is based on a solution scheme of successively subdivising X and estimating lower bounds by solving the continuous relaxation problems. It is clear that branch and bound methods may fail to find the optimal solution of (P) in the absence of convexity. When the objective function and constraint functions are separable and the number of constraints is no more than two, dynamic programming method (Cooper (1981) and Cooper (1982)) can be used as an efficient solution algorithm to solve (P). Lagrangian method (Fisher (1981) and Geoffirion (1974)) deals with (P) in an indirect way: By incorporating constraints into the objective function, a relaxation problem is formed and a dual search is performed to find an optimal solution of the primal problem. In contrast to a guarantee of its success in continuous convex optimization, Largranian method may fail to identify an optimal solution of integer programming problem (P) even when the problem is of a linear or convex structure.

This paper aims to analyze the reasons behind possible failures of the classical Lagrangian dual method and to summarize the recent development in nonlinear Lagrangian duality theory. In section 2, we illustrate the possible failure of the classical Lagrangian method and show geometrically the distinct characterization of the perturbation function and the lower envelope function of (P). In section 3, a p-norm surrogate constraint method (Li (1999)) is introduced that can be adopted to convert a multi-constraint integer programming into an equivalent integer programming problem with a single surrogate constraint. A p-th power transformation method (Li (1997b)) is discussed in Section 4 to overcome the nonconvexity of the lower envelope function and thus to guarantee the success of a dual search. A minimax-type Lagrangian function (Yang and Goh (1999)) that achieves a zero duality gap is briefly summarized in section 5. The logarithmic-cxponential dual formulation (Sun and Li (1998)) is reported in section 6 with an illustrative numerical example. Finally, we give our concluding remarks in section 7.

2 DUALITY GAP IN THE CLASSICAL LAGRANGIAN METHOD

Just as in the duality theory for continuous optimization, we associate with i-th inequality constraint a nonnegative dual variable λ_i and form a Lagrangian function:

$$L(x,\lambda) = f(x) + \sum_{i=1}^{m} \lambda_i g_i(x).$$

The resulting Lagrangian relaxation problem is

$$\psi(\lambda) = \min_{x \in X} L(x,\lambda).$$

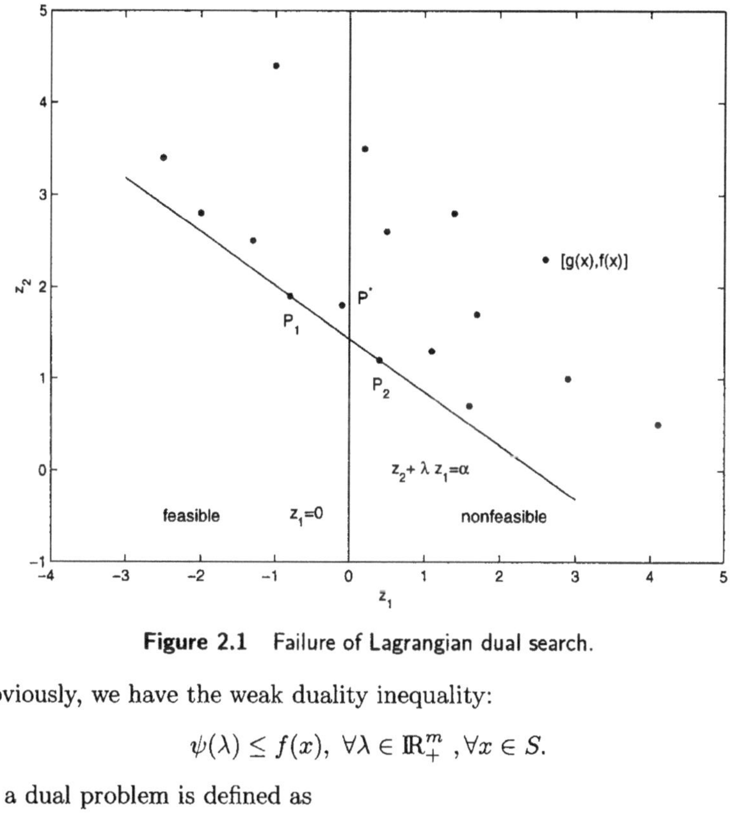

Figure 2.1 Failure of Lagrangian dual search.

Obviously, we have the weak duality inequality:

$$\psi(\lambda) \le f(x),\ \forall \lambda \in \mathbb{R}^m_+ , \forall x \in S.$$

Then a dual problem is defined as

$$\sup_{\lambda \in \mathbb{R}^m_+} \psi(\lambda)$$

The following example shows the possibility to witness a failure of the conventional dual search and an existence of the duality gap.

Example 2.1

$$\begin{aligned}\min\ & f(x) = 0.2(x_1-3)^2 + 0.1(x_2-5)^2 + 0.1 \\ \text{s. t. } & g(x) = x_1^2 - 2.5x_1 + 1.2x_2 - 1 \le 0, \\ & x \in X = \{x \in [0,3]^2 |\ \ x_1, x_2 \text{ integer}\}.\end{aligned}$$

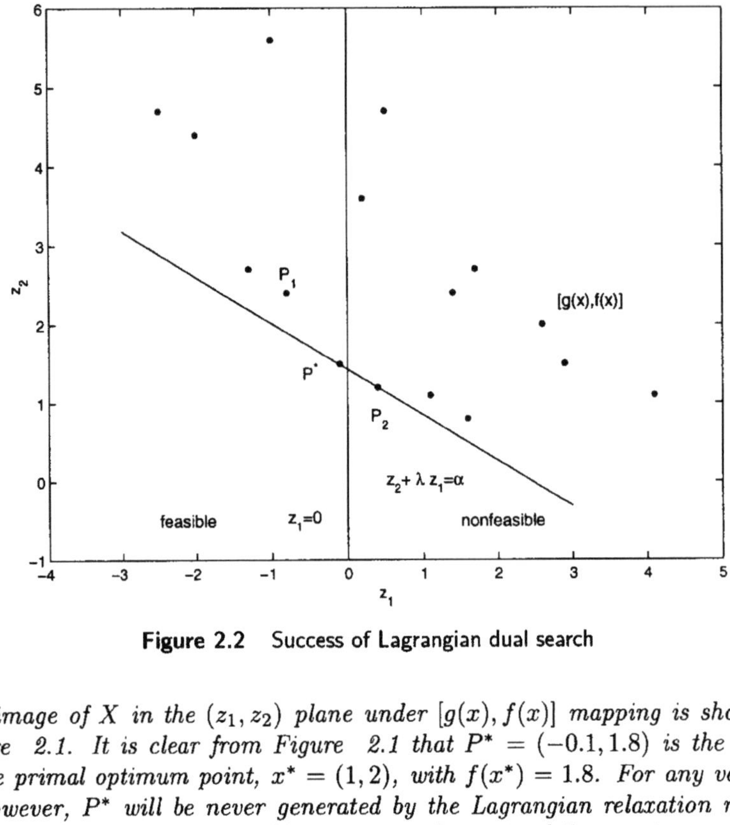

Figure 2.2 Success of Lagrangian dual search

The image of X in the (z_1, z_2) plane under $[g(x), f(x)]$ mapping is shown in Figure 2.1. It is clear from Figure 2.1 that $P^* = (-0.1, 1.8)$ is the image of the primal optimum point, $x^* = (1, 2)$, with $f(x^*) = 1.8$. For any value of λ, however, P^* will be never generated by the Lagrangian relaxation method as an optimal solution. In addition, since the optimal Lagrangian multiplier is $\lambda^* = 7/12$ with $\psi(\lambda^*) = 43/30 \approx 1.433 < 1.8$, a duality gap exists for the Lagrangian dual formulation.

Let us examine another example problem.

Example 2.2

$$\begin{aligned}\min\ & f(x) = 0.3(x_1 - 2)^2 + 0.4(x_2 - 3)^2 + 0.8 \\ \text{s. t. } & g(x) = x_1^2 - 2.5x_1 + 1.2x_2 - 1 \le 0, \\ & x \in X = \{x \in [0,3]^2 |\ x_1, x_2 \text{ integer}\}.\end{aligned}$$

In this example, as shown in Figure 2.2, the dual search happens to be able to identify the primal optimal solution P^*. A crucial observation is that there does not exist a supporting plane at P^* in Figure 2.1 while there does exist such a supporting plane at P^* in Figure 2.2.

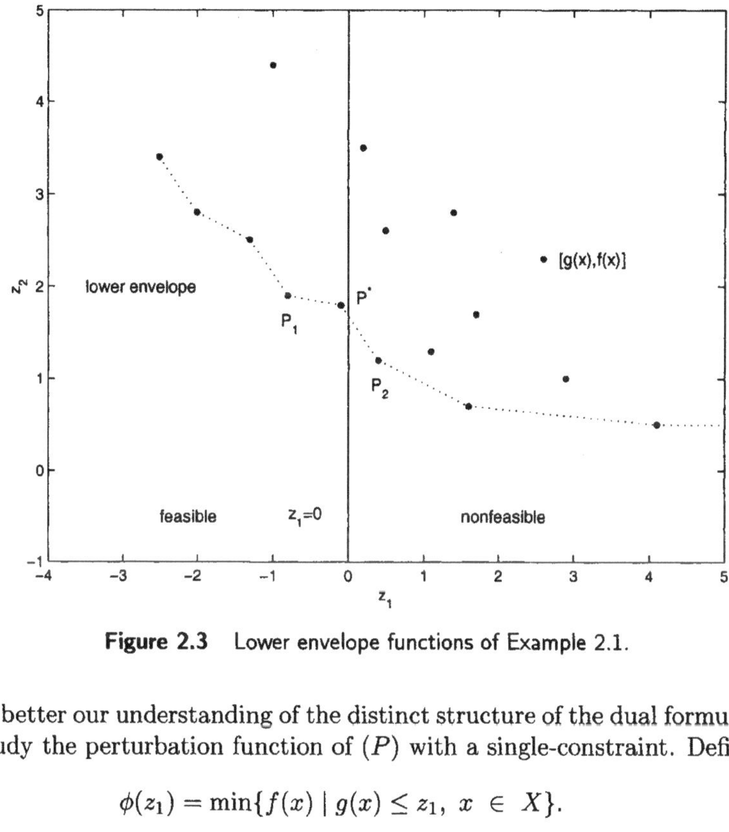

Figure 2.3 Lower envelope functions of Example 2.1.

To better our understanding of the distinct structure of the dual formulation, we study the perturbation function of (P) with a single-constraint. Define

$$\phi(z_1) = \min\{f(x) \mid g(x) \leq z_1,\ x\ \in\ X\}.$$

It is obvious that $\phi(z_1)$ is a nonincreasing piecewise-constant function of z_1 and it is continuous from the right. The domain of $\phi(\cdot)$ is

$$Y = \{z_1 \in \mathrm{I\!R} \mid \text{there exists } x \in X \text{ such that } g(x) \leq z_1\}.$$

Let Ω be the finite set of discontinuous points of $\phi(z_1)$ in (z_1, z_2) plane,

$$\Omega = \{(a_1, b_1), \ldots, (a_N, b_N)\}$$

with $a_1 > a_2 > \ldots > a_N$ and $b_1 < b_2 < \ldots < b_N$. Define $l(z_1)$ to be the lower envelope function of (P),

$$l(z_1) = \begin{cases} b_1 + \xi_1(z_1 - a_1), & a_2 \leq z_1 \leq a_1 \\ b_2 + \xi_2(z_1 - a_2), & a_3 \leq z_1 \leq a_2 \\ \ldots & \ldots \\ b_{N-1} + \xi_{N-1}(z_1 - a_{N-1}), & a_N \leq z_1 \leq a_{N-1} \end{cases}$$

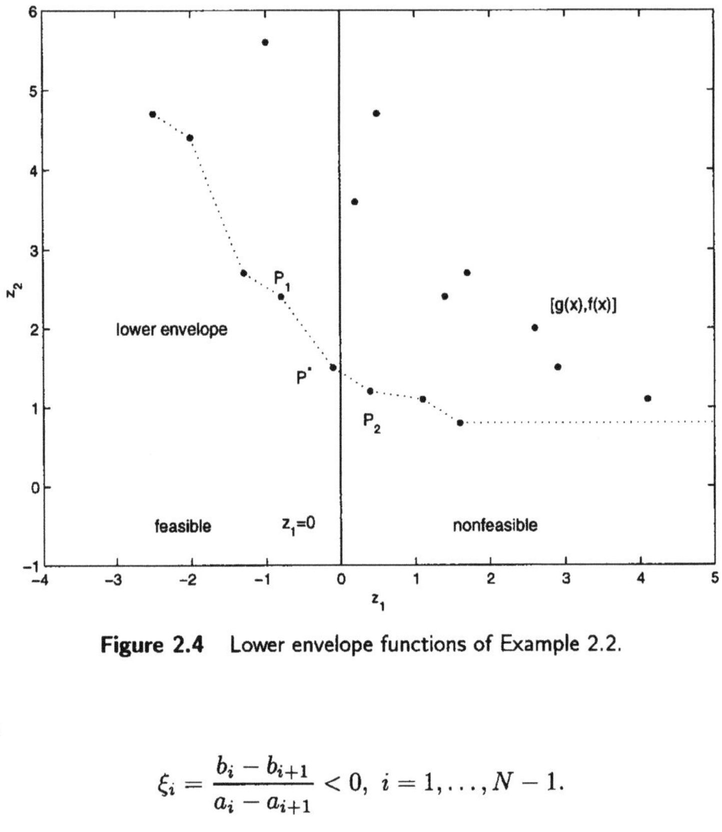

Figure 2.4 Lower envelope functions of Example 2.2.

where

$$\xi_i = \frac{b_i - b_{i+1}}{a_i - a_{i+1}} < 0, \ i = 1, \dots, N-1.$$

Figures 2.3 and 2.4 show the lower envelope functions of Examples 2.1 and 2.2, respectively. We observe that the lower envelope functions are nonconvex in both examples. The lower envelope function in Example 2.2, however, is locally convex at P^*.

It should become apparent now that the following two approaches could be adopted to guarantee the success of a dual search: (a) to convexify the lower envelope function by a suitable transformation, or (b) to replace the (linear) Lagrangian function by a nonlinear Lagrangian function so that the contour of the dual function can be tangent to an optimal point.

3 P-NORM SURROGATE CONSTRAINT METHOD

Certain results of some new dual formulations are derived for integer programming problems with a single constraint. For multi-constraint cases of (P), the

p-norm surrogate technique (Li (1999)) can be used to convert problem (P) into an equivalent problem with a single surrogate constraint.

Rewrite problem (P) into the following equivalent form,

$$\begin{aligned} &\min\ f(x) \\ &\text{s.t. } g_i(x) + b \le b,\ i = 1, \dots, m, \\ &\quad\ x \in X, \end{aligned}$$

where b is a strictly positive number that is chosen to make $g_i(x) + b > 0$ for all $x \in X$ and for all $i = 1, 2, \dots, m$.

The p-norm surrogate constraint formulation of (P) is formed as follows for $p \ge 1$,

$$\begin{aligned} &\min\ f(x) && && (3.1) \\ &\text{s.t. } g_s(x) &=& \left\{\sum_{i=1}^{m}[g_i(x) + b]^p\right\}^{1/p} \le m^{1/p}b \\ &x \in X \end{aligned}$$

Define S_t to be the feasible region of (3.1)

$$S_p = \left\{ x \mid \left\{\sum_{i=1}^{m}[g_i(x) + b]^p\right\}^{1/p} \le m^{1/p}b \text{ and } x \in X \right\}$$

The following results state that a single surrogate constraint can be always constructed using p-norm such that the feasible sets in a surrogate relaxation and the primal problem are identical.

Theorem 3.1 (Li (1999)) *There exists a finite positive p^* such that*

$$S = S_p$$

for all $p > p^*$.

Corollary 3.1 (Sun and Li (1998)) *Suppose that all g_i, $i = 1, 2, \dots, m$, are integer-valued functions, e.g., polynomial functions with integer coefficients. Then*

$$S = S_p$$

when $p > \ln(m)/\ln(\frac{b+1}{b})$.

4 P-TH POWER LAGRANGIAN METHOD

Consider the following single-constraint integer programming problem:

$$\min\ f(x) \tag{4.1}$$
$$\text{s. t. } x \in S = \{x \in X \mid g(x) \le b\}, \tag{4.2}$$

where $f(x) > 0$, $g(x) > 0$ for all $x \in X$, and $b > 0$. In order to convexify the lower envelope function of (4.1)-(4.2) and thus to guarantee the identification of a primal optimal solution via dual search, the following p-th power transformation is introduced in Li (1997b) with $p > 0$,

$$\min \ [f(x)]^p \tag{4.3}$$

$$\text{s. t. } [g(x)]^p \le b^p, \ x \in X. \tag{4.4}$$

The associated Lagrangian relaxation is:

$$v(\lambda, p) := \min_{x \in X} L_p(x, \lambda) = [f(x)]^p + \lambda\{[g(x)]^p - b^p\}, \tag{4.5}$$

and the corresponding dual problem is:

$$\sup_{\lambda \ge 0} v(\lambda, p). \tag{4.6}$$

The following results indicate that if p is sufficiently large, then a successful dual search is guaranteed.

Theorem 4.1 (Li (1997b)) *There exists a finite $q > 0$ such that the lower envelope function $l(z_1)$ of the p-th power-transformed problem (4.3)-(4.4) is a convex function of z_1 for any $p > q$.*

Theorem 4.2 (Li (1997b)) *If x^* is an optimal solution of (4.1)-(4.2) that results in a discontinuous point on the perturbation function ϕ, then there exists a $\lambda^* \ge 0$ such that (x^*, λ^*) is an optimal primal-dual pair when $p > q$ (defined in Theorem 4.1), i.e., x^* solves* (4.5) *with $\lambda = \lambda^*$ and λ^* solves* (4.6).

Remark 4.1 Although Theorem 4.1 assures the existence of a finite q, no practical procedure is reported in Li (1997b) to estimate a lower bound of q for general situations.

5 MINIMAX-TYPE LAGRANGIAN FUNCTION

The following minimax-type Lagrangian function is introduced in Yang and Goh (1999) for (P):

$$N(x, \lambda) = \max\{f(x), g_1(x)/\lambda_1, \ldots, g_m(x)/\lambda_m\}, \ \lambda_i > 0 \ , i = 1, 2, \ldots, m. \tag{5.1}$$

The corresponding relaxation problem is:

$$v(\lambda) := \min_{x \in X} N(x, \lambda). \tag{5.2}$$

The following weak duality holds:

$$v(\lambda) \le f(x), \ \forall x \in S, \ \forall \lambda_i > 0.$$

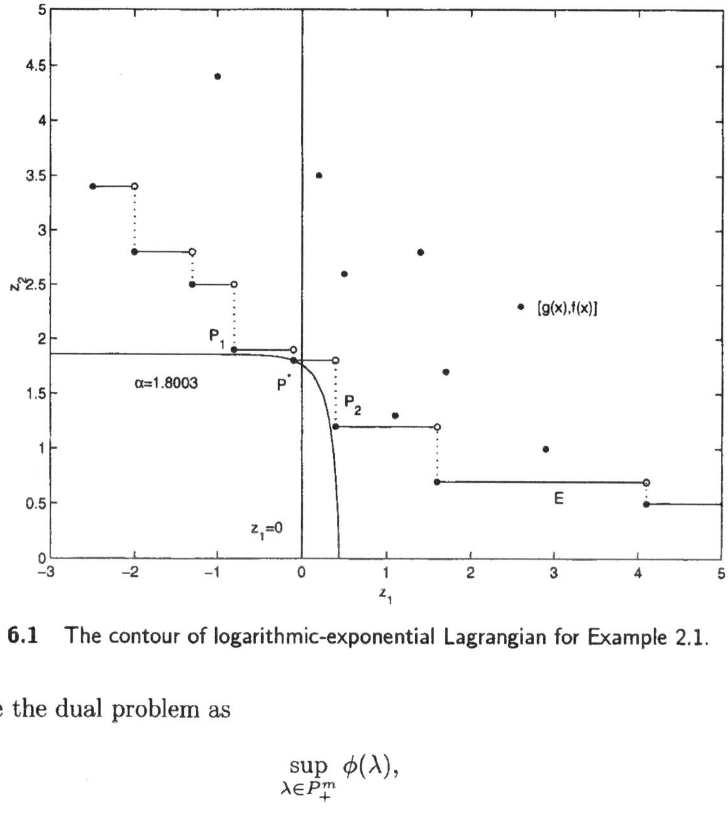

Figure 6.1 The contour of logarithmic-exponential Lagrangian for Example 2.1.

Define the dual problem as

$$\sup_{\lambda \in P_+^m} \phi(\lambda), \tag{5.3}$$

where $P_+^m = \{x \in \mathbb{R}^m \mid x_i > 0, i = 1, \ldots, m\}$.

It can be shown that the minimax-type dual formulation (5.1)-(5.3) possesses the following strong duality property.

Theorem 5.1 (Yang and Goh (1999)) $\min_{x \in S} f(x) = \sup_{\lambda \in P_+^m} v(\lambda)$.

Remark 5.1 The minimax-type Lagrangian is of a nonsmooth structure. The relaxation problem (5.2) is difficult to solve, even harder than (P) in most situations.

6 LOGARITHMIC-EXPONENTIAL DUAL FORMULATION

We will report in this section a dual formulation that possesses the strong duality and at the same time is of a computationally tractable structure in the resulting relaxation problem. We introduce the following logarithmic-exponential

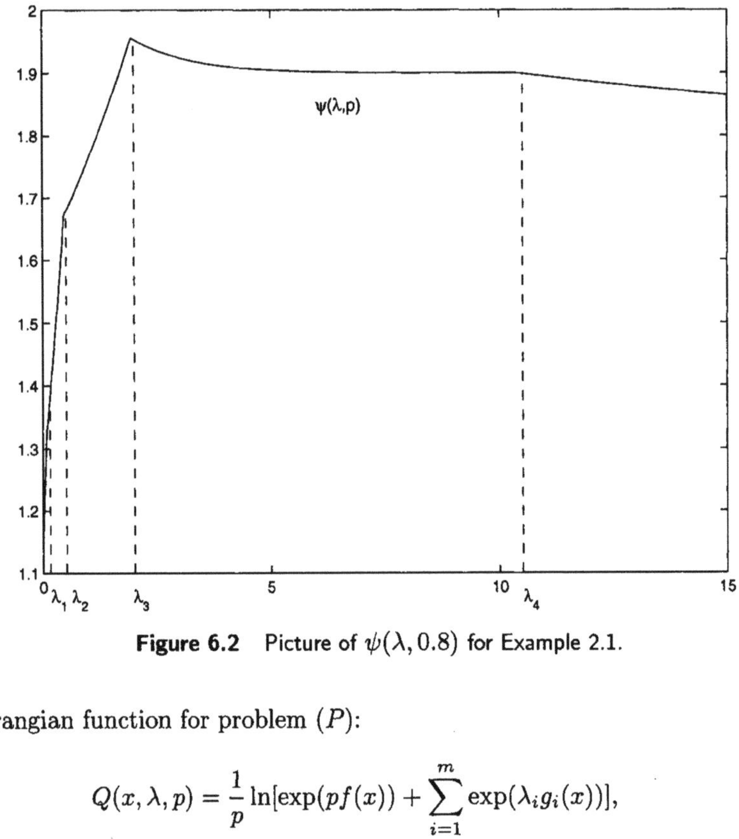

Figure 6.2 Picture of $\psi(\lambda, 0.8)$ for Example 2.1.

Lagrangian function for problem (P):

$$Q(x, \lambda, p) = \frac{1}{p} \ln[\exp(pf(x)) + \sum_{i=1}^{m} \exp(\lambda_i g_i(x))], \tag{6.1}$$

where $p > 0$ is a parameter and $\lambda \in \mathbb{R}_+^m$ are dual variables. Geometrically, the contour $Q(x, \lambda, p) = \alpha$ in $[g(x), f(x)]$ plane can be tangent to any "hidden" point on the curve of perturbation function. See Figure 6.1 for a demonstration of a single-constraint case.

Define the relaxation problem as

$$\psi(\lambda, p) := \min_{x \in X} Q(x, \lambda, p). \tag{6.2}$$

The dual problem then becomes

$$\theta(p) := \sup_{\lambda \in \mathbb{R}_+^m} \psi(\lambda, p). \tag{6.3}$$

Some basic properties of $Q(x, \lambda, p)$ are

(a) $Q(x, \lambda, p)$ is a smooth function of x if f and g_i are smooth;

(b) $Q(x, \lambda, p)$ is a convex function of x if f and g_i are convex.
An asymptotic strong duality of dual formulation (6.1)-(6.3) holds.

Theorem 6.1 (Sun and Li (1998)) $\lim_{p\to\infty} \theta(p) = \min_{x\in S} f(x)$.

We will consider in the following the single-constraint case of (P): $S = \{x \in X \mid g(x) \le 0\}$. For multi-constraint cases, the p-norm surrogate technique (Li (1999)) can be first used to convert the problem into an equivalent problem with a single surrogate constraint.

Theorem 6.2 (Sun and Li (1998)) *For any $p > 0$, dual function $\psi(\lambda, p)$ is a monotonically increasing function on $[0, \lambda_p^*]$ and a monotonically decreasing function on $[\lambda_p^*, +\infty)$, where λ_p^* is the optimum point of dual problem (6.3).*

Remark 6.1 Although the dual function $\psi(\lambda, p)$ is not a concave function, we can see from Theorem 6.2 that it is a unimodular function, that will greatly facilitate the dual search. See Figure 6.2.

Another prominent property of dual formulation (6.1)-(6.3) is that the primal feasibility can be achieved for $\lambda > \lambda_p^*$.

Theorem 6.3 (Sun and Li (1998)) *For any $p > 0$, if $\lambda > \lambda_p^*$, then any optimal solution x^* of (6.2) is primal feasible, this is, $x^* \in S$.*

Let

$$\delta = \min\{f(x) \mid x \in S \setminus S^*\} - f^*,$$

where $f^* = \min_{x\in S} f(x)$, $S^* = \{x \in S \mid f(x) = f^*\}$. Denote

$$\begin{aligned} \bar{f} &= \max_{x\in X} f(x), \\ \underline{g} &= \min_{x\in X\setminus S} g(x). \end{aligned}$$

Theorem 6.4 (Sun and Li (1998)) *If $p > \ln 2/\delta$ and $\lambda > p\bar{f}/\underline{g}$, then any optimal solution of (6.2) is also an optimal solution of (P).*

Remark 6.2 Theorem 6.4 shows that for single-constraint cases of (P), it is not necessary to find the exact dual optimal solution $\lambda^*(p)$. If δ, $\bar{f}$ and $\underline{g}$ are known, we can guarantee the success of dual search by setting p and λ to be greater than the threshold values indicated in the above theorem.

Example 6.1

$$\begin{aligned} \min\ f(x) &= -5x_1x_2x_3x_4 - x_1x_2x_3 + x_2x_3x_4 + x_3x_4x_5 - 4x_1x_2 \\ &\quad + x_2x_3 + x_3x_4 - x_4x_5 + x_1 + x_2 + 6 \\ \text{s. t. } g_1(x) &= x_1x_2 + x_2x_3 + x_3x_4 + 2 \le 3, \\ g_2(x) &= -x_1x_3 + x_3x_2 + 2x_2x_5 + 2 \le 3, \\ g_3(x) &= x_1x_4 - x_4x_5 + 2x_5x_3 + 2 \le 3, \\ x &\in X = \{0,1\}^5. \end{aligned}$$

There are two optimal solutions: $x_1^* = (1,1,0,0,0)$ *and* $x_2^* = (1,1,0,1,0)$ *with* $f^* = 4$. *First, we convert the problem into an equivalent problem with a single surrogate constraint by means of the p-norm surrogate method.*

$$\begin{aligned} &\min\ f(x) \\ &\text{s. t. } g_s(x) = (g_1(x))^4 + (g_2(x))^4 + (g_3(x))^4 - 3 \times 3^4 \leq 0, \\ &\qquad x \in X = \{0,1\}^5. \end{aligned}$$

It is easy to see that: $\delta \geq 1$, $\underline{g_s} \geq 1$ *and* $\bar{f} \leq 12$. *So* $\ln 2/\delta \leq \ln 2 \approx 0.6931$ *and* $p\bar{f}/\underline{g_s} \leq 12p$. *Take* $p = 0.7$ *and* $\lambda = 8.5 > 8.4 = 12p$, *we obtain the following relaxation problem:*

$$\min_{x \in X} Q(x, 8.5, 0.7) = \min_{x \in \{0,1\}^5} 1/0.7 \ln[\exp(0.7f(x)) + \exp(8.5g_s(x))].$$

Its optimal solution is attained at a primal optimal solution $x_1^* = (1,1,0,0,0)$.

7 CONCLUSIONS

In this paper, we have briefly surveyed several new dual formulations recently reported to solve general integer programming problems. We have shown that the success guarantee of a dual search can be obtained by adopting p-th power transformation or some nonlinear Lagrangian formulations. These dual formulations provide us new insights into the distinct features in integer optimization. In particular, the logarithmic-exponential dual formulation holds a promise to deal with general integer programming problems from both the theoretical and computational aspects. It can be observed from our illustrative numerical example that the parameter p and the dual variable λ can be easily determined in certain special situations, e.g., 0-1 polynomial integer programming. It is worth to point out that the efficiency of the dual search also depends on the structure of the resulting relaxation problem at the lower level.

Acknowledgments

This research was partially supported by a grant from the Research Grants Council of the Hong Kong Special Administrative Region, China (RGC Ref. No. CUHK358/96P).

References

Cooper, M. W. (1981), A survey of methods for pure nonlinear programming, *Management Science*, Vol, 27, pp. 353-361.

Cooper, M. W. (1982), Nonlinear integer programming for various forms of constraints, *Naval Research Logistics Quarterly*, Vol. 29, pp. 585-592.

Fisher, M. L. (1981), The Lagrangian relaxation method for solving integer programming problems, *Management Science*, Vol. 27 pp. 1-18.

Geoffirion, A. M. (1974), Lagrangian relaxation for integer programming, *Mathematical Programming Study*, Vol. 2, pp. 82-114.

Gupta, O. K., Ravindran, A. (1985), Branch and bound experiments in convex nonlinear integer programming, *Management Science*, Vol. 31, pp. 1533-1546.

Körner, F. (1988), A new branching rule for the branch and bound algorithm for solving nonlinear integer programming problems, *BIT*, Vol. 28, pp. 701-708.

Li, D. (1999), Zero duality gap in integer programming: p-th norm surrogate constraint method, *Operations Research Letters*, Vol. 25, pp. 89-96.

Li, D. (1997b), Success guarantee of dual search in integer programming: p-th power Lagrangian method, technical report, SEEM97-004, Department of System Engineering and Engineering Management, The Chinese University of Hong Kong.

Sun, X. L. and Li, D. (1998), Logarithmic-exponential dual formulation for integer programming, technical report, SEEM98-002, Department of System Engineering and Engineering Management, The Chinese University of Hong Kong.

Yang, X. Q. and Goh, C. J. (1999), A nonlinear Lagrangian function for discrete optimization problems, in *From Local to Global Optimization*, edited by Migdalas, A., Pardalos, A., Varbrand, P. and Holmqvist, K., Kluwer, (to appear).

6 SIMULATED ANNEALING AND PENALTY METHODS FOR BINARY MULTICOMMODITY FLOW PROBLEMS

X.Q. Yang,

Department of Applied Mathematics,
The Hong Kong Polytechnic University,
Kowloon, Hong Kong

A.I. Mees and K. Campbell

Centre for Applied Dynamics and Optimization,
Department of Mathematics and Statistics,
The University of Western Australia,
Nedlands, WA 6907, Australia

Abstract: Multicommodity Flow (MCF) problems are an important class of combinatorial optimisation problems which can be used to model practical situations such as computer networks, traffic systems and warehouse allocation. The discrete Multicommodity Flow problem is NP-Hard. Many methods have been presented which attempt to find solutions to combinatorial problems such as the discrete MCF problem within a reasonable time frame. In this paper we look at the binary Multicommodity Flow problem and its representation using linear and nonlinear penalty methods. We implement the Simulated Annealing algorithm to find approximate solutions to the minimum cost problem, and compare the performance of variants of the algorithm on a set of test networks. Simulated Annealing requires the definition of a neighborhood of a solution: to enable this, we introduce the Painted Multi-Path Algorithm.

Key words: multicommodity flow, network optimization, penalty function, simulated annealing.

X. Yang et al. (eds.), Progress in Optimization, 93–105.

1 INTRODUCTION

The multicommodity flow problem (MCF) can be used to model a range of systems, including traffic systems: see Farvolden et al (1993), and LeBlanc et al (1975), and routing on a computer network: see Cantor and Gerla (1974). Previous solution methods for continuous MCF problems include decomposing the network into several single commodity problems in Kennington and Shalaby (1977), and applying a modified version of the simplex algorithm in Farvolden et al (1993). See Assad (1978) and Farvolden et al (1993) for a review of various techniques. The discrete multicommodity flow problem is NP-Hard, see Garey and Johnson (1979). This means that there is no known algorithm that will solve the discrete multicommodity flow problem in polynomial time.

The simulated annealing method is based on the physical process used to make spin glasses. The physical system operates so as to minimise the potential energy. An important parameter in this process is the temperature. If the temperature is allowed to cool slowly enough, an equilibrium state of glasses will always be reached. This idea was first applied to optimization problems by Kirkpatrick et al (1983). It was shown that simulated annealing method will converge to an optimal solution, see Lundy and Mees (1986). This convergence is not, however, guaranteed to happen in a 'reasonable' time, and for any given application, we cannot be sure of having found the optimum. In this sense, simulated annealing acts like a heuristic method and amounts to finding approximate solutions.

In this paper, the simulated annealing algorithm is developed for the binary multicommodity flow problem. Multi-path and random path algorithms are developed for defining a neighborhood of solutions for a given solution. The multi-path algorithm finds all possible paths between the source node and the sink node of a commodity, while the random path algorithm finds a specific path among them. The difficulty of solving the MCF is due to the existence of shared capacity constraints. We handle shared capacity constraints by penalty methods, so we are able to work with solutions that may not satisfy them, but we ensure that the divergence constraints are satisfied at all times. The shared capacity constraints are penalized by both linear and nonlinear penalty functions, developed recently in Tovey and Mees (1995) and Yang and Goh (1999). Numerical testing shows the simulated annealing method developed is efficient for small networks with a few commodities; more extensive tests on larger problems will be reported elsewhere.

In section 2, the binary multicommodity flow problem is formulated. We show that the Lagrangian dual method may be useful, but may fail to find the optimal solution of the binary multicommodity flow problem. In section 3, linear and nonlinear penalty functions are investigated. In section 4, a simulated annealing algorithm is developed using a painted multi-path algorithm and a random path algorithm to define a neighborhood structure. In section 5, numerical testing is given for networks with different commodities and densities.

2 THE BINARY MULTICOMMODITY FLOW PROBLEM

Assume that a directed network $\mathcal{G} = (\mathcal{N}, \mathcal{A})$ is given with incidence matrix E. A binary MCF can be defined on the network $\mathcal{G}$. Since we are looking at a binary problem, the flow of a commodity through an arc can be either 0 or 1.

We describe a system where we want to minimise the cost of the flow passing through a network whilst satisfying the capacity constraint on each arc.

Definition 2.1 Binary Multicommodity Flow Problem (BMCF) *Given a network with incidence matrix E, the Binary Multicommodity Flow Problem is to*

$$\textit{minimize} \quad f(X) = \sum_{j \in \mathcal{A}} c_j X_j, \tag{2.1}$$

subject to

$$EX = B, \tag{2.2}$$

$$X_j = \sum_{k=1}^{K} x_{j,k} \leq 1, \tag{2.3}$$

$$x_{j,k} \in \{0,1\}, \quad \forall\, j \in \mathcal{A},\ \forall\, k = 1, 2, ..., K. \tag{2.4}$$

Here,

(i) The flow $x_{j,k}$ of commodity $k = 1, \ldots, K$ in arc $j \in \mathcal{A}$ is either 0 or 1.

(ii) X is the multicommodity flow through the network. This is a $|\mathcal{A}| \times |K|$ matrix with jth row representing flows of all commodities on arc j of which the sum is X_j.

(iii) B is the supply-demand matrix in which each column has only two nonzero entries, 1 and -1 corresponding to the source and sink nodes for the corresponding commodity.

(iv) c_j is the cost of flow on arc j.

(v) As the capacity on each arc is assumed to be 1, this means that at most one commodity can flow through an arc.

Essentially, we want to minimize the cost f of flow X through the network $\mathcal{G} = (\mathcal{N}, \mathcal{A})$, where the cost function is defined in equation (2.1). The restrictions that are placed on X are that it satisfies the divergence constraint (2.2) and the capacity constraint (2.3).

The following example shows how the Lagrangian dual method may sometimes be applied to find solutions of the BMCF.

Example 2.1 *Consider a network with the following data*

$$c = \begin{pmatrix} 1 \\ 2 \\ 5 \\ 4 \\ 1 \\ 2 \end{pmatrix}; \quad E = \begin{pmatrix} 1 & 0 & 0 & 1 & -1 & 0 \\ -1 & 1 & 1 & 0 & 0 & 0 \\ 0 & 0 & -1 & 0 & 1 & -1 \\ 0 & -1 & 0 & -1 & 0 & 1 \end{pmatrix}; \quad B = \begin{pmatrix} 1 & 0 \\ 0 & -1 \\ 0 & 1 \\ -1 & 0 \end{pmatrix}.$$

By enumeration, the optimal value for (BMCF) is 6 *and the optimal solution is*

$$X = \begin{pmatrix} 0 & 1 \\ 0 & 0 \\ 0 & 0 \\ 1 & 0 \\ 0 & 1 \\ 0 & 0 \end{pmatrix}.$$

Lagrangian relaxation produces the following subproblem for each $\lambda = (\lambda_1, \cdots, \lambda_6) \geq 0$*:*

$$\begin{array}{ll} \textit{minimize} & \sum_{j=1}^{6} c_j X_j + \sum_{j=1}^{6} \lambda_j (\sum_{k=1}^{2} x_{j,k} - 1), \\ \textit{subject to} & EX = B, \end{array}$$

The optimal value for this subproblem is denoted by $\phi(\lambda)$*. The dual problem is*

$$\textit{maximize} \quad \phi(\lambda), \quad \textit{subject to} \quad \lambda \geq 0.$$

The optimal value for this problem is also 6 *which is attained at* $\lambda = (1, 0, 0, 0, 0, 0)$*. Hence there is no gap in this example.*

However, it is well-known that Lagrangian dual methods are only useful for continuous convex optimization problems, but normally fail to solve discrete optimization problems. Several such examples have been given in Bazaraa and Shetty (1979), and Sun and Li (1998). In this paper, we apply penalty methods and simulated annealing to solve the BCMF.

3 PENALTY FUNCTIONS FOR DISCRETE PROBLEMS

Penalty function methods have been developed for solving discrete optimization problems. In Tovey and Mees (1995), penalty functions were used to embed discrete optimization problems in continuous optimization probems and applied to the BCMF and to other problems. Several nonlinear penalty schemes were used for solving discrete optimization problems, such as minimax, p-th and logarithmic-exponential types in Sun and Li (1998) and Yang and Goh (1999). In this section, a discrete exact penalization is derived for nonlinear penalty functions, which is a counterpart of a similar result for linear penalty functions in Tovey and Mees (1995).

Consider the following discrete inequality constrained optimization problem P:

$$\begin{array}{ll}\text{minimize} & f_0(x) \\ \text{subject to} & x \in \mathcal{X}, \quad f_i(x) \le 0, \quad i \in I,\end{array}$$

where $\mathcal{X} \subset \mathbb{R}^n$ is a discrete set of finitely many elements, $I = \{1, \cdots, m\}$, and $f_i : \mathcal{X} \to \mathbb{R}$, $i \in I$ are real-valued functions.

Let the set of feasible solutions of problem P be

$$\mathcal{X}_0 = \{x \in \mathcal{X} : f_i(x) \le 0, \quad i \in I\}.$$

The following defines a penalty cost function for the discrete optimization problem, see e.g. Tovey and Mees (1995).

Definition 3.1 *A function ϕ that satisfies the following properties is said to be a penalty cost function of P:*
(a) $\phi(x) > 0$ if $x \notin \mathcal{X}_0$.
(b) $\phi(x) = 0$ if $x \in \mathcal{X}_0$.

A typical penalty function is the quadratic penalty function: (see Luenberger (1984))

$$\phi(x) = \sum_{i \in I} [\max\{0, f_i(x)\}]^2, \quad \text{(quadratic penalty function)} \qquad (3.1)$$

The following is a result for the penalty cost function of P (which is actually an exact penalty cost function.)

Theorem 3.1 *(Discrete exact penalization) If ϕ is a penalty cost function, then there is some $\hat{\alpha}$ such that any solution of the discrete penalty problem*

$$\textit{minimize} \quad f_0(x) + \alpha\phi(x), \quad \textit{subject to } x \in \mathcal{X} \qquad (3.2)$$

is a solution of problem P for all $\alpha \ge \hat{\alpha}$. Furthermore if x^* is an optimal solution to P, then $\hat{\alpha}$ can be chosen as

$$\hat{\alpha} = max_{x \in \mathcal{X} \setminus \mathcal{X}_0} \frac{f(x^*) - f(x)}{\phi(x)}.$$

It is clear that the cost function of the above discrete penalty problem is linear in the cost function of the original inequality constraint optimization problem. So we often refer it to as a linear penalty function.

Recently a new type of penalty function for the discrete problem P was given in Yang and Goh (1999) with multiple penalty parameters. As these penalty functions may be nonlinear in the cost function of the original discrete inequality constraint optimization problem, we refer to the corresponding penalty functions as nonlinear penalty functions. Here we present a generalization of the result in Yang and Goh (1999). From the computational point of view, we present the result with a single parameter.

Without loss of generality, assume that $f_0(x) > 0, \forall x \in \mathcal{X}$.

Definition 3.2 *Let $\gamma > 0$. The function $\varphi : \mathbb{R}^n \times \mathbb{R}^m \times \mathbb{R} \to \mathbb{R}$ is said to be a nonlinear penalty cost function for the discrete problem P if it satisfies the following properties:*
(a) $\varphi(x, f_0, f_1, \cdots, f_m, \gamma) \geq \gamma max_{1\leq i\leq m} f_i(x)$ if $x \notin \mathcal{X}_0$.
(b) $\varphi(x, f_0, f_1, \cdots, f_m, \gamma) = f_0(x)$ if $x \in \mathcal{X}_0$.
The nonlinear penalty problem is defined as

$$minimize \quad \varphi(x, f_0, f_1, \cdots, f_m, \gamma), \quad subject\ to\ x \in \mathcal{X} \tag{3.3}$$

Example 3.1 *The following functions satisfy the properties of a nonlinear penalty function: for $\gamma > 0$, $x \in \mathcal{X}$.*

$$\varphi_\infty(x, f_0, f_1, \cdots, f_m, \gamma) = max\{f_0(x), \gamma max_{1\leq i\leq m}\{0, f_i(x)\}\}, \tag{3.4}$$

$$\varphi_p(x, f_0, f_1, \cdots, f_m, \gamma) = \left(f_0^p(x) + \gamma^p \sum_{i=1}^{m} [max\{0, f_i(x)\}]^p\right)^{\frac{1}{p}}, \quad p > 0.$$

The following is a counterpart of Theorem 3.1.

Theorem 3.2 *(Discrete nonlinear exact penalization) Suppose that $\mathcal{X}$ is a finite set. If φ is a nonlinear penalty cost function for P then there is some $\hat{\gamma} > 0$ such that any solution of the nonlinear "unconstrained" discrete problem*

$$minimize \quad \varphi(x, f_0, f_1, \cdots, f_m, \gamma), \quad subject\ to\ x \in \mathcal{X}$$

is a solution of problem P for all $\gamma \geq \hat{\gamma}$. More precisely, if x^* is an optimal solution to P, then $\hat{\alpha}$ can be chosen as

$$\hat{\gamma} = max\left\{\frac{f_0(x^*)}{f_i(\hat{x})} : \hat{x} \in \mathcal{X} \setminus \mathcal{X}_0, f_i(\hat{x}) > 0\right\}.$$

Proof: Let $\mathcal{X}^*$ be the optimal solution set to P and let $x^* \in \mathcal{X}^*$. Assume that $\gamma \geq \hat{\gamma}$.

Let $\hat{x}$ be any solution to the nonlinear "unconstrained" discrete problem. If $\hat{x} \in \mathcal{X}_0$, then $\hat{x}$ solves P since $\varphi(x, f_0, f_1, \cdots, f_m, \gamma) = f_0(x)$ if $x \in \mathcal{X}_0$.

Assume for contradiction that $\hat{x} \notin \mathcal{X}^*$, then $\hat{x} \in \mathcal{X} \setminus \mathcal{X}_0$ by taking the contrapositive of the previous statement. Therefore $\varphi(\hat{x}, f_0, f_1, \cdots, f_m, \gamma) \geq \gamma f_i(\hat{x})$ for some i.

Since $\hat{x}$ is an optimal solution for (3.3) it follows that

$$\varphi(\hat{x}, f_0, f_1, \cdots, f_m, \gamma) \leq \varphi(x^*, f_0, f_1, \cdots, f_m, \gamma) = f_0(x^*).$$

Then by $\hat{x} \notin \mathcal{X}_0$ there exists i such that $f_i(\hat{x}) > 0$, thus

$$\gamma f_i(\hat{x}) \leq \varphi(\hat{x}, f_0, f_1, \cdots, f_m, \gamma) \leq \varphi(x^*, f_0, f_1, \cdots, f_m, \gamma) = f_0(x^*).$$

So

$$\gamma \leq \frac{f_0(x^*)}{f_i(\hat{x})}$$

a contradiction. Hence $\hat{x} \in \mathcal{X}^*$.

Discussions In some algorithms, discrete penalty problems are solved using branch and bound methods to implicitly enumerate and compare their continuous counterparts. In this case, the nonlinear penalty cost function $\varphi_p(x, f_0, f_1, \cdots, f_m, \gamma)$ has the potential to convexify a nonconvex function f_0. In fact, let $f_0(x) = \sqrt{|x|}$ and $p = 4$. Then the nonlinear penalty cost function

$$\bar{\varphi}_4(x, f_0, f_1, \cdots, f_m, \gamma) = f_0^4(x) + \gamma^4 \sum_{i=1}^{m} [\max\{0, f_i(x)\}]^4,$$

is convex on the feasible region. Note that this is obtained by removing the power $\frac{1}{4}$ of the the nonlinear penalty cost function

$$\varphi_4(x, f_0, f_1, \cdots, f_m, \gamma) = \left(f_0^4(x) + \gamma^4 \sum_{i=1}^{m} [\max\{0, f_i(x)\}]^4 \right)^{\frac{1}{4}},$$

as this does not matter when solving the nonlinear penalty problems.

However the linear penalty cost function $f_0(x)+\alpha\phi(x)$ is never convex on the feasible region. This difference does not usually matter when heuristic discrete algorithms are applied for solving penalty problems, but it does matter for any algorithm which solves continuous penalty problems.

The difficulties of solving BMCF is, in general, due to the existence of the capacity constraints (2.3). Thus these constraints are penalized. As applications of linear and nonlinear penalty functions to BMCF, we obtain the following relaxed problems of BMCF.

Linear Penalty Problem

$$\begin{aligned} \text{minimize} \quad f_{linear}(X) \quad &= \quad \sum_{j \in \mathcal{A}} c_j X_j + \alpha \sum_{j \in \mathcal{A}} (\sum_{k=1}^{K} X_{j,k} - 1)^2 \\ \text{subject to} \quad & \quad EX = B, \\ & \quad x_{j,k} \in \{0,1\}, \quad \forall\, j \in \mathcal{A},\ \forall\, k = 1, 2, ..., K. \end{aligned}$$

Nonlinear Penalty Problem

$$\begin{aligned} \text{minimize} \quad f_{nonlinear}(X) \quad &= \quad \max(\sum_{j \in \mathcal{A}} c_j X_j, \gamma \max_{j \in \mathcal{A}} \{0, \sum_{k=1}^{K} X_{j,k} - 1\}) \\ \text{subject to} \quad & \quad EX = B, \\ & \quad x_{j,k} \in \{0,1\}, \quad \forall\, j \in \mathcal{A},\ \forall\, k = 1, 2, ..., K. \end{aligned}$$

Note that these problems only have divergence constraints and can be solved much easier than BMCF. These problems will be solved using the simulated annealing algorithm to be developed in the next section.

4 SIMULATED ANNEALING FOR DISCRETE MULTICOMMODITY FLOWS

The simulated annealing algorithm uses the temperature parameter t to regulate the length of time taken by the algorithm before it terminates. It is therefore important to select the initial temperature t_i and the final temperature t_f with reference to the cooling schedule to ensure that the algorithm performs sufficient iterations, and ceases after a 'reasonable' amount of time. Moreover, the probability of accepting an uphill move decreases as the temperature decreases, hence at low temperatures, once the algorithm is in the region of a local minimum it is unlikely to escape. It was suggested by Lundy and Mees (1986) that the algorithm should be terminated when $t \leq \epsilon / \ln[(|\mathcal{X}| - 1)/\Theta]$, where $\mathcal{X}$ is the set of feasible flows for BMCF and Θ is the error probability of the solution from the optimal solution. Lundy and Mees (1986) also suggested that the algorithm cease at an earlier time, and a steepest descent search performed on the current solution. If the solution is within the region of local minima, and the probability of accepting uphill moves is negligible, then the descent search will find this minimum without attempting to perform uphill moves as would occur if simulated annealing was used. The starting temperature needs to be sufficiently high to ensure that all states are equally probable, and mobility of the spins is high. This will ensure that there is a large range of possible solutions.

The simulated annealing algorithm (SAA) for BMCF is given below.

Simulated Annealing Algorithm

```
Initialise search:

    choose initial solution X_0 ∈ 𝒳
    set X = X_0
    set X* = X_0
    set t_i and t_f, let t = t_i

while ¬(stopping condition) do

    for i=1..p do
        randomly select X' ∈ N_m(X), and let δf = f(X') − f(X)
        if δf < 0 then set X = X'
        else set X = X' with probability e^{−δf/t}
        if f(X) < f(X*), set X* = X'
    end for
```

$t = \alpha(t)$

```
end while
```

Remark 4.1: (i) If SAA is run with a temperature of zero, then effectively a random descent search is used.

(ii) p is preselected which decides the number of iterations required in the local search.

(iii) If $\delta f \geq 0$ then set $X = X'$ with probability $e^{-\delta f/t}$. To do this, generate a random number r within the interval [0,1]. If $r < e^{-\delta f/t}$ then set $X = X'$. If not, then do not change X.

(iv) As $t \to 0$, the probability of accepting uphill moves approaches zero.

(v) The stop condition is set as $t < t_f$ where t_f is the preselected final temperature.

(vi) The value $\alpha(t)$ is the cooling rate. It is suggested in Lundy and Mees (1986) that

$$\alpha(t) = \frac{t}{1 + \beta t}$$

where β is a parameter. The larger the β is, the faster the cooling is. This cooling scheme will be used in numerical tests in section 5.

SAA starts with a solution X_0 that satisfies the divergence constraint $EX = B$. Such a X_0 can be found by applying the Painted Multi-Path Algorithm and the Random Path Algorithm given below.

The Painted Multi-Path Algorithm is an adaptation of the Painted Path Algorithm, see Rockafellar (1984). The objective of the Painted Multi-Path Algorithm is to make it possible to find all paths between the source node $F(k)$ and the sink node $T(k)$ of a selected arc k while maintaining efficiency. The number of possible paths increases very quickly as the size of the network increases, but the algorithm is linear with the number of arcs because it does not explicitly generate the paths; instead it provides the information needed to generate the paths later in an efficient manner. Rather than using the set *label* in the Painted Path Algorithm, the matrix *NodeHistory* is used to record the preceding nodes.

The output of the algorithm is the matrix *NodeHistory*. *NodeHistory* can be used to generate all possible paths between any two nodes for a network $\mathcal{G}$ given the direction and painting of its arcs. The matrix *NodeHistory* records each node that can be used to immediately precede a node i in a path.

The Painted Multi-Path Algorithm

Initialisation:

set $S = \{F(k)\}$ and $NodeHistory = [0]_{|N|,|N|}$

```
repeat
```

```
Find the set M where
    M = {j : (F(j) ∈ S and T(j) ∈ 𝒩 \ S and j is green or
     white) or (T(j) ∈ S and F(j) ∈ 𝒩 \ S and j is
     green or black)}
If M ≠ {} then ∀ j ∈ M
    set i = { F(j),  if p = T(j) ∈ S
            { T(j),  if p = F(j) ∈ S
    set S = S ∪ {i}
    set NodeHistory_{i,p} = 1
until M = ∅
```

Remark 4.2: (i) That the arc j is 'green', 'white', or 'black' means that on the arc j the flow is permitted 'backwards or forwards', 'forwards', or 'backwards', respectively.

(ii) If the destination node $T(k)$ is not an element of S, then no path exists. This implies that all of the elements of the row of *NodeHistory* corresponding to the destination $T(k)$ are zero. If $T(k)$ is in S, then the Random Path Algorithm described below can be applied to *NodeHistory* to determine a path. The matrix $[0]_{|N|,|N|}$ is the $|N| \times |N|$ zero matrix.

Note that the solution obtained by the Random Path Algorithm does not necessarily satisfy the capacity constraint $X_j = \sum_{k=1}^{K} x_{j,k} \leq 1$. The implementation of SAA begins with an initial solution, and performs a search of neighbouring states to try to find a solution that satisfies the capacity constraint and has a low cost. We select a change ΔX to the flow X so that the divergence constraint remains satisfied. That is,

$$E[X + \Delta X] = EX + 0 = B$$

In order to find an alternative path for the flow through an arc k, we use the Multi Path Algorithm and Random Path Algorithm. The Multi Path Algorithm produces a list of all possible paths between the source node $F(k)$ and the sink node $T(k)$. A path is randomly selected from this list, and flow is pushed through this alternative route. Such a path is selected by the following Random Path Algorithm.

Random Path Algorithm

```
Initialisation:

    set pathnodes = {T(k)}
    set nodes = {}
    set visited = [0]_{|N|,|N|}
    set i_c = T(k)
```

```
while (F(k) ∉ pathnodes) do
    nodes = {i : i ∈ N and NodeHistory_{i_c,i} = 1
    and i ∉ pathnodes and visited_{i_c,i} = 0}
    if (nodes ≠ {}) then
        randomly select i_prev from nodes
        set pathnodes = {i_prev} ∪ pathnodes
        set visited_{i_c,i_prev} = 1
        set i_c = i_prev
    else
        set elements of i_c^th row of visited to 0
        set pathnodes = pathnodes \ {i_c}
        set i_c = pathnodes(1)
    end if
end while
```

5 RESULTS

The tests were run on several sets of sample networks. There were nine different types of networks, consisting of combinations of 2, 3 and 4 commodities with densities of 0.4, 0.6 and 0.8; every network has 10 nodes. The density of the network refers to the ratio between the number of arcs in the network and the total number of possible arcs, $\binom{10}{2} = 45$. The cost for each arc and each commodity ranged from 1 to 10 and the source-sink pair for each commodity was different. All these were generated randomly.

Table 5.1 shows the percentages of runs on which SAA terminated successfully with the best cost found. Each instance of the network was run 10 times with each of 4 different cooling schedule parameter $\beta = 0.05, 0.1, 0.15$ and 0.5. So there were 40 runs for each network instance. The stopping condition for each run was that the temperature became less than a given value which was calculated by

$$5/\log((|\mathcal{N}| * |\mathcal{A}| * K - 1)/\rho), \quad \rho = 0.85.$$

The best cost was the smallest cost found among these 40 runs. Of course, the best cost may be not the true optimal value but determining the true optimum was in general not feasible because it required enumerating of all possible feasible solutions which was outside the capacity of our workstations. The time for each SAA run ranged from a few seconds to a few minutes on a SGI O2 Workstation.

The reason the tests were performed for different β parameter values was to determine the effect that varying the parameter had on the performance of the algorithm. It is easy to see from Table 5.1 that small cooling parameter results in a better rate to find the best cost.

It is also noticeable that the success rate on low density networks is higher than that on high density networks, as we would expect since the sparse networks will have fewer local minima.

The average time for both linear and nonlinear penalty functions for a run is about a few seconds. The smaller cooling rate takes longer to terminate. Moreover, the time increases significantly as the size of the network increases. Some initial test shows that more than 1000 seconds are needed to find a solution for a network with 4 commodities, 20 nodes and 0.4 density.

In Table 5.1, the columns labelled l-p and n-p correspond to the success rate using linear and nonlinear penalty functions. (For example, 0.92 indicates a 92% success rate.) It can be seen that there is not much difference between the success percentages for linear and nonlinear penalty functions.

Table 5.1 Success Rates of SA

β	0.05		0.1		0.15		0.5	
(K,dens)	l-p	n-p	l-p	n-p	l-p	n-p	l-p	n-p
(2, 0.4)	0.92	0.93	0.86	0.87	0.79	0.74	0.54	0.47
(3, 0.4)	0.91	0.88	0.81	0.77	0.74	0.76	0.44	0.45
(4, 0.4)	0.89	0.85	0.75	0.75	0.68	0.67	0.31	0.34
(2, 0.6)	0.71	0.72	0.60	0.64	0.41	0.55	0.23	0.19
(3, 0.6)	0.45	0.57	0.30	0.28	0.28	0.24	0.05	0.05
(4, 0.6)	0.50	0.57	0.35	0.32	0.24	0.22	0.00	0.04
(2, 0.8)	0.35	0.40	0.23	0.26	0.13	0.19	0.04	0.06
(3, 0.8)	0.22	0.19	0.11	0.06	0.04	0.02	0.00	0.00
(4, 0.8)	0.16	0.15	0.04	0.02	0.03	0.00	0.00	0.00

6 CONCLUSIONS

In this paper, the binary multicommodity flow problem was represented using both linear and nonlinear penalty functions. The simulated annealing algorithm was used to solve the resulting optimization problems. Tests on some small networks showed that the algorithm can find the best cost in a few seconds with a reliable success rate and that there is little difference between the performance of linear and nonlinear penalty functions. Tests of the algorithm on large networks will be carried out in the future.

Acknowledgments

This research was partially supported by the Australian Research Council.

References

Assad, A. (1978), Multicommodity network flow - a survey, *Networks*, Vol. 8, pp. 37-91.

Bazaraa, M.S. and Shetty, C.M. (1979), *Nonlinear Programming*, John Wiley & Sons, New York, Singapore.

Cantor, D., and Gerla, M. (1974), Optimal routing in a packet-switched computer network, *IEEE Transactions on Computers*, Vol. C-23, pp. 1062-1068.

Farvolden, J., Powell, W, and Lustig, I., (1993), A primal partitioning solution for the arc-chain formulation of a multicommodity network flow problem, *Operations Research* Vol. 41, pp. 669-693.

Garey, M. and Johnson, D. (1979), *Computers and Intractability: A Guide to the Theory of NP-Complete*, W.H. Freeman. New York.

Kennington, J. and Shalaby, M. (1977), An effective subgradient procedure for minimal cost multicommodity flow problem, *Management Science*, Vol. 23, pp. 994-1004.

Kirkpatrick, S., Gelatt, C. and Vecchi, M. (1983), Optimization by simulated annealing, *Science*, Vol. 220, pp. 671-680.

LeBlanc, L., Morlok, E., and Pierskalla, W. (1975), An efficient approach to solving the road network equilibrium traffic assignment problem, *Transportation Research*, Vol. 9, pp. 309-318.

Lundy, M and Mees, A. (1986), Convergence of an annealing algorithm, *Mathematical Programming*, Vol. 34, pp. 111-124.

Luenberger, D.G. (1984), *Linear and Nonlinear Programming, second ed.*, Addison Wesley, Reading, MA.

Reeves, R. (eds) (1993) *Modern Heuristic Techniques for Combinatorial Problems*, Halsted Press, New York.

Rockafcllar, R.T. (1984), *Network Flows and Monotropic Optimization*, John Wiley & Sons, New York, Singapore.

Sun, X.L. and Li, D. (1998), New dual formulations in constrained integer programming, in *Progress in Optimization II: Contributions From Australasia*, edited by Yang, X.Q., Mees, A.I., Fisher, M. and Jennings, L.S., Kluwer, pp. 219-231.

Tovey, C. and Mees, A.I. (1995), A general approach to deterministic annealing, in *Recent Trends in Optimization Theory and Applications*, edited by Agarwal, R.P., World Scientific Publishing Co., pp. 409-421.

Yang, X. Q. and Goh, C. J. (1999), A nonlinear Lagrangian function for discrete optimization problems, in *From Local to Global Optimization*, edited by Migdalas, A., Pardalos, A., Varbrand, P. and Holmqvist, K., Kluwer, (to appear).

II NONSMOOTH OPTIMIZATION

7 A QUADRATIC RECOURSE FUNCTION FOR THE TWO-STAGE STOCHASTIC PROGRAM

John R. Birge,

School of Engineering and Applied Science,
Northwestern University,
Evanston, Illinois 60208, USA

Stephen M. Pollock

Department of Industrial and Operations Engineering,
University of Michigan,
Ann Arbor, MI 48109, USA

and Liqun Qi

School of Mathematics,
The University of New South Wales,
Sydney, New South Wales 2052, Australia

Abstract: We present a quadratic recourse representation of the two-stage stochastic linear problem. Unlike the usual linear recourse model, it is differentiable with respect to the first stage decision variables. This offers the possibility of applying high convergence rate methods to solve the two-stage problem. We show that the quadratic recourse function approximates the linear recourse function (and the corresponding solution of the two-stage problem with quadratic recourse converges to the solution of the two-stage problem with linear recourse) as a parameter $k \to \infty$ and another parameter $\epsilon_k \to 0$. We also give a bound for this approximation.

Key words: stochastic linear problem, quadratic recourse.

X. Yang et al. (eds.), Progress in Optimization, 109–121.

1 INTRODUCTION

One of the main paradigms of stochastic programming is represented by the two-stage stochastic linear program formulated as a master problem and a recourse problem (see Kall (1976); Kall and Wallace (1994)). The *master problem* is

$$\begin{array}{ll} \min & f(x) = c^t x + \Phi(x) \\ \text{s.t.} & Ax = b \\ & x \geq 0, \end{array} \tag{1.1}$$

where $x \in \mathbb{R}^n$ is the first-stage decision vector, $c \in \mathbb{R}^n$ is the cost coefficient vector for x, $Ax = b$ and $x \geq 0$ are the linear constraints on x, with $b \in \mathbb{R}^m$ and $A \in \mathbb{R}^{m \times n}$.

$\Phi(x)$ is the expected value of the linear recourse function, $\Phi(x) = E(\phi(x, \xi))$, where $\phi(x, \xi)$ is defined by the *recourse problem*

$$\begin{array}{rlrcl} \phi(x,\xi) = & \min & (q(\xi))^t y & & \\ & \text{s.t.} & W(\xi)y & = & T(\xi)x - h(\xi) \\ & & y & \geq & 0. \end{array} \tag{1.2}$$

Here ξ is a random variable of dimension r with distribution $P(\cdot)$, so by definition,

$$\Phi(x) = E(\phi(x,\xi)) = \int \phi(x,\xi) P(d\xi). \tag{1.3}$$

The vector, $y \in \mathbb{R}^s$, is the second-stage decision vector, $h \in \mathbb{R}^r$ is the demand or resource vector, and $T \in \mathbb{R}^{r \times n}$ is a technology matrix.

The usual origin of the constraint equations in (1.2) is the desire to satisfy the condition $T(\xi)x - h(\xi) = 0$. However, because of the random nature of both T and h, $T(\xi)x - h(\xi)$ does not equal zero in general. Thus Wy is introduced to represent the "discrepancy" (where $W \in \mathbb{R}^{r \times s}$) and $q \in \mathbb{R}^s$, $q > 0$, is the associated cost coefficient vector for non-zero y.

To simplify our discussion, we only discuss in detail the *fixed* recourse problem: when W does not depend upon ξ. We also assume that T and q are independent of ξ and $h = \xi$, conditions corresponding to uncertain resource levels but known prices and technologies. Equation (1.2) then becomes

$$\begin{array}{rlrcl} \phi(x,\xi) = & \min & q^t y & & \\ & \text{s.t.} & Wy & = & Tx - \xi \\ & & y & \geq & 0. \end{array} \tag{1.4}$$

For convenience, we let ξ be a discrete random vector, so that

$$\Phi(x) = \sum_{j=1}^{l} \phi(x, \xi_j) p_j, \tag{1.5}$$

where $p_j \geq 0$ and $\sum_{j=1}^{l} p_j = 1$. Our results can be readily extended to continuous ξ.

There are two disadvantages associated with the two-stage stochastic program with fixed recourse represented by equations (1.1), (1.5) and (1.4). First, generally Φ is nondifferentiable with respect to x, which prevents the use of algorithms with high rates of convergence. Secondly, (1.4) may not be feasible for some x. (A problem with (1.4) feasible for all x is called a problem with *complete recourse.* If (1.4) is feasible for all x satisfying the linear constraints of (1.1), then it is called a problem with *relatively complete* recourse.)

To address these difficulties we consider below a problem with a quadratic recourse function $\phi_k(x, \xi)$, where $k > 0$ is a parameter. This quadratic recourse function is always continuously differentiable, which paves the way for using algorithms with high rates of convergence to solve the two-stage stochastic program.

Two-stage quadratic recourse models have been proposed before. In particular, Rockafellar and Wets (1986) discuss the extended linear-quadratic problem (ELQP). ELQP introduces *additional* coefficient matrices but there are no direct links between its solutions and the solutions of the two-stage stochastic program with linear recourse. On the other hand, our quadratic recourse function will have a direct link with (1.4) in that it is proposed as an alternative way to model the original linear recourse problem. We show that the quadratic recourse function $\phi_k(x, \xi)$ converges to the linear recourse function $\phi(x, \xi)$, and the solution of the two-stage stochastic program with quadratic recourse converges to the solution of the two-stage stochastic program with linear recourse respectively, as $k \to \infty$. An error bound is also given for this convergence.

The remaining part of this paper is as follows. In Section 2 we present the quadratic recourse function $\phi_k(x, \xi)$, discuss its meanings and point out its differences from the ELQP model. We discuss its differentiability properties and algorithmic meanings in Section 3. In Section 4 we prove that it converges to the linear recourse function as $k \to \infty$. An error bound is given for this convergence and a simple example is presented to illustrate the use of this bound. We also show that the solution of the two-stage stochastic program with recourse $\phi_k(x, \xi)$ converges to the solution of the two-stage stochastic program with recourse $\phi(x, \xi)$. In Section 5 we make some concluding remarks.

2 THE RECOURSE FUNCTION

We now consider an alternative to (1.1), (1.4) and (1.5) that has the same second-stage optimal solution under our nonnegative objective assumptions. We do this by minimizing the square of the recourse function; i.e., we solve

$$\begin{array}{rllcl} \phi^2(x, \xi) = & \min & (q^t y)^2 & & \\ & \text{s.t.} & Wy & = & Tx - \xi \\ & & y & \geq & 0, \end{array} \tag{2.1}$$

for $q > 0$.

We could use this function directly in an optimization procedure, but more than first-order differentiability is needed for high convergence rate methods. We therefore approximate the problem represented in (2.1) by the following parametrized quadratic recourse function:

$$\begin{array}{rll}\phi_k^2(x,\xi) = & \min & (q^t y)^2 + k\|Wy - Tx + \xi\|^2 + \epsilon_k \\ & \text{s.t.} & y \geq 0,\end{array} \tag{2.2}$$

where $\|v\|^2 = v^t v$ is the 2-norm of the vector v, and k and ϵ_k are two positive parameters. We also assume $\phi_k(x,\xi) \geq 0$. Thus, there is a clearly defined ϕ_k that satisfies equation (2.2). Correspondingly, (1.1) and (1.5) are replaced by

$$\begin{array}{ll}\min & f_k(x) = c^t x + \Phi_k(x) \\ \text{s.t.} & Ax = b \\ & x \geq 0,\end{array} \tag{2.3}$$

where

$$\Phi_k(x) \equiv E(\phi_k(x,\xi)) = \sum_{j=1}^{l} \phi_k(x,\xi_j) p_j. \tag{2.4}$$

We may think of (2.2) as an alternative representation of the situation expressed by the equations of (2.1). Instead of minimizing $(q^t y)^2$ with the constraint $Wy = Tx - \xi$, we minimize the weighted sum of $(q^t y)^2$ and $\|Wy - Tx + \xi\|^2$, with k reflecting the relative importance of satisfying the $Tx - \xi = Wy$ constraints compared to minimizing $(q^t y)^2$. The additional parameter ϵ_k is added to ensure that $\phi_k^2 \geq \epsilon_k > 0$. This will be useful for establishing differentiability properties of ϕ_k. [We could also replace k by an $n \times n$ diagonal matrix K such that each diagonal element of K "weights" a component of $Wy - Tx + \xi$. In this case, (2.2) would have the form

$$\begin{array}{rll}\phi_k^2(x,\xi) = & \min & (q^t y)^2 + (Wy - Tx + \xi)^t K (Wy - Tx + \xi) + \epsilon_k \\ & \text{s.t.} & y \geq 0.\end{array} \tag{2.5}$$

Again, for simplicity, we restrict our discussion to the form (2.2), although the results below can be shown to hold under the conditions $\|K\| \to \infty$ and $\epsilon_k \to 0$.]

The idea of converting the objective function of (1.2) by a quadratic form is not new. For example, the ELQP model considers the dual problem of (1.4):

$$\begin{array}{rll}\phi(x,\xi) = & \max & (Tx - \xi)^t u \\ & \text{s. t.} & W^T u \leq q,\end{array} \tag{2.6}$$

where u is the dual variable vector. ELQP adds a quadratic term $\frac{1}{2}u^t H u$ to (2.6), to obtain

$$\begin{array}{rll}\phi_H(x,\xi) = & \max & (Tx - \xi)^t u + \frac{1}{2}u^t H u \\ & \text{s. t.} & W^T u \leq q,\end{array} \tag{2.7}$$

where H is an $n \times n$ positive definite symmetric monitor matrix.

As a result of formulations such as (2.7), superlinearly convergent algorithms have been developed by Qi and Womersley (1995), Chen, Qi, and Womersley (1995), and Birge, Chen and Qi (1994). However, the model represented by

(2.7) has no immediate relation to the original linear recourse model (1.4), and, in particular, it is hard to interpret the "meaning" of the matrix H. Moreover, although we may let $\|H\| \to 0$ so that $\phi_H \to \phi$, this results in computational instability in the second-order algorithms used to solve the two-stage recourse problem with recourse ϕ_H.

On the other hand, as we show below, when we solve the quadratic version of the original problem, not only does $\phi_k \to \phi$ as $k \to \infty$, but the algorithms used to solve the problem represented by (2.3), (2.4) and (2.2) are stable as $k \to \infty$.

3 DIFFERENTIABILITY OF THE QUADRATIC RECOURSE FUNCTION

By the theory of linear programming, $\phi(x, \xi)$, defined by (1.4), is not in general differentiable with respect to x. This makes it impossible to apply superlinearly convergent methods, such as the Newton method, to solve the stochastic program defined by (1.1), (1.4) and (1.5). Classically, superlinear convergence of a Newton method for solving a nonlinear optimization problem requires that the objective and the constraint functions of the nonlinear optimization problem are twice continuously differentiable. Based upon the superlinear convergence theory for nonsmooth equations in Pang and Qi (1993); Qi (1993); Qi and Sun (1994), Qi (1994) developed superlinearly convergent generalized Newton methods for solving a nonlinear optimization problem, whose objective and constraint functions are SC^1, (i.e., they are continuously differentiable and their derivatives are semismooth, see Pang and Qi (1995)).

In general, an SC^1 function is not twice differentiable. A nonlinear optimization problem with an SC^1 objective function and linear constraints is called an SC^1 problem in Pang and Qi (1995). It has been shown that the ELQP is an SC^1 problem (Qi (1994); Qi and Womersley (1995)). The superlinearly convergent generalized Newton method proposed in Qi (1994) was globalized by using a line search and the trust region strategy in Pang and Qi (1995) and Jiang and Qi (1996) respectively. These methods were applied to the ELQP in Birge, Chen and Qi (1994); Chen, Qi, and Womersley (1995); Qi and Womersley (1995).

In this section, we will show that the two-stage stochastic program defined by (2.3), (2.4) and (2.2) is also an SC^1 problem. This opens the way to apply the superlinearly and globally convergent generalized Newton method in Qi (1994); Pang and Qi (1995); Jiang and Qi (1996); Birge, Chen and Qi (1994) to solve this problem. Before doing this, we briefly review the definition of semismoothness of a vector function and related concepts of generalized Jacobians of vector functions.

Suppose that $F : \mathbb{R}^n \to \mathbb{R}^m$ is a Lipschitz vector function. By Rademacher's theorem, F is differentiable almost everywhere. Let D_F be the set where F is differentiable. At any point $x \in \mathbb{R}^n$, the B-differential (see Pang and Qi

(1993); Qi (1993)) of F at x is defined by

$$\partial_B F(x) = \{ \lim_{\substack{y \to x \\ y \in D_F}} \nabla F(y) \},$$

which is a nonempty set. The Clarke subdifferential (Clarke (1983)) of F at x is

$$\partial F(x) = \text{conv } \partial_B F(x),$$

which is a nonempty convex set. If for any $h \in \mathbb{R}^n$,

$$\lim_{\substack{V \in \partial F(x+th) \\ t \downarrow 0}} \{Vh\} \tag{3.1}$$

exists, then we say that F is semismooth at x. In this case, F is also directionally differentiable at x and $F'(x;h)$ equals the limit in (3.1). The concept of semismooth functions was introduced for functionals by Mifflin (1977) and extended to vector functions by Qi and Sun (1994).

The following theorem establishes the SC^1 property of the stochastic two-stage problem with quadratic recourse.

Theorem 3.1 *The stochastic program (2.3), where Φ_k is defined by (2.4) and (2.2), is an SC^1 problem.*

Proof. Since $\phi_k^2 \geq \epsilon_k > 0$, it suffices to prove that ϕ_k^2 is an SC^1 function with respect to x. Define $z \equiv Tx - \xi$ and $\psi_k(z) \equiv \phi_k^2(x, \xi)$. Then rewriting (2.2) gives

$$\begin{aligned} \psi_k(z) = \quad \min \quad & (q^t y)^2 + k\|Wy - z\|^2 + \epsilon_k \\ \text{s.t.} \quad & y \geq 0. \end{aligned} \tag{3.2}$$

If ψ_k is differentiable with respect to z, then ϕ_k^2 is differentiable with respect to x and

$$\nabla_x \phi_k^2(x, \xi) = T^t \nabla_z \psi_k(z).$$

Let $\bar{z} = (2z^t, 0)^t \in \mathbb{R}^{r+1}$, $\overline{W} = \binom{kW}{q^t}$, and $u = \overline{W}y \in \mathbb{R}^{r+1}$. Then $U = \{u = \overline{W}y : y \geq 0\}$ is a polyhedron in $\mathbb{R}^{r+1}$.

Define $g : \mathbb{R}^{r+1} \to \mathbb{R} \cup \{+\infty\}$ by

$$g(u) = \begin{cases} u^t u, & \text{if } u \in U, \\ +\infty, & \text{otherwise.} \end{cases}$$

Then g is a closed proper extended-valued strongly convex function.

We now can write (3.2) as

$$\begin{aligned} \psi_k(z) &= kz^t z - \max_{u \in U}\{\bar{z}^t u - u^t u\} + \epsilon_k & (3.3) \\ &= kz^t z - g^*(2z, 0) + \epsilon_k, & (3.4) \end{aligned}$$

where g^* is the conjugate function of g. By Theorem 23.5 of Rockafellar (1970), since g is strongly convex, g^* is finite and continuously differentiable everywhere and its derivative at $\bar{z}$ is the unique maximum point in the maximum operation in (3.3). Furthermore, the derivative of g^* is Lipschitz. Actually, it is not difficult to see that the unique maximum of the maximum operation in (3.3) is piecewise linear with respect to $\tilde{z}$. Hence, the derivative mapping of g^* is semismooth (Qi and Sun (1994)). This shows that ψ_k, hence Φ_k, is an SC^1 function. Therefore, the stochastic programming problem (2.3), (2.4) and (2.2) is an SC^1 optimization problem. □

Now we can apply the generalized Newton (SQP) method proposed in Qi (1994); Pang and Qi (1995); Jiang and Qi (1996) to solve (2.3). With an adequate nonsingularity condition, this method is superlinearly and globally convergent. In the next section, we will show that $\psi_k(z)$ converges to $\psi(z) \equiv \phi^2(x,\xi)$ as $k \to \infty$, and give an error bound. This shows that the generalized Newton method is also stable for this problem.

It is noted that, although g^* is convex, ψ_k and Φ_k are not convex in general. In fact, by (3.3), if W is a nonnegative matrix, then ψ_k is the difference of two convex functions of $\tilde{z}$.

If P is continuous, with an argument similar to that in Birge and Qi (1995) we can show that Φ_k is twice differentiable. Then superlinear convergence can be established for quasi-Newton methods solving (2.3).

4 APPROXIMATION TO THE LINEAR RECOURSE FUNCTION

In this section, we show that $\psi_k(z)$ approximates $\psi(z)$ monotonically from below when $k \to \infty$, and give an error bound for this approximation.

Theorem 4.1 *Suppose that (1.4) is feasible for $z = Tx - \xi$. Then for any $0 < k \le K$, we have*

$$\sqrt{\phi_k^2(x,\xi) - \epsilon_k} \le \sqrt{\phi_K^2(x,\xi) - \epsilon_K} \le \phi(x,\xi).$$

Proof. Let y_k and y_K be solutions to the minimum operations in the definitions of $\psi_k(z)$ and $\psi_K(z)$ by (3.2) respectively. Let y^* be any feasible solution of (1.4). Then we have

$$\begin{aligned}
\psi_k(z) - \epsilon_k &= (q^t y_k)^2 + k\|Wy_k - z\|^2 \\
&\le (q^t y_K)^2 + k\|Wy_K - z\|^2 \text{ by optimality of } y_k \\
&\le (q^t y_K)^2 + K\|Wy_K - z\|^2 \text{ since } k \le K \\
&= \psi_K(z) - \epsilon_K \\
&\le (q^t y^*)^2 + K\|Wy^* - z\|^2 \text{ by optimality of } y_K \\
&= (q^t y^*)^2 \text{ by feasibility of } y^* \\
&= \psi(z).
\end{aligned} \tag{4.1}$$

The conclusion of the theorem follows by taking square roots (and noting that $q^t y > 0$). □

Corollary 4.1 *In Theorem 4.1, as $k \to \infty$,*

$$\|Wy_k - z\| \to 0,$$

where y_k is the unique solution of the minimum operation in the definition of $\psi_k(z)$.

Proof. By step 2 of the proof of Theorem 4.1,

$$0 \le k\|Wy_k - z\|^2 \le \psi(z) - (q^t y_k)^2 \le \psi(z) < \infty.$$

The conclusion follows since $\psi(z)$ is finite. □

The next theorem shows that as $k \to \infty$, $\phi_k(x,\xi)$ converges to $\phi(x,\xi)$ and gives an error bound.

Theorem 4.2 *Suppose that (1.4) is feasible for $z = Tx - \xi$ and α is the maximum value of 2-norms of optimal dual solutions of (1.4). Then, $\sqrt{\phi_k^2(x,\xi) - \epsilon_k}$ monotonically converges to $\phi(x,\xi)$ from below and, for k large enough,*

$$0 \le \frac{\phi(x,\xi) - \sqrt{\phi_k^2(x,\xi) - \epsilon_k}}{\phi(x,\xi)} \le \frac{\alpha}{\sqrt{k}}. \tag{4.2}$$

Proof: Let y_k and y^* be the same as in Theorem 4.1. Then, y_k solves

$$\begin{array}{rlrcl} \sqrt{\phi_k^2(x,\xi) - \epsilon_k} = & \min & q^t y & & \\ & \text{s.t.} & Wy & = & Wy_k \\ & & y & \ge & 0, \end{array} \tag{4.3}$$

while y^* solves

$$\begin{array}{rlrcl} \phi(x,\xi) = & \min & q^t y & & \\ & \text{s.t.} & Wy & = & z \\ & & y & \ge & 0. \end{array} \tag{4.4}$$

By Corollary 4.1, (4.3) can be regarded as a perturbation of (4.4) with a small change of $Wy_k - z$ on the right hand side of the constraint of the linear program (4.4). By the perturbation theory of linear programming, we have

$$|\sqrt{\phi_k^2(x,\xi) - \epsilon_k} - \phi(x,\xi)| \le \alpha\|Wy_k - z\|.$$

By the proof of Corollary 4.1,

$$\|Wy_k - z\| \le \frac{q^t y^*}{\sqrt{k}}.$$

By Theorem 4.1,

$$\sqrt{\phi_k^2(x,\xi) - \epsilon_k} \le \phi(x,\xi).$$

Combining these three inequalities, we have (4.2). The last conclusion follows. □

Corollary 4.2 *If $\epsilon_k \to 0$, then $\phi_k(x,\xi)$ converges to $\phi(x,\xi)$.*

Proof: The conclusion follows mfrom (4.2) and Theorem 4.1. □

Figure 4.1 Convergence of ϕ_k to ϕ for $z = (1, 0.5)$.

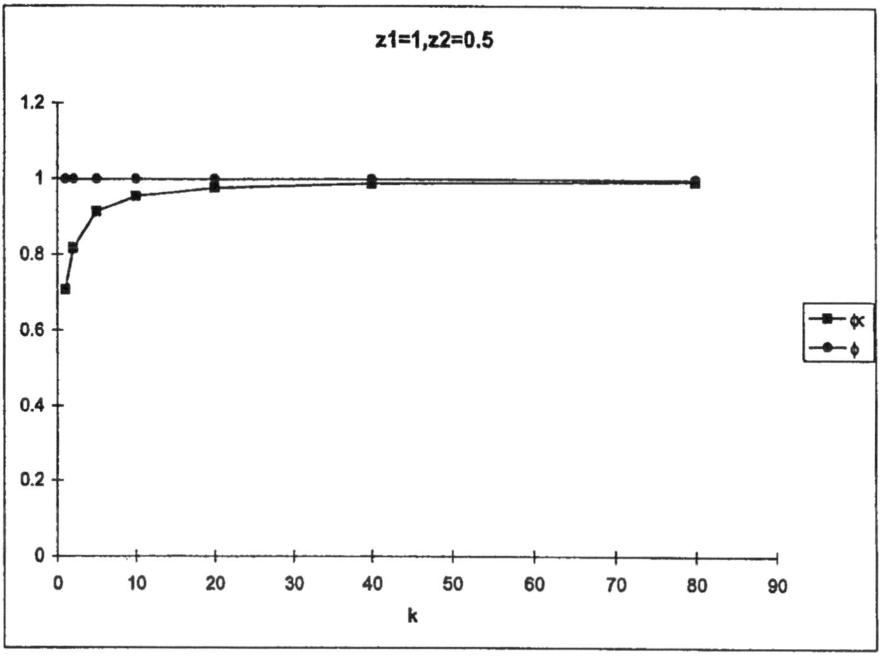

Example. To see how these bounds appear in practice, consider the following simple example (where $\epsilon_k = 0$ for illustration).

$$\begin{array}{llll} \phi(x,\xi) = \min & y_1 + y_2 + y_3 & & \\ \text{s.t.} & y_1 + y_3 & = & z_1 \\ & y_2 + y_3 & = & z_2 \\ & y & \geq & 0. \end{array} \tag{4.5}$$

Figures 1 and 2 show convergence of ϕ_k for $z = (1, 0.5)$ and $z = (1, 1)$. Note that the convergence is somewhat faster for $z = (1, 1)$. In this case, the dual solution to (4.5) is not unique and has the same maximum norm, $\alpha = 1$, as for $z = (1, 0.5)$. The convergence behavior for $z = (1, 1)$ may benefit from a smaller norm for the average of the norms of dual solutions.

Figure 4.2 Convergence of ϕ_k to ϕ for $z = (1, 1)$.

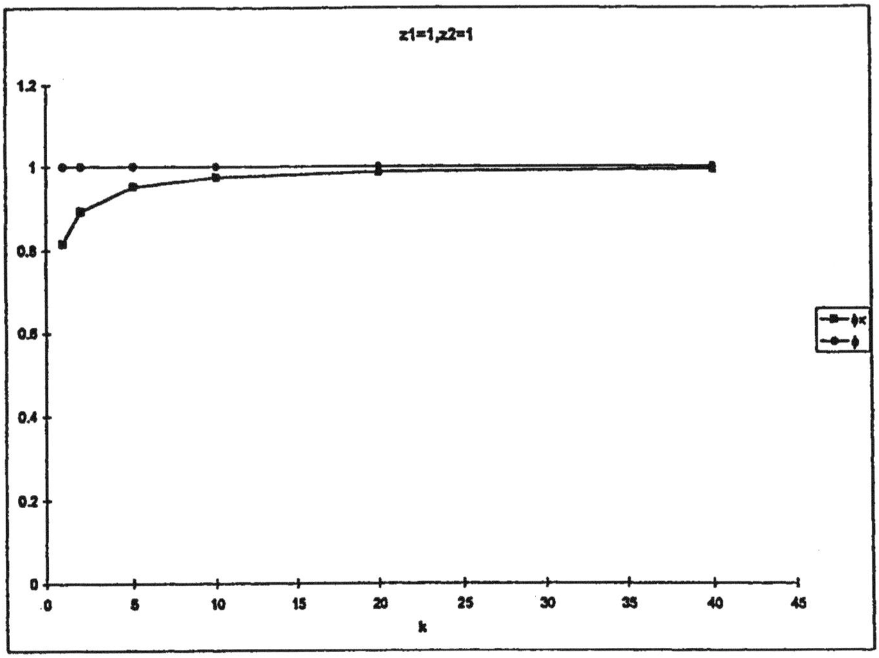

Theorem 4.2 also gives us the condition for the optimal solutions of (2.2) to converge to an optimal solution of (1.2). We use the following two results from the theory of *epi-convergence* (Attouch and Wets (1981)). A sequences of functions, $\{f^\nu\}$, *epi-converges* to f, if the epi-graphs of f^ν, converge as sets to the epi-graph of f.

Theorem 4.3 *(Birge and Wets (1986), Theorem 2.3.) Suppose a sequence of functions,* $\{g^\nu, \nu = 1, \ldots, \}$, *epi-converges to* g, *then* $\limsup_{\nu\to\infty}(\mathit{inf} g^\nu) \leq \mathit{inf} g$, *and, if* $x^k \in \operatorname{argmin} g^{\nu_k}$ *for some subsequence,* $\{g^\nu\}$, *of functions from* $\{g^{\nu_k}\}$, *and* $x = \lim_k x^k$, *then*

$$x \in \operatorname{argmin} g, \text{ and } \lim_k(\mathit{inf} g^{\nu_k}) = \mathit{inf} g. \tag{4.6}$$

The next result shows that the sequence $\{\phi_k\}$ converges to ϕ.

Theorem 4.4 *(Attouch and Wets (1981), Proposition 3.12). Suppose a sequence of functions,* $\{g^\nu, \nu = 1, \ldots, \}$, *pointwise converge everywhere to* g, *i.e.,*

$\lim_\nu g^\nu(x) = g(x)$. *If the* g^ν *are monotone increasing, or monotone decreasing, and* g *is lower semi-continuous, then* g^ν *epi-converges to* g.

The following theorem follows immediately from these observations.

Theorem 4.5 *(Birge and Wets (1986), Theorem 2.7.) Suppose* $\{g, g^\nu, \nu = 1, \ldots\}$ *form a collection of functions defined on* $\mathbb{R}^n \times \Xi$, *with value in* $\mathbb{R} \cup \infty$, *for all* x, $\xi \to g(x,\xi)$ *is measurable, if* $P[\xi | g(x,\xi)) < \infty] = 1$, *then* $\int_\Xi g(x,\xi)P(d\xi) < \infty$, *and* $g^\nu(x,\xi)$ *epi-converges to* $g(x,\xi)$ *for all all* $\xi \in \Xi$, *and all* g^ν *are absolutely bounded by integrable functions, then the expectation functionals,* $\int_\Xi g^\nu(x,\xi)P(d\xi) < \infty$, *epi- and pointwise converge to* $\int_\Xi g(x,\xi)P(d\xi) < \infty$.

Now, we can state our result for convergence of the optimal solutions.

Theorem 4.6 *Suppose that* ξ *has finite second moments, let* $\{x^k, \pi^k\}$ *be a sequence of optimal primal-dual solution pairs to (2.3), where the multipliers* π^k *are associated with the linear constraints,* $Ax = b$. *If* $\{x^k, \pi^k\}$ *has a limit point,* (x^*, π^*), *with* $\epsilon_k \to 0$, *then* (x^*, π^*) *is an optimal primal-dual solution pair of (1.1).*

Proof. If ξ has second moments, then it can be shown easily that ϕ is absolutely integrable (see, for example, (Kall (1976))) and that, if $P[\xi|\phi(x,\xi)) < \infty] = 1$, then $\int_\Xi \phi(x,\xi)P(d\xi) < \infty$. The extensions of these results to ϕ_k follow, for example, from Theorem 4.2.

To invoke Theorem 4.5, let $\tilde{\phi}_k(x,\xi) = \sqrt{\phi_k^2(x,\xi) - \epsilon_k}$ and define $g^k(x,\xi) = \phi_k(x,\xi) + \delta_{\{Ax=b, x\geq 0\}}(x)$, where the indicator function, $\delta_S(x) = 0$, if $x \in S$, and $+\infty$ if $x \notin S$. Using the bound in Theorem 4.2 and the result from Theorem 4.4, we then meet the conditions for Theorem 4.5. If we replace f_k in (2.3) with $\tilde{f}_k$, where ϕ_k is replaced by $\tilde{\phi}_k$, then the expectation functionals $\tilde{f}_k$ converge. Using Theorem 4.3, we have that limit points of minimizers of $\tilde{f}_k$ are optimal in (1.1).

Next, we must relate $\tilde{f}_k$ to f_k. Note that the optimality conditions for $\tilde{f}_k$ yield $\tilde{x}_k$ and $\tilde{\pi}_k$ such that

$$c + \nabla\tilde{\Phi}_k(\tilde{x}_k) - \tilde{\pi}_k^t A = \tilde{\rho}_k \geq 0, \tilde{\rho}_k^t \tilde{x}_k = 0. \tag{4.7}$$

For f_k, the optimality conditions are:

$$c + \nabla\Phi_k(x_k) - \pi_k^t A = \rho_k \geq 0, \rho_k^t x_k = 0. \tag{4.8}$$

Note that for any x such that ϕ_k is finite for all ξ, we have $\nabla\phi_k$ continuous and, thus,

$$\begin{aligned} \nabla\Phi_k(x) &= \int \nabla\phi_k(x,\xi)P(d\xi) \\ &= \int \frac{\tilde{\phi}_k(x,\xi)}{\phi_k(x,\xi)} \nabla\tilde{\phi}_k(x,\xi)P(d\xi) \\ &= \Delta_k \nabla\tilde{\Phi}_k(x), \end{aligned} \tag{4.9}$$

where $\Delta_k \to 1$ as $k \to \infty$. For any limit point, $(\bar{x}, \bar{\pi})$ of (x_k, π_k), we must have, from (4.8):

$$c + \nabla\tilde{\Phi}(\bar{x}) - \bar{\pi}^t A = \bar{\rho} \geq 0, \bar{\rho}^t \bar{x} = 0, \tag{4.10}$$

where $\tilde{\Phi}(x) = \lim_k \tilde{\Phi}_k(x)$. Thus $(\bar{x}, \bar{\pi})$ forms an optimal primal-dual pair in (2.3) with $\tilde{\Phi}$ replacing Φ_k, and $(\bar{x}, \bar{\pi})$ must be optimal in (1.1) by the argument above. □

5 CONCLUSIONS

We present a quadratic approximation of the stochastic program with linear recourse. The approximation uses a linear-quadratic loss function similar to that used in decision theory. It also allows the use of superlinearly convergent methods that cannot be directly to the linear recourse problem. We showed that the approximation converges and that solutions of the approximation also converge to those of the original problem. A possible further research topic is to use this quadratic approximation and the recently developed smoothing Newton method (Chen, Qi and Sun (1998)) to construct a globally and superlinearly convergent algorithm to solve the stochastic program with linear recourse.

Our interest in this form of quadratic approximation extends beyond an advantage for fast computational methods. We believe that the approach may also enable further approximations using known moments of the relevant random variables.

Acknowledgments

The first author's work was supported in part by the National Science Foundation under Award Number DDM-9215921. The third author's work was supported in part by the Australian Research Council.

References

Attouch, H. and Wets, R. J-B (1981), Approximation and convergence in nonlinear optimization, in *Nonlinear programming, 4*, edited by Mangasarian, O.L., Meyer, R.R. and Robinson, S.M. Academic Press, New York-London.

Birge, J.R., Chen, X. and Qi, L. (1994), A stochastic Newton method for solving stochastic quadratic programs with recourse, AMR 94/9, Applied Mathematics Report, University of New South Wales.

Birge, J.R. and Qi, L. (1995), Subdifferentials in approximation for stochastic programming, *SIAM Journal on Optimization*, Vol. 5, pp. 436-453.

Birge, J.R. and Wets, R. J–B. (1986), Designing approximation schemes for stochastic optimization problems, in particular, for stochastic programs with recourse, *Mathematical Programming Study*, Vol. 27, pp. 54-102.

Chen, X., Qi, L. and Sun, D. (1998), Global and superlinear convergence of the smoothing Newton method and its application to general box constrained variational inequalities, *Mathematics of Computation*, Vol. 67, pp. 519-540.

Chen, X., Qi, L. and Womersley, R. (1995), Newton's method for quadratic stochastic programs with recourse, *Journal of Computational and Applied Mathematics*, Vol. 60, pp. 29-46.

Clarke, F.H. (1983), *Optimization and Nonsmooth Analysis* Wiley, New York.

Jiang, H. and Qi, L. (1996), A globally and superlinearly convergent trust region algorithm for convex SC^1 minimization problems and its application to stochastic programs, *Journal of Optimization Theory and Applications*, Vol. 90, pp. 649-670.

Kall, P. (1976), *Stochastic Linear Programming*, Springer Verlag, Berlin.

Kall, P. and Wallace, S.T. (1994), *Stochastic Programming*, Wiley, New York.

Mifflin, R. (1977), Semismooth and semiconvex functions in constrained optimization, *SIAM Journal on Control and Optimization*, Vol. 15, pp. 957-972.

Pang, J.-S. and Qi, L. (1993), Nonsmooth equations: Motivation and algorithms, *SIAM Journal on Optimization*, Vol. 3, pp. 443–465.

Pang, J.-S. and Qi, L. (1995), A globally convergent Newton method for convex SC^1 minimization problems, *Journal of Optimization Theory and Applications*, Vol. 85, pp. 633-648.

Qi, L. (1993), Convergence analysis of some algorithms for solving nonsmooth equations, *Mathematics of Operations Research*, Vol. 18, pp. 227–244.

Qi, L. (1994), Superlinearly convergent approximate Newton methods for LC^1 optimization problems, *Mathematical Programming*, Vol. 64, pp. 277–294.

Qi, L. and Sun, J. (1994), A nonsmooth version of Newton's method, *Mathematical Programming*, Vol. 58, pp. 353–367.

Qi, L. and Womersley, R.S. (1995), An SQP algorithm for extended linear problems in stochastic programming, *Annals of Operations Research*, Vol. 56, pp. 251-285.

Rockafellar, R.T. (1970), *Convex Analysis*, Princeton University Press, Princeton, N.J.

Rockafellar, R.T. and Wets, R.J.-B. (1986), A Lagrangian finite-generation technique for solving linear-quadratic problems in stochastic programming, *Mathematical Programming Study*, Vol. 28, pp. 63-93.

8 LAGRANGE MULTIPLIERS FOR NONCONVEX OPTIMIZATION

B. D. Craven

Department of Mathematics and Statistics,
The University of Melbourne,
Parkville, Victoria 3052, Australia

Abstract: Consider a constrained minimization problem, in which directional derivatives exist, but are not convex functions of direction. Necessary conditions are obtained for optimality, in which the Lagrange multipliers are generally functions of direction. Sufficient conditions, also an analog of Wolfe's dual, are obtained under invex-type hypotheses.

Key words: nonconvex optimization, necessary conditions, Lagrange multipliers, sufficient conditions.

X. Yang et al. (eds.), Progress in Optimization, 123–128.

1 INTRODUCTION

Assume that the constrained minimization problem:

$$\min \quad f(x) \ \text{ subject to } g(x) \leq 0, \tag{1.1}$$

in which $f : \mathbb{R}^n \rightarrow \mathbb{R}$ and $g : \mathbb{R}^n \rightarrow \mathbb{R}^m$ are nonconvex Lipschitz functions, reaches a local minimum at $x = p$. The minimum can be described using Clarke generalized directional derivatives, see Clarke (1983), however these derivatives represent local convex approximations to nonconvex functions, so may not describe the local behaviour at all closely. The following simple examples show this, in cases when the functions are nonconvex and directionally differentiable.

Example 1.1

$$\min_{x \in \mathbb{R}} f(x) \quad \textit{subject to} \quad x \leq 0,$$

where

$$f(x) := \begin{cases} -x & \textit{if } x < 0, \\ -cx & \textit{if } x \geq 0 \end{cases}$$

with $c > 1$. The minimum is at $x = 0$. Since the gradients of f near 0 are -1 and $-c$, the Clarke subdifferential $\partial^C f(0) = \ell co\{-1, -c\} = [-c, -1]$ (ℓco denotes the closed convex hull). Hence the Clarke directional derivatives are

$$f^c(0; -1) = sup_{v \in [-c,-1]} \langle v, -1 \rangle \ = c$$

$$f^c(0, +1) = sup_{v \in [-c,-1]} \langle v, +1 \rangle = -1.$$

These (as is well known) do not relate to the directional derivatives, which are 1 and $-c$ respectively. Although Karush-Kuhn-Tucker (KKT) conditions using $\partial^C f(0)$ are well known:

$$c + \lambda(-1) \geq 0 \quad \textit{and} \quad -1 + \lambda(1) \geq 0$$

(for negative and positive directions), and are satisfied with multiplier $\lambda = 1$, one may ask whether any modification of KKT will hold for directional derivatives? (Obviously one cannot satisfy both $1 + \lambda(-1) \geq 0$ and $-c + \lambda(1) \geq 0$ with the same λ.)

Example 1.2 *Consider the (maximizing) linear program indicated by Figure 1 as follows:*

$$\max_{x \in \mathbb{R}^2} x_1 + x_2 \textit{ subject to } x_1 \geq 0, \ x_2 \geq 0, \ -\alpha x_1 + x_2 \leq 1, \ x_1 + \alpha x_2 \leq 1 + \alpha,$$

with $0 < \alpha < 1/2$. A nonsingular affine transformation of the domain can transform it equivalently to a program with active constraints $x_1 \leq 1, x_2 \leq 1$. See Figure 2. So it suffices to consider the linear program:

$$\max_{x \in \mathbb{R}^2} c^T x \quad \textit{subject to} \quad x_1 \leq 1, x_2 \leq 1$$

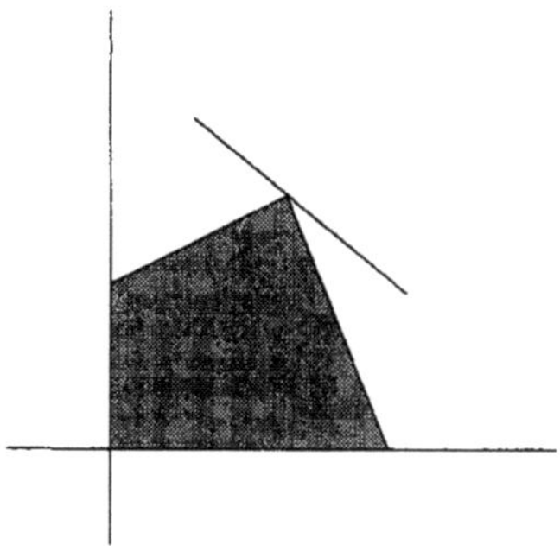

Figure 1

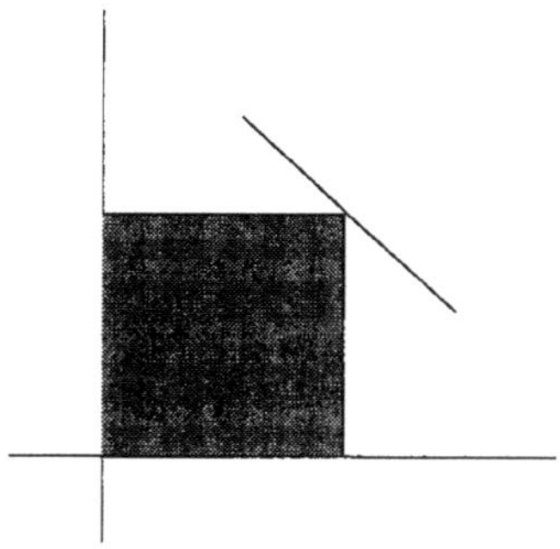

Figure 2

with some $c = [c_1, c_2]^T > 0$, *omitting constraints inactive at the optimum point. The optimum at* $(1,1)$ *is unique, and necessary KKT conditions require*

$$-[c_1, c_2] + [\lambda_1, \lambda_2] = [0, 0],$$

hence $\lambda^T := [\lambda_1, \lambda_2] = c^T$. *Consider a Lagrangian* $L(x, \mu) := -c^T x + \mu^T (x - e)$, *with* μ *to be chosen and* $e = [1, 1]^T$. *For a direction* d *pointing into the feasible region from the optimum point, thus for* $d \leq 0$, *the directional derivative of* $L(., \mu)$ *in direction* d *is* $(-c^T + \mu^T)d$, *and this is* ≤ 0 *provided that* $-c^T + \mu^T \geq 0$.

These examples motivate the question, whether necessary conditions can be found for an optimum of a nonconvex directionally differentiable function, using the directional derivatives and Lagrange multipliers that may depend on the direction.

2 NECESSARY CONDITIONS

Consider the problem (1.1) with f and g also directionally differentiable at the minimum point $p \in \mathbb{R}^n$. It is *not* assumed that the directional derivatives $f'(p; d)$ and $g'(p; d)$ are convex functions of direction d. Denote by $\mathbf{D}$ the set of feasible directions from p, allowed by the constraint $g(x) \leq 0$. Let $\mathbf{M} := \{(f'(p; d), g'(p; d)) : d \in \mathbf{D}\}$.

Theorem 2.1 *(Necessary conditions) Let* p *be a local minimum for (1). For each* $d \in \mathbf{D}$, *there are multipliers* $\tau \geq 0$ *and* $\sigma \geq 0$, *not both zero and depending on* d, *such that the following condition of Fritz John type holds:*

$$\tau f'(p; d) + \sigma g'(p; d) \geq 0, \sigma^T g(p) = 0. \tag{2.1}$$

If also $g'(p; d) < 0$ *then* $\tau \neq 0$.

Proof: Fix $d \in \mathbf{D}$, with $\|d\| = 1$. The minimum is unaffected by omitting constraints inactive at p. Then, for some $\delta > 0$, $\phi(\alpha) := f(p+\alpha d)$ is minimized at $\alpha = 0$ over $[0, \delta)$, subject to $\rho(\alpha) := g(p + \alpha d) \leq 0$. Hence necessary Fritz John conditions hold. Thus, there exist $\tau \geq 0$ and $\sigma \geq 0$, not both zero and depending on d, such that $\tau\phi'(0) + \sigma\rho'(0) \geq 0$ Since $\phi'(0) = f'(p; d)$ and $\rho'(0) = g'(p : d)$, and σ may be taken as 0 whenever $g_j(p) < 0$, (2.1) follows. If $\tau = 0$ then $\sigma g'(p; d) \geq 0$ and $\sigma > 0$, contradicting $g'(p; d) < 0$. □

Proposition 2.1 *(See Craven and Glover (1996), Lemma 10) Let A be compact and convex; let B be a closed convex cone; then*

$$(\exists a \in A)(\forall b \in B)\ a^T b \geq 0 \ \Leftrightarrow (\forall b \in B)(\exists a \in A)\ a^T b \geq 0.$$

Theorem 2.2 *Let Ξ be any closed convex cone contained in $\mathbf{M}$. Denote $q := (f, g)$. Assume that some d^* satisfies $q'(p; d^*) \in \Xi$ and $g'(p; d^*) < 0$. Then there exists a multiplier $\lambda \in \mathbb{R}^m_+$, dependent on Ξ but not on $d \in \Xi$, such that*

$$(\forall d \in \Xi) f'(p; d) + \lambda^T g'(p; d) \geq 0,\ \lambda^T g(p) = 0. \tag{2.2}$$

Proof: From Theorem 2.1, for each $d \in \mathbf{D}$, there exist τ and σ satisfying (2.1). Denote $\theta^T := (\tau, \sigma^T)$; let $K := \{\rho \in \mathbb{R}^{m+1}_+ : \rho^T e = 1\}$, where e denotes a vector of ones. Then $0 \neq \theta \in \mathbb{R}^{m+1}_+$, and $\theta \in K$ can be assumed. Then

$$(\forall d \in \mathbf{D})(\exists \theta \in K)\ \theta^T q'(p; d) \geq 0.$$

Let Ξ be any closed convex cone contained in $\{q'(p; d) : d \in \mathbf{D}\}$. Substituting $w := q'(p : d)$,

$$(\forall w \in \Xi)(\exists \theta \in K)\ \theta^T w \geq 0. \tag{2.3}$$

Since Ξ is a closed convex cone and K is compact convex, Proposition 2.1 shows that

$$(\exists \theta \in K)(\forall w \in \Xi)\ \theta^T w \geq 0.$$

Thus there exist $\beta \geq 0$ and $\lambda \in \mathbb{R}^m_+$, not both zero, such that

$$q'(p; d) \in \Xi \implies \beta f'(p; d) + \lambda^T g'(p; d) \geq 0,\ \lambda^T g(p) = 0.$$

If $\beta = 0$, and $(\exists d^*) q'(p; d^*) \in \Xi$, $g'(p; d^*) < 0$, then $\lambda^T g'(p; d^*) \geq 0$ and < 0, a contradiction; hence $\beta \neq 0$, and (2.2) follows. □

Remarks 2.1 From Example 1.1, the same λ cannot generally be found for all feasible directions d.

If (2.3) holds, then it still holds when Ξ is enlarged to $\Xi + \mathbb{R}^{m+1}_+$. Under some generalized convexity hypotheses, $\Xi + \mathbb{R}^{m+1}_+$ may be a convex set, although Ξ is not.

These necessary conditions consider only directions $d \in \mathbf{D}$, so do not include all directions $x - p$ when $g(x) \leq 0$, as for example when p is at a cusp of a nonconvex feasible set.

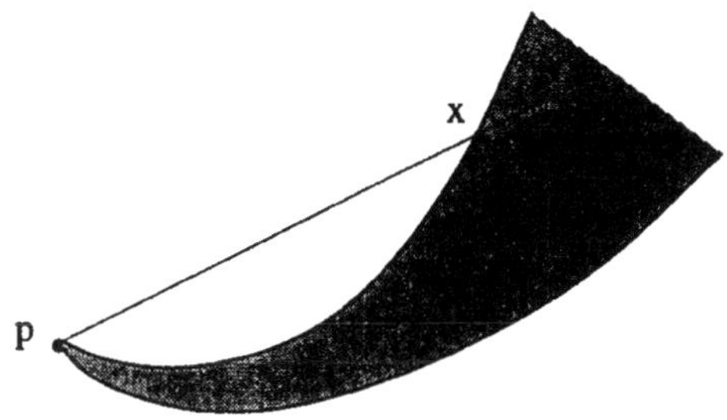

Figure 3

3 SUFFICIENT CONDITIONS

Denote by S^n the unit sphere in $\mathbb{R}^n$. Consider the hypotheses:

$$(\exists\mu : S^n \to \mathbb{R}^m_+)(\forall d \in \mathbf{D}) f'(p;d) + \mu(d)^T g'(p;d) \geq 0;\ \mu(d)^T g(p) = 0; \quad (3.1)$$

and the following *directional invex* hypotheses on functions f and g near p:

$$\begin{aligned}(\exists\omega : \mathbb{R}^n \times \mathbb{R}^n \to \mathbf{D})(\forall z) \quad & f(p+z) - f(p) \geq f'(p;\omega(z,p)); \\ & g(p+z) - g(p) \geq g'(p;\omega(z,p)). \end{aligned} \quad (3.2)$$

Theorem 3.1 *(Sufficient conditions) If (3.1) and (3.2) hold, then p is a minimum for (1.1).*

Proof: Let $g(p+z) \leq 0$. From (3.2) with $\lambda(.) \geq 0$,

$$[f(p+z)+\lambda(z)^T g(p+z)] - [f(p)+\lambda(z)^T g(p)] \geq f'(p;\omega(z))+\lambda(z)^T g'(p;\omega(z)) \geq 0$$

from (3.1), given $\omega(z) \in \mathbf{D}$. Since $\lambda(z)^T g(p+z) \leq 0$, and $\lambda(z)^T g(p) = 0$ from (2.3), it follows that $f(p+z) \geq f(p)$, proving the minimum. □

4 DUALITY

Consider the following possible Wolfe-type dual problem to (1.1):

$$\begin{aligned} \max_{u,v} \quad & f(u) + v^T g(u) \\ \text{subject to} \quad & v \geq 0,\ f'(u;.) + v^T g'(u;.) \geq 0 \text{ on } \Xi. \end{aligned} \quad (4.1)$$

where Ξ, as in Theorem 2.2, is a closed convex cone contained in $\mathbf{M}$.

Theorem 4.1 *(Duality) Assume the hypotheses of Theorem 2.2, and the directional invex hypothesis (3.2), with* $\omega(x-u,u) \in \Xi$. *Then problem (4.1) is a strong dual to problem (1.1).*

Proof: If x is feasible for (1.1) and (u,v) is feasible for (4.1), then

$$[f(x)+v^T g(x)] - [f(u)+v^T g(u)] \geq f'(u;\omega(x-u,u)) + g'(u;\omega(x-u,u)) \geq 0. \tag{4.2}$$

Since also $v^T g(x) \geq 0$, (4.2) proves weak duality for (1.1) and (4.1).

Under the hypotheses of Theorem 2.2, there exists $\lambda \geq 0$ such that $(u,v) = (p,\lambda)$ is a feasible point for (4.1). With weak duality, this shows that there is zero duality gap between (1.1) and (4.1). □

References

Clarke, F.H. (1983), *Optimization and Nonsmooth Analysis*, Wiley-Interscience, New York.

Craven, B.D. and Glover, B.M. (1996), An approach to vector subdifferentials, *Optimization*, Vol. 38, pp.237-251.

9 CLASS–INCLUSION PROPERTIES FOR CONVEX FUNCTIONS

Andrew Eberhard

Department of Mathematics,
Royal Melbourne Institute of Technology,
124 Latrobe St, Melbourne, Victoria 3000,
Australia

and Charles E. M. Pearce

Department of Applied Mathematics,
The University of Adelaide,
South Australia 5005, Australia

Abstract: Class–inclusion properties are derived for some classes of generalized convex functions. It is shown that some of these classes coincide under a restriction to lower semicontinuity.

Key words: convexity, quasi–convexity, lower semicontinuity, class inclusions.

X. Yang et al. (eds.), Progress in Optimization, 129–133.

1 INTRODUCTION

Convexity is a key concept in optimization and a variety of generalizations exist (see, for example, Ponstein (1967)). Some relations between these are known. Thus suppose $C(I)$, $W(I)$, $J(I)$ denote respectively the classes of convex, Wright–convex and Jensen–convex functions on an interval $I \subset \mathbb{R}$. Wright (1954) conjectured that

$$C(I) \subset W(I) \subset J(I)$$

and that both inclusions are proper. This result was subsequently proved by Kenyon (1956) and Klee (1956). Recently their result has been extended by Dragomir and Pearce (1998) to cover quasiconvex, Wright–quasiconvex and Jensen–quasiconvex functions.

The above suggests the existence of further results. In this paper we address this question. The following definitions will prove useful.

Definition 1.1 *Suppose $f : I \to \mathbb{R}$, where I is an interval, and let $g(f, \cdot, \cdot, \cdot)$ be a real-valued function on $I \times I \times (0,1)$. We say that f is g–convex on I if, for each $x, y \in I$ and $\lambda \in (0,1)$,*

$$f(\lambda x + (1-\lambda)y) \le g(f, x, y, \lambda). \tag{1.1}$$

Similarly f is Wright g–convex if

$$f(\lambda x + (1-\lambda)y) + f(\lambda y + (1-\lambda)x) \le g(f, x, y, \lambda) + g(f, x, y, 1-\lambda) \tag{1.2}$$

and Jensen g–convex if

$$f((x+y)/2) \le g(f, x, y, 1/2). \tag{1.3}$$

We denote the classes of such functions by $gC(I)$, $WgC(I)$, $JgC(I)$ respectively.

With the choice

$$g(f, x, y, \lambda) := \lambda f(x) + (1-\lambda) f(y), \tag{1.4}$$

the above classes are respectively those of the familiar convex, Wright–convex and Jensen–convex functions. For

$$g(f, x, y, \lambda) := \max[f(x), f(y)], \tag{1.5}$$

we obtain quasiconvex, Wright–quasiconvex and Jensen–quasiconvex functions respectively.

Related and broader classes of convex function are of importance in connection with Hadamard's inequality. Thus Dragomir, Pečarić and Persson (1995) treat functions of P type, which correspond to g–convexity for f nonnegative and

$$g(f,x,y,\lambda) := f(x) + f(y), \tag{1.6}$$

as well as functions of Godunova–Levin type, which are g–convex functions for f nonnegative and

$$g(f,x,y,\lambda) := \frac{f(x)}{\lambda} + \frac{f(y)}{1-\lambda} \quad (\lambda \in (0,1)). \tag{1.7}$$

We remark that if g is given by (1.6) or (1.7), then setting $y = x$ in (1.3) shows that $f \in JgC(I)$ implies f has to be nonnegative.

It is immediate from the definitions that

$$gC(I) \subset WgC(I) \subset JgC(I). \tag{1.8}$$

The content of Kenyon (1956), Klee (1956) and Dragomir and Pearce (1998) is in showing that the inclusions are in fact proper for g given by (1.4) and (1.5).

In the next section we show that this also holds for the second inclusion when g is given by (1.6) or (1.7). We were unable to show the corresponding result for the first inclusion.

In Section 3 we show that the restrictions of $gC(I)$, $WgC(I)$, $JgC(I)$ to lower–semicontinuous functions f coincide for g given by (1.4) and (1.5). This strengthens the result also obtained in Dragomir and Pearce (1998) that the restrictions to continuous functions coincide for g given by (1.5). The result for continuous–function restrictions and g given by (1.4) is classical.

2 PROPER CLASS INCLUSIONS

Theorem 2.1 *The second inclusion in (1.8) is proper for g given by (1.6) or (1.7).*

Proof: Consider the Dirichlet map $f : \mathbb{R} \to \mathbb{R}$ defined by

$$f(u) = \begin{cases} 1 & \text{for } u \text{ irrational} \\ 0 & \text{for } u \text{ rational.} \end{cases}$$

If x and y are both rational, then so is $(x+y)/2$, so that in this case

$$f((x+y)/2) = f(x) + f(y). \tag{2.1}$$

If one of x, y is rational and the other irrational, then $(x+y)/2$ is irrational and so again (2.1) holds. If both x and y are irrational, then

$$f((x+y)/2) < 2 = f(x) + f(y).$$

Hence since f is nonnegative we have

$$f((x+y)/2) \leq f(x) + f(y) \leq \frac{f(x)}{1/2} + \frac{f(y)}{1/2}, \tag{2.2}$$

so that $f \in JgC(\mathbb{R})$ for g given by (1.6) or (1.7).

On the other hand, if x and y are distinct rationals, there are uncountably many values of $\lambda \in (0,1)$ for which $\lambda x + (1-\lambda)y$ and $(1-\lambda)x + \lambda y$ are both irrational. For each such λ,

$$f(\lambda x+(1-\lambda)y)+f((1-\lambda)x+\lambda y) > [f(x)+f(y)]\left(\frac{1}{\lambda}+\frac{1}{1-\lambda}\right) \geq f(x)+f(y)$$

so that $f \notin WgC(I)$ with g given by (1.6) or (1.7). This completes the proof.□

3 LOWER SEMICONTINUITY

Theorem 3.1 *If f is lower semicontinuous on I, then the classes $gC(I)$, $WgC(I)$, $JgC(I)$ coincide for g given by (1.4) or (1.5).*

Proof: First we deal with g given by (1.4) with f lower semicontinuous. Suppose $f \in JgC(I)$, so that $f((x+y)/2) \leq [f(x)+f(y)]/2$ for all $x, y \in I$. A straightforward induction provides

$$f(\beta x + (1-\beta)y) \leq \beta f(x) + (1-\beta)f(y) \tag{3.1}$$

for all $\beta = i.2^{-n}$ with $i = 0, 1, \ldots, 2^n$. For any $\lambda \in (0,1)$, there exists a sequence (β_n) of such values β with $\beta_n \to \lambda$. We have from (3.1) that

$$\operatorname{liminf} f(\beta_n x + (1-\beta_n)y) \leq \lambda f(x) + (1-\lambda)f(y),$$

and so by lower semicontinuity

$$f(\lambda x + (1-\lambda)y) \leq \lambda f(x) + (1-\lambda)f(y).$$

Thus $f \in gC(I)$, and by (1.8) the required result is established.

The proof is parallel when g is given by (1.5), starting from

$$f((x+y)/2) \leq \max[f(x), f(y)]$$

for $f \in JgC(I)$. □

References

Dragomir, S.S., Pečarić, J.E. and Persson, L.E. (1995), Some inequalities of Hadamard type, *Soochow Journal of Mathematics*, Vol. 21, pp. 335–341.

Dragomir, S.S. and Pearce, C.E.M. (1998), Quasi–convex functions and Hadamard's inequality, *Bulletin of the Australian Mathematical Society*, Vol. 57, pp. 377–385.

Kenyon, H. (1956), Note on convex functions, *American Mathematical Monthly*, Vol. 63, pp. 107.

Klee, V.L. (1956), Solution of a problem of E.M. Wright on convex functions, *American Mathematical Monthly*, Vol. 63, pp. 106–107.

Ponstein, J. (1967), Seven types of convexity, *SIAM Review*, Vol. 9, pp. 115–119.
Wright, E.M. (1954), An inequality for convex functions, *American Mathematical Monthly*, Vol. 61, pp. 620–622.

10 ON GENERIC LOCALLY CONVEX VECTOR FUNCTIONS

V. Gershkovich, B.D. Craven and D. Ralph,

Department of Mathematics and Statistics,
The University of Melbourne,
Parkville, Vic. 3052, Australia.

Abstract: Local convexifiability, LC, is studied for generic smooth vector functions and vector functions with small codimension singularities. We discuss its connection with Pareto optima and describe the global structure of manifolds admitting generic locally convex vector functions. We describe also the spaces of LC vector functions.

Key words: vector functions on manifolds, invariant convexity, Pareto minima, multicriteria optimization.

X. Yang et al. (eds.), Progress in Optimization, 135–152.

1 INTRODUCTION

We study "locally convexifiable", LC, smooth vector functions on smooth manifolds. By way of background, we first give a very brief discussion of the scalar-valued case.

Convex functions have been investigated in analysis, Rockafellar (1970), optimization, Hiriart-Urruty et al (1993), geometry, Greene et al (1981a), Greene et al (1981b), Yau (1974), and economics, Debreu (1974), to mention only the briefest selection from the vast literature.

The global aspect of optimization problems is usually the most difficult. The class of convex problems is a nice exception — a local minimum is a global one and convexity itself is a local property. Recently in optimization there have been numerous attempts, see references in Craven et al (1996) which date back to Hanson (1980) and Craven (1981), to extend the notion of convexity to a wider class of (scalar) functions with the same properties. A key point is that a Morse (scalar) function, under some additional assumptions, has a unique local minimum when it does not have saddle points, see Craven et al (1996) for details. In Craven et al (1996) it was also shown that this class of functions, i.e. Morse functions without saddle points, covers the main generalizations proposed earlier and in a sense all possible generalizations of classes of functions with either no mimima or exactly one minimizer (including generalization to "tractable" classes of non-smooth functions, see e.g. Gershkovich et al (1997)).

The aim of this paper is to generalize these results to vector functions, where "generalized convexity" is formulated naturally in terms of Pareto critical points and then appears to be closely connected with multicriteria optimization and the research of S. Smale, namely Smale (1973), Smale (1975); and his students, see e.g. Wan (1975), Wan (1977), van Gedrop (1980).

Definition 1.1 (Locally Convex Vector Function) *A smooth vector function $F = (f_1, f_2, \ldots, f_k) : \mathbb{R}^n \to \mathbb{R}^k$ is said to be* locally convex *at $x \in \mathbb{R}^n$ if there exists a neighborhood $U(x)$ of x such that*

$$(\forall y \in U(x) \setminus \{x\}) \qquad F(y) - F(x) - d_x F(y - x) \in \bar{\mathbb{R}}^k_+. \qquad (1.1)$$

F *is* locally strictly convex *at x if (1.1) is satisfied after replacing $\bar{\mathbb{R}}^k_+$ by its interior $\mathbb{R}^k_+$.*

Definition 1.2 (Locally Convexifiable Vector Function) $F = (f_1, \ldots, f_k) : M_n \to \mathbb{R}^k$ *is said to be* locally convexifiable, *LC, or an LC-*function *at x if there exists a local coordinate system $\phi : U(x) \to \mathbb{R}^n$, such that $F\phi^{-1} : \phi(U(x)) \to \mathbb{R}^k$ is convex. F is* strictly LC *or SLC, if $F\phi^{-1}$ is strictly convex.*

In this paper we investigate the problem of local convexity for generic vector functions $M_n \to \mathbb{R}^k$ and for vector functions with small (codimension) *singularities.* We more deeply explore the case $2k < n$, which can be analysed by analogy with the case of scalar functions because any generic map $M_n \to \mathbb{R}^k$ is a $(k-1)$-mersion when $2k < n$, Golubitsky et al (1973). The case $2k > n$

(more precisely $2k > n+4$) is far more difficult and will be discussed in detail in elsewhere.

By analogy with scalar functions, the LC-property means that any Pareto critical point (of the multicriteria problem with criteria $f_1, \ldots, f_k$) is a local Pareto minimum. This provides connections with the pioneering research of Smale (1973), Smale (1975), and subsequent work along these lines, see e.g. Wan (1975), Wan (1977), van Gedrop (1980), related to the structure of the Pareto critical set. Another topic of this paper is the structure of the space of LC and SLC functions (and relations between them). We present also some results related to LC vector functions with small singularities.

We use some concepts related to singularity theory. Any one of the books Bröcker et al (1975), Golubitsky et al (1973), Arnold et al (1985) contains all the definitions and results we use in this paper; Smale (1973) is a nice short introduction in problems related to Pareto optima.

1.1 Notation

We collect here the notation to be used throughout the paper. Further "local" notation will be introduced in corresponding sections.

M_n refers to a smooth n-dimensional manifold, $\mathbb{R}^k$ to the k-dimensional Euclidean space. Let $\mathbb{R}^k_+$ be the open positive orthant in $\mathbb{R}^k$ and $\bar{\mathbb{R}}^k_+$ be its closure. More generally, let $\bar{X}$ be the closure of any subset X of a topological space. Let O be the origin in $\mathbb{R}^k$.

We write $M_n \sim \tilde{M}_n$ to denote diffeomorphic manifolds. We denote by $C^r(M_n)$ the space of C^r functions on M_n, where $0 \leq r \leq \infty$. We denote by $C^r_k(M_n)$ the space of C^r vector functions $M_n \to \mathbb{R}^k$.

All manifolds, coordinate systems, maps, (vector) functions, perturbations and vector fields are supposed to be C^∞ (or sufficiently) smooth.

For $x \in M_n$ we denote by $T_x M_n$ the tangent space at x. The differential of a smooth map $F: M \to N$ at a point $x \in M$ is denoted by $d_x F$; we denote by $\text{rk}_x dF$ the rank of $d_x F$, and by $\ker_x dF$ the kernel of $d_x F$, i.e. the inverse image $d_x F^{-1}(0)$.

Let $f: M_n \to \mathbb{R}^1$ be a scalar function. Recall a point x is said to be a critical point of f if $d_x f = 0$. We denote by $\text{Hess}_x f = \left\{ \frac{\partial^2 f(x)}{\partial x_i \partial x_j} \right\}$ the Hessian of f at a critical point x of f. The index of f at x, $\text{ind}_x f = \text{ind}(\text{Hess}_x f)$, is the number of negative eigenvalues of the Hessian.

2 REGULAR AND CRITICAL POINTS OF VECTOR FUNCTIONS

In this section we include preliminary concepts and results on generic vector functions. We do not assume $2k < n$.

2.1 Conical hull

A study of local convexity for vector functions F employs the concept of cones $\bar{\mathcal{K}}_F$ of (scalar) functions (generated by the coordinate functions) and study of singularities of functions from $\bar{\mathcal{K}}_F$.

Definition 2.1 (Conical hull of coordinate functions) *The* closed conical hull *of a vector function* $F = (f_1, \ldots, f_k) : \mathbb{R}^n \to \mathbb{R}^k$ *is the convex (polyhedral) cone* $\bar{\mathcal{K}}_F \subset C^\infty(M_n)$ *defined by*

$$\bar{\mathcal{K}}_F = \{f = \alpha_1 f_1 + \ldots + \alpha_k f_k \mid \alpha_i \in \bar{\mathbb{R}}^1_+\}.$$

For the open cone $\mathcal{K}_F$*, replace* $\bar{\mathbb{R}}^1_+$ *by* $\mathbb{R}^1_+$.

Remark 2.1 *The coefficients of* $\Lambda_F = (\alpha_1, \ldots, \alpha_k)$ *are defined uniquely up to a common nonzero factor, hence* Λ_F *can be considered as a point* $(\alpha_1 : \ldots : \alpha_k)$ *of the* $(k-1)$*-dimensional projective space* $P_{k-1}\mathbb{R}$.

The following statements are obvious.

Proposition 2.1 *A vector function* $F : \mathbb{R}^n \to \mathbb{R}^k$ *is locally convex if and only if each function in* $\bar{\mathcal{K}}_F$ *is locally convex.*

As was mentioned above, see also Craven et al (1996), local convexity is not an invariant notion, and depends on the choice of the linear structure in the tangent spaces (local coordinate system); one way to construct an invariant notion is to replace local convexity by (local) convexifiability. The following reformulation of the definition is evident.

Proposition 2.2 *A vector function* $F : M_n \to \mathbb{R}^k$ *is LC at* x *if and only if there exists a local coordinate system* $\phi : U(x) \subset M_n \to \mathbb{R}^n$ *such that* $f\phi^{-1}$ *is locally convex for each function* $f \in \bar{\mathcal{K}}_F$.

The following result is stronger; it follows from Smale-Wan theorem (Section 2.6), see also Section 2.5, in conjunction with the results of Section 3.7.

Theorem 2.1 *A vector function* $F : M_n \to \mathbb{R}^k$ *is LC at* x *if and only if for any* $f \in \bar{\mathcal{K}}_F$ *there exists a local coordinate system* $\phi : U(x) \subset M_n \to \mathbb{R}^n$ *such that* $f\phi^{-1}$ *is locally convex.*

2.2 F-regular points.

We consider the case $k \leq n$.

Definition 2.2 (F-regular and F-singular points) *Let* $F : M_n \to \mathbb{R}^k$ *be a vector function. A point* $x \in M_n$ *is said to be* F-regular *if* $\text{rk}_x dF = k$*, and* F-singular *(or* F-critical) *otherwise. Denote by* Ω_F *the set of* F-singular *points.*

Proposition 2.3 *A vector function* $F : M_n \to \mathbb{R}^k$ *is SLC at all* F*-regular points.*

Proof In accordance with the implicit function theorem, there exists a local coordinate system $\{x_i\}_{i=1}^n$ at any F-regular point x such that $f_i = x_i + (x_i)^2$ near x for each i. Hence F is SLC at any F-regular point.

A simple "globalization" of this result sounds as follows, see Craven et al (1996). Recall that a map $\phi : M_n \to N_k$ is called *proper* if $\phi^{-1}K$ is compact for any compact $K \subset N_k$.

Proposition 2.4 *Let $k \leq n$ and $F : M_n \to \mathbb{R}^k$ be a proper submersion, that is $\mathrm{rk}_x dF \equiv k$. Then there exists a compact manifold M_{n-k} and a diffeomorphism $\phi : M_{n-k} \times \mathbb{R}^k \to M_n$ such that $F\phi : M_{n-k} \times \mathbb{R}^k \to \mathbb{R}^k$ is the projection on the second factor.*

Remark 2.2 *A description of the set of proper vector functions without critical points is an interesting and non-trivial problem which will be considered elsewhere. The initial part of the problem is to describe, for a given M_n, all M_{n-k} such that $M_{n-k} \times \mathbb{R}^k \sim M_n$.*

2.3 Cones of functions and their critical points

A point $x \in M_n$ is F-critical when $d_x f_1, d_x f_2, \ldots, d_x f_k$ are linearly dependent. The set of points $\lambda_F^x = (\alpha_1, \ldots, \alpha_k) \in \mathbb{R}^k$ such that $d_x(\alpha_1 f_1 + \ldots + \alpha_k f_k) = 0$ is a linear subspace of $\mathbb{R}^k$ of dimension $k - \mathrm{rk}_x dF$. Denote $\Lambda_F^x = \langle \lambda_F, F \rangle = \{\sum \alpha_i f_i,\ \{\alpha_i\} \in \Lambda_F\}$. To study points of Ω_F we have to distinguish two essentially different cases of configurations of linearly dependent vectors. Given any set $\Xi = \{\xi_1, \ldots, \xi_l\} \subset \mathbb{R}^n$, denote by conv Ξ the convex hull of vectors $\{\xi_1, \ldots, \xi_l\}$.

Definition 2.3 (Critical and regular configurations of vectors) *A set $\Xi = \{\xi_1, \ldots, \xi_k\} \subset \mathbb{R}^n$ is said to be a* critical configuration *if $0 \in \mathrm{conv}\ (\Xi)$; and a* regular configuration *otherwise.*

Notice any configuration of linearly independent vectors is regular.

Definition 2.4 ($\bar{\mathcal{K}}_F$-regular and $\bar{\mathcal{K}}_F$-critical points) *Let $F = (f_1, \ldots, f_k) : M_n \to \mathbb{R}^k$ be a vector function. A point $x \in \Omega_F$ is said to be* $\bar{\mathcal{K}}_F$-regular *if $\{d_x f_1, \ldots, d_x f_k\}$ is a regular configuration and* $\bar{\mathcal{K}}_F$-critical *otherwise. Denote by C_F the set of $\bar{\mathcal{K}}_F$-critical points.*

Evidently x is $\bar{\mathcal{K}}_F$-regular if and only if x is regular for any function in $\bar{\mathcal{K}}_F$.

Definition 2.5 (The cone of critical functions) *Let $F = (f_1, \ldots, f_k) : M_n \to \mathbb{R}^k$ be a vector function and $x \in C_F$. The cone $\bar{\mathcal{K}}_F^x \in \bar{\mathcal{K}}_F$ of functions which have a critical point at x is called a* critical cone *at x.*

Denote $\lambda_F^x(+) = \lambda_F^x \cap \mathbb{R}_+^k$ and $\Lambda_F^x(+) = \langle \lambda_F(+), F \rangle$. The following lemmata are evident.

Lemma 2.1 $\bar{\mathcal{K}}_F^x = \bar{\mathcal{K}}_F \cap \Lambda_F^x = \bar{\mathcal{K}}_F \cap \Lambda_F^x(+)$.

Lemma 2.2 *$\Lambda_F^x(+) \subset T_x M_n$ is a convex cone in $T_x M_n$.*

2.4 Points of $\Omega_F \setminus C_F$

Theorem 2.2 *A vector function $F : M_n \to \mathbb{R}^k$ is $(S)LC$ at any point $x \notin C_F$.*

The theorem follows immediately from the next evident lemma.

Lemma 2.3 *Let x be a $\bar{\mathcal{K}}_F$-regular point of a vector function $F = (f_1, \ldots, f_k) : M_n \to \mathbb{R}^k$. Then there exists a local coordinate system $\{x_1, \ldots, x_k\}$ such that*

$$f_i = a_i x_1 + \sum_{j=2}^{n} c_j^i x_j + \sum_{p,q} c_{p,q}^i x_p x_q + \ldots, \quad i = 1, \ldots, k,$$

where all a_i are strictly positive.

2.5 Points of C_F

Proposition 2.5 *Let $k < n$ and $F : M_n \to \mathbb{R}^k$ be a generic vector function, $x \in M_n$. Then for all $x \in M_n$ the set of Morse functions in $\bar{\mathcal{K}}_F^x$ is open and everywhere dense in $\bar{\mathcal{K}}_F^x$.*

Theorem 2.3 *$F : M_n \to \mathbb{R}^k$ is LC if and only if for any $x \in C_F$ all Morse functions from $\bar{\mathcal{K}}_F^x$ have zero index at x.*

Proof The proof follows from Smale-Wan theorem, see Section 2.6.

As a corollary we obtain a proof of Theorem 2.1.

Remark 2.3 *There are several interesting problems on relation of LC and SLC functions. (A natural question is how to describe the closure of SLC in LC and to point out conditions when these classes coincide.) We discuss some of these problems in Section 3.4.*

2.6 Local Pareto minima

We give another interpretation of C_F as a set of Pareto critical points. Recall that the concept of Pareto critical points relates to the multicriteria optimization, see Debreu (1974), Smale (1973). Suppose we have k functions (criteria) $f_1, \ldots, f_k$ on M_n.

Definition 2.6 (Pareto regular and Pareto critical points) *A point $q \in M_n$ is said to be* Pareto critical *if $d_q F^{-1}(\mathbb{R}_+^k) = \emptyset$ and* Pareto regular *otherwise.*

Definition 2.7 (Pareto optimal points) *A point $q \in M_n$ is said to be* strict Pareto maximum *if $f_i(p) \geq f_i(q)$ for all i implies $p = q$. (Pareto minimum and local Pareto minima and maxima can be defined in a similar way).*

Note that the set of strict local Pareto minima does not generally consist of isolated points, nor is it generally connected. Local Pareto minima are described by the following theorem, obtained by Smale (1975) and Wan (1975); (the proofs were simplified later in van Gedrop (1980)), see also Wan (1977).

Theorem 2.4 (Smale-Wan)

1. *Suppose* $x \in M_n$ *is a local Pareto minimum for functions* $f_1, \ldots, f_k$ *and* $k < n$. *Then the following condition is satisfied. Suppose* $\alpha_1, \ldots, \alpha_k > 0$ *and* $\sum_{i=1}^{k} \alpha_i \, d_x f_i = 0$. *Then* $\mathrm{Hess}_x \sum_{i=1}^{k} \alpha_i f_i$ *is positive semidefinite quadratic form on* $\ker_x dF$.

2. *If in the last condition* $\mathrm{Hess}_x \sum_{i=1}^{k} \alpha_i f_i$ *is positive definite then* x *is a strict local Pareto minimum.*

Note that the first condition of the theorem relates to the cone of critical functions (Definition 2.5). Evidently we have the following statements.

Proposition 2.6 *Let* $F : M_n \to \mathbb{R}^k$ *be a vector function. Then* C_F *is the set of Pareto critical points.*

Denote by $C_F^{lc} \subset C_F$ the subset of points where F is LC.

Proposition 2.7 C_F^{lc} *is the set of local Pareto minima.*

Corollary 2.1 *Let* $F : M_n \to \mathbb{R}^k$ *be a LC vector function. Then* C_F *is the set of local Pareto minima. (That is any local Pareto critical point is a local Pareto minimum).*

2.7 Restrictions on submanifolds

A restriction of a convex function to any submanifold is convex. This statement does not extend to LC functions. (One of the simplest counterexamples is the scalar function $f : \mathbb{R}^2 \to \mathbb{R}^1$ defined as $f(x,y) = x - y^2$ and the submanifold defined by the equality $x = 0$ or $x =$ any constant.)

Nevertheless the following statement holds.

Proposition 2.8 *Let* $F : M_n \to \mathbb{R}^k$ *be a generic LC function. Let* $S \subset M_n$ *be a submanifold and* $x \in C_F \cap S$. *Then* $F_{|S}$ *is LC at* x.

In this paper we prove this result only under the condition $2k < n$, in which case it follows from the results of Sections 3.7 and 3.8.

3 GENERIC LC VECTOR FUNCTIONS $F : M_N \to \mathbb{R}^K$ WITH $2K - 4 < N$

The condition on the dimensions ($2k - 4 < n$) simplifies the situation strongly. The reason is that the rank of the generic map $F = (f_1, \ldots, f_k) : M_n \to \mathbb{R}^k$ is bounded from below by $k - 1$; see results realted to m-mersions in Feit (1969) and Gromov (1986), and also results related to Pareto optima in Smale (1973). As a result one can choose a local coordinate system such that all functions f_i

but one are coordinate functions. The properties of F (e.g. LC-property) are determined by just one scalar function. Another reason to present a detailed study of the simpler case $2k < n$ is that, in applications, one needs not only generic functions but also functions with small singularities, see Igusa (1984). We study in detail a case with a slightly stronger inequality $2k < n$. Under this condition one can completely describe vector functions with small singularities.

3.1 Rank

Definition 3.1 (Feit (1969) (m-mersion)) *A map $F : M_n \to N_k$ is said to be an m-*mersion *if $\text{rk}_x\, dF \geq m$ for all $x \in M_n$. (When $n \geq k$, k-mersions are called* submersions. *When $n \leq k$, n-mersions are called* immersions*).*

Proposition 3.1 (Golubitsky et al (1973)) *Suppose $2k < n$, then the set of $(k-1)$-mersions $F : M_n \to \mathbb{R}^k$ is open and everywhere dense in the set of C^2-smooth vector functions on M_n.*

Then for a generic vector function $F : M_n \to \mathbb{R}^k$, the cone $\bar{\mathcal{K}}_F$ is either empty or one-dimensional, and in the latter case there is a a unique critical function up to a constant positive factor:

$$\bar{\mathcal{K}}_F^x = \{\alpha f_{\bar{\mathcal{K}}}^x; \alpha \geq 0\}$$

for some nonzero $f_{\bar{\mathcal{K}}}^x \in \bar{\mathcal{K}}_F$. As a result the classification of singularities of generic cones of functions coincides with the classification of singularities for scalar functions. This allows us to give the following definitions.

Definition 3.2 (Morse cone of functions) *The closed cone of functions $\bar{\mathcal{K}}_F$ is said to be* Morse *if all nonzero functions in this cone are Morse.*

Definition 3.3 (Index for Morse cones of functions) *Let $\bar{\mathcal{K}}_F$ be Morse and x be $\bar{\mathcal{K}}_F$-critical. Then the index* $\text{ind}_x\bar{\mathcal{K}}_F$ *of $\bar{\mathcal{K}}_F$ at x is defined as the index of $f_{\bar{\mathcal{K}}}^x$ at x.*

So the Pareto minima x of F, see Corollary 2.1, are indexed by the critical functions $f_{\bar{\mathcal{K}}}^x$.

3.2 Isolated singularities

It is well known that the comnplement of the set of scalar functions with isolated singularities has infinite codimension in $C^\infty(M_n)$, see e.g. Bröcker et al (1975). This can easily be extended to vector functions. Denote by $I_k(M_n)$ the set of vector functions $M_n \to \mathbb{R}^k$ such that all functions from $\bar{\mathcal{K}}_F$ have isolated singularities.

Proposition 3.2 *The complement $C_k^\infty(M_n)\setminus I_k(M_n)$ has infinite codimension in $C_k^\infty(M_n)$.*

Assumption: *Hereafter all vector functions $F : M_n \to \mathbb{R}^k$ are $(k-1)$-mersions and have only isolated singularities.*

3.3 Degeneracy

Generic cones of vector functions include degenerate scalar functions but, under our assumptions, only with codimension-one (so called birth-death) and codimension-two singularities, see Igusa (1984), Arnold et al (1985). We define them in terms of normal forms.

Definition 3.4 (Birth-death singularities of functions) *A function $f : \mathbb{R}^n \to \mathbb{R}^1$ is said to have a* birth-death singularity *at x if there exist k and a local coordinate system around x such that*

$$f \sim x_n^3 - \sum_{i=1}^{k} x_i^2 + \sum_{i=k+1}^{n-1} x_i^2.$$

Birth-death singularities are the simplest, that is a codimension-one, singularities (they do not occur for generic functions but occur for generic one parametric families of functions, see e.g. Arnold et al (1985)). We recall the origination of the term birth-death. A small perturbation $f \to f + \epsilon x_n$ with $\epsilon < 0$ decomposes the birth-death singularity into non-degenerate critical points with indexes k and $k+1$, while for $\epsilon > 0$ the critical point disappears.

Codimension-two singularities also can be defined by their normal forms. (A more formal treatment would first define codimension-two singularities by the usual definition of codimension; and then prove that functions with codimension-two singularities are characterized by the normal form given next.)

Definition 3.5 (Codimension-two singularities of functions) *A function $f : \mathbb{R}^n \to \mathbb{R}^1$ with a codimension-two singularity at x has a normal form*

$$f \sim x_n^4 - \sum_{i=1}^{k} x_i^2 + \sum_{i=k+1}^{n-1} x_i^2.$$

3.4 Restriction of F to Ω_F.

To discuss normal forms (in the sense of Arnold et al (1985)) of vector functions we need a little notation related to local rings. Let $N \subset M_n$ be a submanifold. Denote by $\mathcal{R}_x(N)$ the local ring of smooth functions on N at x; by $\mathcal{M}_x(N)$ the unique maximal ideal in $\mathcal{R}_x(N)$, the family of smooth functions vanishing at x, see Bröcker et al (1975); and by $\mathcal{M}_x^i(N)$ its i-th power, that is the set of functions vanishing up to i-order. When $N = M_n$, we condense the corresponding notation to $\mathbb{R}_x$ and $\mathcal{M}_x^i$ respectively.

Let $F : M_n \to \mathbb{R}^k$ be a $(k-1)$-mersion and $x \in \Omega_F$. Then there exists a local coordinate system $(x_1, \ldots, x_n)$ such that (after renumbering the coordinates in $\mathbb{R}^k$)

$$f_i = x_i,\ i = 1, \ldots, k-1, \quad \text{and } f_k = \sum_{j=1}^{k-1} c_i x_i + R_F(x_1, \ldots, x_n), \tag{3.1}$$

where $R_F \in \mathcal{M}_x^2$. Recall C_F is the set of $\bar{\mathcal{K}}_F$-critical point and $x \in C_F$ if and only if $c_i \leq 0$ in (3.1) for all i.

Remark 3.1 *For a generic vector function F the set C_F includes, in particular, a finite number of points where all $c_i = 0$, which are precisely the points where one of the coordinate functions has a non-degenerate critical point.*

Lemma 3.1 *Suppose $x \notin C_F$. Then F is $(S)LC$ at x.*

A proof for a more general case has been presented in Section 2.4. We give another one here, based on the formula (3.1), which is useful in the following study of generic vector functions.

Proof There exists j such that $c_j > 0$ in (3.1). One can chose local coordinates $\{y_i\}$ such that $x_j = y_j + Z\sum_{i=1}^n y_i^2$ and $x_i = y_i + \epsilon y_i^2$ for $i \neq j$, where Z is a large positive number and ϵ is a small positive number. Then $f_i = y_i + \epsilon y_i^2$, $i = 1, \ldots, k-1$, is strictly convex; also

$$f_k = \sum_{i=1}^{k} c_i y_i + Z c_j \sum_{i=1}^{n} y_i^2 + \epsilon \sum_{i \neq j} c_i y_i^2 + R_F$$

is strictly convex for large enough Z.

When all $c_i \leq 0$ in formula (3.1) then F is LC at x if and only if R_F is locally convex. We reformulate this condition in invariant terms and obtain a simple LC criterion in the next section.

3.5 The Quadratic Part of R_F

When $x \in C_F$, the function R_F in formula (3.1) is defined uniquely up to a positive constant factor for a generic vector function F. One can define the rank $\mathrm{rk}_x R_F = \mathrm{rk}\,\mathrm{Hess}_x R_F$ and $\mathrm{ind}_x R_F = \mathrm{ind}_x \mathrm{Hess}_x R_F$ at x. The following result is evident from formula (3.1) (and also this is a partial case of Smale-Wan theorem, see Section 2.6).

Theorem 3.1 (Necessary and sufficient conditions for LC) *Let $F : M_n \to \mathbb{R}^k$ be a $(k-1)$-mersion and $x \in C_F$.*

1. *If $F : M_n \to \mathbb{R}^k$ is LC at x then $\mathrm{ind}_x R_F = 0$.*

2. *If $\mathrm{rk}_x R_F = n$ and $\mathrm{ind}_x R_f = 0$ then F is SLC at x*

In the next three sections we prove that $\mathrm{rk} R_F = n$ at all critical points of generic scalar functions and then (S)LC is completely determined by the index (as for scalar functions).

The rank $\mathrm{rk}_x R_F$ ($\mathrm{ind}_x R_F$) can be interpreted as the sum $\mathrm{rk}_x d_x^2 F + \mathrm{rk}_x f_{\bar{\mathcal{K}}_F^x}$ (resp. $\mathrm{ind}_x d_x^2 F + \mathrm{ind}_x f_{\bar{\mathcal{K}}_F^x}$), where $d_x^2 F$ is the *quadratic differential* of F and $f_{\bar{\mathcal{K}}_F^x}$ is the critical function of $\bar{\mathcal{K}}_F$. We review the definition of the quadratic differential in the next section.

3.6 Quadratic Differential and its Index

For a generic vector function the quadratic differential $d_x^2F : \ker_x Df \to \operatorname{coker}_x D_F$ can be identified with a quadratic form on $\ker_x F$, see Bröcker et al (1975); its rank and index are properly defined; we denote them respectively by $\operatorname{ind}_x F$ and $\operatorname{rk}_x^2 F$. structure on $\mathbb{R}^k$ (which takes place when we consider vector functions). Denote by $\operatorname{rk}_x^2 F$ the rank and by $\operatorname{ind}_x F$ the index of this quadratic form. (It is an exercise to check that they are properly defined). Let us call a (non-unique) local coordinate system satisfying the formula (3.1) *F-adapted.* Define the germ of a smooth submanifold N_F at x by the equations $f_1 = \ldots = f_{k-1} = 0$, that is $x_1 = \ldots = x_{k-1} = 0$ in an F-adapted local coordinate system $\{x_1, \ldots, x_n\}$.

The set $N_F(x)$ cannot be chosen in a canonical way. At points of C_F where any $k-1$ coordinate functions have linearly independent gradients we have k choices for N_F namely we have $N_F^i = \{z \mid f_{j_1}(z) = f_{j_2}(z) \text{ for } j_1, j_2 = 1, \ldots, k\}$, where $i = 1, \ldots, n$. Nevertheless we have the following statement (which follows as in the case of linear mappings).

Proposition 3.3 *Let $x \in C_F$. Then all manifolds $N_F^i(x)$ have the same tangent space at x.*

Denote $\operatorname{Hess}_x F = \operatorname{Hess}_x(f_{k|N_F^i})$.

Corollary 3.1 *The index and rank of* $\operatorname{Hess}_x F$ *at a point $x \in C_F$ do not depend on the choice of an F-adapted coordinate system and the choice of $N_F = N_F^i$.*

We have $\operatorname{rk}_x^2 F = \operatorname{rk}(\operatorname{Hess}_x(f_{k|N_F}))$ and $\operatorname{ind}_x F = \operatorname{ind}(\operatorname{Hess}_x(f_{k|N_F}))$.

3.7 Vector functions near full rank points

Proposition 3.4 *Let $F : M_n \to \mathbb{R}^k$ be a generic vector function, $x \in M_n$. Suppose* $\operatorname{rk}_x R_F = n$. *Then there exists a local coordinate system $\{x_i\}$ at x satisfying the following properties.*

1. *The germ of Ω_F at x is a germ of a smooth $(n-k+1)$-dimensional submanifold defined by the equations $x_k = x_{k+1} = \ldots = x_n = 0$.*

2. *The germ of N_F at x is a germ of a smooth $(k-1)$-dimensional submanifold defined by the equations $x_1 = x_2 = \ldots = x_{k-1} = 0$.*

3. *$F_{|\Omega_F}$ is an immersion; in local coordinates $F_{|\Omega_F}$ is defined by the equations*

$$
\begin{aligned}
& f_i - x_i,\ i = 1, \ldots, k \quad 1, \\
& f_k = \sum_{i=1}^{k-1} c_i x_i + H_F(x_1, \ldots, x_{k-1}) + P_F(x_1, \ldots, x_{k-1}) + Q_F(x_k, \ldots, x_n)
\end{aligned}
\tag{3.2}
$$

where H_F, Q_F are a non-degenerate quadratic forms, and $P_F \in \mathcal{M}_x^3(\Omega_F)$.

Proof The formula (3.1) can be transformed to (3.2) form under assumption of the proposition, see Milnor (1963), p. 7, which also gives the other statements.

Corollary 3.2 *At a point x where* rk $R_F = n$ *the germ of* Ω_F *is the germ of a submanifold transversal to* N_F *and of the complementary dimension.*

Corollary 3.3 *The decomposition* $TM_n = TC_F \oplus TN_F$ *does not depend on the choice* $N_F = N_F^i$.

Under the assumptions of the Proposition, a local coordinate system can be chosen in such a way as to make Q_F diagonal.

Corollary 3.4 *Suppose* $F : M_n \to \mathbb{R}^k$ *is a generic vector function,* $x \in C_F$, $\text{ind}_x F = 0$. *Then there exists a local coordinate system* $\{x_i\}$ *at* x *satisfying the following properties.*

$F_{|\Omega_F}$ *is an immersion; in local coordinates* F *is defined by the equations*

$$\begin{aligned} f_i &= x_i, i = 1, \dots, k-1, \\ f_k &= \sum_{i=1}^{k-1} c_i x_i + H_F(x_1, \dots, x_{k-1}) + \sum_{j=k}^{n} x_j^2 + P_F(x_1, \dots, x_{k-1}) \end{aligned} \tag{3.3}$$

where H_F *is a positive definite quadratic form, and* $P_F \in \mathcal{M}_x^3(\Omega_F)$.

3.8 On the rank of R_F for a generic vector function F.

We prove the following theorem.

Theorem 3.2 rk$R_F \equiv n$ *on* C_F *for generic* F.

The theorem follows from the next two propositions. The first one related to the quadratic differential and the second one to the restriction of the vector function F on the critical set Ω_F.

Denote by $\Omega_F^d \subset N_F$ the set of points of Ω_F where rk $Q_F < \text{dim}N_F = n - k + 1$ and $C_F^d = \Omega_F^d \cap C_F$.

Proposition 3.5 *Let* F *be a generic vector function satisfying the condition* $\text{ind}_x Q_F = 0$ *for all points of* C_F. *Then* $C_F^d = \emptyset$.

Proposition 3.6 *The restriction of a generic Morse vector function* F *to* Ω_F *is a Morse function.*

The proof of both propositions follows from the following three lemmata.

Lemma 3.2

1. S_m^+ *is contractible.*

2. *There exists a smooth homotopy* $\phi_t : S_m \to S_m$, $t \in [0,1]$ *such that* ϕ_0 *is the identity map,* $\phi_\epsilon(S_m^+) \subset S_{m,m}^+$ *for all positive* ϵ, *and* ϕ_1 *is a constant map.*

Proof (of Lemma 3.2) Let $A \in S_{m,m}^+$. Define $\phi_t(B) = (1-t)B + tA$ for $t \in [0,1]$.

The following lemma is evident.

Lemma 3.3 *Let $F : M_n \to \mathbb{R}^k$ be a generic vector function, $x \in C_F$ and $f_{\bar{K}}^{x}$ be a critical function at x. Then $f_{\bar{K}}^{x}$ has only isolated singularities.*

Lemma 3.4 *Let $f : \mathbb{R}^n \to \mathbb{R}^1$ be a convex function with an isolated singularity at the origin. Then there exists an ϵ-perturbation $f \to f_\epsilon$ such that f_ϵ is a Morse LC function with a strict local minima at the origin.*

We omit the proof for brevity.

3.9 Generic LC vector functions

We obtain the following corollaries of theorem 3.2.

Proposition 3.7 *A generic LC vector function is SLC.*

We present two more corollaries which are useful in study of the structure of $C_{\mathcal{F}}$.

Proposition 3.8 *Let $F : M_n \to \mathbb{R}^k$ be a generic vector function. Then the cone $\bar{\mathcal{K}}_F$ is Morse.*

Proposition 3.9 *Let $F : M_n \to \mathbb{R}^k$ be a generic vector function. Then the restriction of the cone $\bar{\mathcal{K}}_F$ to C_F is Morse.*

Theorem 2.1 in Section 2.1 can be shown using the above results in conjunction with the Smale-Wan Theorem in Section 2.6 or Theorem 2.3 in Section 2.5.

4 ON LOCALLY CONVEX VECTOR FUNCTIONS WITH SMALL SINGULARITIES

We begin with the main result of this section; its proof is provided in the subsections that follow.

Theorem 4.1 *The only degenerate points of SLC vector functions $F : M_n \to \mathbb{R}^k$ with singularities of codimension ≤ 2 at the origin are those that are locally equivalent to one of the following two vector functions:*

1. $f_i = x_i$, $i = 1, \ldots, k-1$, *and*

$$f_k = \sum_{i=1}^{k-1} c_i x_i + H_F(x_1, \ldots, x_{k-1}) + x_k^4 + \sum_{k+1}^{n} x_i^2$$

where $c_i < 0$, and H_F is a positive definite quadratic form.

2. $f_i = x_i$, $i = 1, \ldots, k-1$, *and*

$$f_k = \sum_{i=1}^{k-1} c_i x_i + H_F(x_1, \ldots, x_{k-1}) + \sum_{k}^{n} x_i^2$$

where $c_i < 0$, and H_F is locally equivalent to $\sum_{i=1}^{k-2} y_i^2 + y_{k-1}^4$.

4.1 Generic two parametric families of vector functions.

We are going to study vector functions with small singularities (of codimension one and two). To study such degeneracy of vector functions we use the following result.

Proposition 4.1 *Let $2k < n$. Suppose $\mathcal{F}^{a,b} : M_n \to \mathbb{R}^k$ is a generic two-parameter family of vector functions. Then $F^{a,b}$ is a $(k-1)$-mersion for all a, b.*

Proof A simple corollary of the jet transversality theorem, see Golubitsky et al (1973).

4.2 Possible degeneracy of H_F and Q_F.

We have also to investigate points where Q_F and H_F are degenerate. Let us first discuss results related to quadratic forms.

Lemma 4.1 *$S^+_{m,l}$ is a connected smooth submanifold of S^+_m of codimension $c_{m,l} = (m-l)(m-l+1)/2$.*

Proof The proof is a simple modification of the proof of Lemma 9.3 in Bröcker et al (1975). We denote by $M_{k\times l}$ the set of real $k \times l$ matrices, by S^+_m the set of positive semidefinite symmetric matrices, and by $S^+_{m,l} \subset S_m$ subset of matrices of rank l (in particular $S^+_{m,m}$ is the set of positive definite matrices).

Define a smooth map $\psi : S^+_{l,l} \times M_{l\times(m-l)} \times S^+_{m-l} \to S^+_m$ such that

$$\psi(A, B, C) = \begin{pmatrix} A & A^{1/2}B \\ B^T A^{1/2} & B^T B + C \end{pmatrix},$$

where $A \in S^+_{l,l}$, $B \in M_{l\times(m-l)}$ and $C \in S^+_{m-l}$. This map is injective with an open image in S^+_m; rk $\psi(A,B,C) =$ rk $A +$ rk C. Then rk $\psi(A,B,C) = l$ only when $C = 0$. This gives the necessary result.

This lemma shows that for singularities of codimensions one and two the possible value of rk $d^2_x F =$ rk Q_F is $n-k+1$ and $n-k$; and the only possible values of $\mathrm{rk}_x H_F$ is $k-1$ and $k-2$.

We have four possibilities:

1. $\mathrm{rk} d^2_x F = n-k+1$ and $\mathrm{rk}_x H_F = k-1$; this is the non-degenerate case already considered;

2. $\mathrm{rk} d^2_x F = n-k$ and $\mathrm{rk}_x H_F = k-1$;

3. $\mathrm{rk} d^2_x F = n-k+1$ and $\mathrm{rk}_x H_F = k-2$;

4. $\mathrm{rk} d^2_x F = n-k$ and $\mathrm{rk}_x H_F = k-2$;

In what follows we find LC vector functions corresponding to cases two and three and show that there are no LC vector functions in the last case. In fact, cases two and three are similar and it is sufficient to study only one of them.

4.3 *Restriction of* F *to* Ω_F : $\mathrm{rk}H_F = \dim\Omega_F - 1$ *and* $\mathrm{rk}\, Q_F = n - k + 1$.

Lemma 4.2 *Let* $x \in C_F$. *Suppose* $\mathrm{rk}_x H_F = \dim \Omega_F - 1 = k - 2$ *and* $\mathrm{rk}\, Q_F = n - k + 1$. *Then there exists a local coordinate system* $y_1, \dots, y_{k-1}, y_k, \dots y_n$ *satisfying the following properties.*

a) $\{y_1, \dots, y_{k-1}\}$ *is a local coordinate system on* Ω_F*;*

b) $\{y_k, \dots, y_n\}$ *is a local coordinate system on* N_F*;*

c) $F = (y_1, \dots, y_{k-1}, f_k)$ *where*

$$f_k = \sum_{i=1}^{k-1} c_i y_i + H_F(y_1, \dots, y_{k-1}) + \tilde{P}_F(y_k) + Q_F(y_k, \dots, y_n),$$

where $c_i < 0$, H_F *and* Q_F *are positive definite quadratic forms, and* $\tilde{P}_F \in \mathcal{M}^3(\mathbb{R}^1)$.

Proof See Milnor (1963), p. 7.

This gives the following criterion of LC at x.

Proposition 4.2 (Criterion of local convexifiability for the simplest singularity of H_F) *Let* $x \in C_F$, $\mathrm{ind}_x H_F = 0$. *Assume* $\mathrm{rk}\ H_F = \dim \Omega_F - 1$. *Then* F *is* LC *at* x *if and only if* $\tilde{P}_F$ *is* LC *at* x.

Denote $\Omega_1 = \{x \in \Omega \mid \mathrm{rk}H_F = k - 2\}$.

Corollary 4.1 *At a point* $x \in \Omega_1$ *the function* $\tilde{P}_F(y)$ *is equivalent either to* y^3 *or to* $\pm y^4$. *Then* F *is* LC *at* x *only when* $\tilde{P}_F(y) = y^4$.

4.4 *The case* $\mathrm{rk}H_F = \dim\Omega_F - 1$ *and* $\mathrm{rk}Q_F = \dim N_F - 1$.

Denote by $\Omega^{1,1}$ the set of points where $\mathrm{rk}\ H_F = \dim\ \Omega_F - 1$ and $\mathrm{rk}\ Q_F = \dim\ N_F - 1$.

Proposition 4.3 *For an* LC *vector function with codimension of singularity* ≤ 2 *the set* $\Omega^{1,1}$ *is empty.*

Proof Let $x \in \Omega^{1,1}$. Under the assumptions on the respective ranks of H_F and Q_F and on the codimensions of the singularities of F, one has $F = (y_1, \dots, y_{k-1}, f_k)$ where

$$f_k = \sum_{i=1}^{k-1} c_i y_i + H_F(y_1, \dots, y_{k-2}) + y_{k-1}^3 + Q_F(y_k, \dots, y_{n-1}) + y_n^3,$$

where $c_i < 0$, H_F and Q_F are positive definite quadratic forms. Then F is not LC at x.

4.5 *Codimension one singularities*

The results of Sections 4.2–4.4 give the following result (a simple generalization of results related to convex functions).

Proposition 4.4 *There are no LC functions with codimension-one singularities.*

Corollary 4.2 *The set of generic LC functions is connected.*

We present more precise results in the next section.

4.6 *Singularities of codimension* ≤ 5

Under the stronger condition $n > 2k + 2$ one can investigate singularities up to codimension five rather easily. The study is based on the following simple remarks.

Lemma 4.3 *Let* $n > 2k + 2$ *and* $\mathcal{F}$ *be a generic 5-parametric family of vector functions. Then all vector functions in* $\mathcal{F}$ *are* $(k-1)$*-mersions.*

For vector functions with codimension of singularity ≤ 5 only the following combinations of ranks of H_F and Q_F are possible.

1. $\text{rk}\, H_F = k - 1$ and $n - k + 1 \geq \text{rk}\, Q_F \geq n - k - 1$.
2. $\text{rk}\, H_F = k - 2$ and $n - k + 1 \geq \text{rk}\, Q_F \geq n - k - 1$.
3. $\text{rk}\, H_F = k - 1$ and $n - k + 1 \geq \text{rk}\, Q_F \geq n - k$.

It is now possible to classify these singularities using normal forms of functions up to codimension five. The classification process is similar to the above though the results are somewhat more technical, see Arnold et al (1985).

5 FUTURE RESEARCH

The length of the paper does not allow us to discuss the following research topics:

1. The space of *LC* functions on M_n without critical points. In this case, M_n admits a representation as a direct product $M_n \sim M_{n-k} \times \mathbb{R}^k$; notice that this problem is interesting even for the case of scalar functions, cf. Craven et al (1996).
2. The structure of the Pareto optimal set.
3. The image of the Pareto optimal set under F.
4. The behaviour of F when restricted to the set of its critical points.
5. Applications including nonsmooth vector functions, e.g. those whose component functions are max-type functions (the maxima of finite sets of smooth functions), optimization constraints, etc.

The needs of multicriteria optimization motivate these research topics. For topic 1, F has no Pareto optima. Topics 2–4, for the the case when Pareto optima exist, are relevant to refining one's choice of a Pareto optimal point on M_n by using additional criteria.

Acknowledgements

The authors gratefully acknowledge the support of the Australian Research Council for this research.

References

Arnold, V., Gusein-Zade, S.M. and Varchenko, A.N. (1985–1988), *Singularities of Differentiable Maps*, English translation,Monographs in Mathematics, Vol. 82–83, Birkhauser, Boston.

Th. Bröcker and Lander, L. (1975), *Differentiable maps and catastrophes*, Cambridge University Press, Cambridge.

Craven, B. D. (1981), Duality for generalized convex fractional programs, in *Generalized Concavity in Optimization and Economics*, Academic Press, New York (eds. S. Schaible and W. Ziemba), pp. 473–490.

Craven, B. D., Gershkovich, V. and Ralph, D. (1995), Morse theory and invariant concepts of convexity, *Optimization*, Vol. 35, pp. 1-14.

Craven, B. D., Gershkovich, V. and Ralph, D. (1996), On stable locally convex vector functions, *Proceedings of the Optimization Miniconference III*, University of Melbourne, July 1996, pp. 1-4.

Debreu, G. (1974), Four aspects of the mathematical theory of economic equilibrium, *Proceedings of International Congress of Mathematics*, Vancouver.

Feit, S. (1969), k-mersions of manifolds, *Acta Mathematica*, Vol. 122, pp. 173-195.

van Gedrop, J. H. (1980), A Note on Local Pareto Optima, *J. Mathematical Economics*, Vol. 7, pp. 51-54.

Gershkovich, V. (1997), Singularity theory for distance functions on non-positively curved surfaces, *Proceedings of Pacific Rim Singapore Conference*, Walter de Gruyter.

Gershkovich, V. and Rubinstein, H. (1997), Morse theory for Min-type functions, *Asian Journal of Mathematics*, Vol. 4, pp. 696-715.

Golubitsky, M. and Guillemin, V. (1973), *Stable Mappings and Their Singularities*, Springer-Verlag.

Greene, R.E. and Shiomata, K. (1981), Convex functions on complete noncompact manifolds, *Inventiones Math.*, Vol. 63, pp. 129-157.

Greene, R.E. and Shiomata, K. (1981), Convex functions on non-compact manifolds: differentiable structure, *Ann. Ec. Norm. Sup. 4 ser.*, Vol. 4, pp. 357-367.

Gromov, M. (1986), *Partial Differential Relations*, Springer-Verlag, New York.

Hanson, M. A. (1980), On the sufficiency of the Kuhn-Tucker conditions, *J. Math. Anal. Appl.*, Vol. 80, pp. 545–550.

Hiriart-Urruty, J.-B. and Lemaréchal, C. (1993), *Convex Analysis and Minimization Algorithms*, Vol. 1,2, Springer-Verlag, New York.

Igusa, K. (1984), Higher singularities of smooth functions are not necessary, *Ann. Math.*, Vol. 119, pp. 1-58.

Milnor, J. (1963), *Morse theory*, Princeton University Press, Princeton, New Jersey.

Rockafellar, R.T. (1970), *Convex Analysis*, Princeton University Press, Princeton.

Smale, S. (1973), Global Analysis and Economics I, Pareto optima and Generalization of Morse theory, in *Dynamical Systems*, Academic Press, New York, (ed. M. Peixoto).

Smale, S. (1975), *Sufficient condition for an optimum*, Warwick Dynamical Systems 1974, Lecture Notes in Mathematics, Springer-Verlag, pp. 287-292.

Vasiliev, V. (1997), Asymptotics of exponential integrals, Newton diagrams, and classification of minima, *Functional Analysis and its Applications*, Vol. 11, pp. 1-11.

Wan, Y.-H. (1975), On local Pareto Optima, *Journal of Mathematical Economics*, Vol. 2, pp. 35-42.

Wan, Y.-H., On the Algebraic Criteria for Local Pareto Optima, *I. - Topology*, Vol. 16, (1977), pp. 113-117; *II.- Trans. AMS*, Vol. 245, (1978), pp. 385-397.

Yau, S. T. (1974), Non-existence of continuous convex functions on certain Riemannian manifolds, *Math. Ann.*, Vol. 207, pp. 269-270.

11 ESSENTIAL COMPONENTS AND CONNECTEDNESS OF SOLUTION SET FOR COMPLEMENTARITY PROBLEMS

George Isac

Department of Mathematics and Computer Science,
Royal Military College of Canada,
P.O. Box 17000 STN Forces,
Kingston Ont., Canada K7K 7B4

and George X.Z.Yuan

Department of Mathematics,
The University of Queensland,
Brisbane, 4072, Australia,
and the corresponding author

Abstract: The Complementarity Theory is now recognized as an interesting domain of applied mathematics with deep relations with fundamental mathematics and with many applications (e.g., see Bulavsky et al (1998), Hyers et al (1997), Isac (1992-1998), Isac et al (1998), Isac and Carbone (1999), Karamardian (1971) and related references therein). The *complementarity condition* is strongly related to equilibria not only in the physical sense, but also in economical sense. Because of this relation between complementarity problems and the equilibria it is very important to study the *stability* and the *sensitivity* of *solution set.* The concept of the connectedness is related to sensitivity. The aim of this paper is to stimulate the research in the sense of the study of connectedness and in particular, the study of existence of essential components of solution set. This paper is a survey on this subject which contains some new results and ideas.

Key words: complementarity problem, stability, essential components, connectedness, fixed points.

X. Yang et al. (eds.), Progress in Optimization, 153–165.

1 INTRODUCTION

It is known that Complementarity Theory is an interesting domain of applied mathematics with deep relations with fundamental mathematics and with many applications (e.g., see Bulavsky et al (1998), Hyers et al (1997), Isac (1992-1998), Isac et al (1998), Isac and Carbone (1999), Karamardian (1971) and related references therein). The *complementarity condition* is strongly related to equilibria not only in the physical sense, but also in economical sense. Because of this relation between complementarity problems and the equilibria it is very important to study the *stability* and the *sensitivity* of *solution set.*

We know that in general it is hard to obatin stability results even though the existence results are available. However, by using the concept of the connectedness related to the sensitivity and stability, we are able to establish stability for the solutions of complementarity problems without traditional assumptions such as objective functions should be Lipschitz or differentiable. Hopefully the method to study stability by using topological concept of *connectedness* would be further developed. This topological method belongs to the category of the so-called generic stability approach and interested readers can find more informations and details from Yuan (1999).

The aim of this paper is to stimulate the research in the sense of the study of connectedness and in particular, the study of existence of essential components of solution set. This paper is a survey on this subject which contains some new results and ideas.

2 PRELIMINARIES

We first recall some notions and definitions. Let $\langle E, E^*\rangle$ be a duality of Banach spaces. The dual system $\langle E, E^*\rangle$ can be a Hilbert space $(H, \langle , \rangle)$ or the Euclidean space $(\mathbb{R}^n, \langle , \rangle)$. If $D \subset E$ is a non-empty subset, *the interior*, *closure*, and *the boundary* of D will be denoted by $int(D)$, $\overline{D}$ and $\partial(D)$, respectively. The distance between a point $x \in E$ and the set D is defined by $d(x, D) := \inf\{\|x - y\| : y \in D\}$. Let (X, d) be a metric space. Denote by 2^X the family of all subsets of X and $P_{cl}(X) = \{A \subset X : A \neq \emptyset$ and A is closed $\}$. If $A, C \in P_{cl}(X)$, we define $h^*(A, C) := \sup\{d(a, C) : a \in A\}$ and $h^*(C, A) = \sup\{d(c, A) : c \in C\}$. Then the metric h defined by $h(A, C) := \max\{h^*(A, C), h^*(C, A)\}$ is the Hausdorff metric on X. We note that the Hausdorff metric h can take the value $+\infty$. Then we have the following result (e.g., see Hu and Papageorgiou (1997)).

Theorem 2.1 *If (X, d) is a complete metric space, then $(P_{cl}(X), h)$ is a complete metric space.*

If X, Y are Hausdorff topological spaces and $T : Y \to 2^X$ is a set-valued mapping, then (1): T is upper semicontinuous at $y_0 \in Y$ if for each open set U in X with $T(y_0) \subset U$, there exists an open neighborhood $O(y_0)$ of y_0 such that $T(y) \subset U$ for any $y \in O(y_0)$; (2): T is upper semicontinuous (in short, USC) if T is upper semicontinuous at every $y \in Y$.

Let E be a vector space. We say that $K \subset E$ is a closed pointed convex cone of E if (1) K is a closed subset in E; (2) $K + K \subseteq K$; (3) $\lambda K \subseteq K$, for all $\lambda \in R_+$; and (4): $K \cap (-K) = \{0\}$. The dual of K is, by definition, $K^* = \{y \in E^* : \langle x, y \rangle \geq 0 \text{ for all } x \in K\}$. If $(H, \langle , \rangle)$ is a Hilbert space and $K \subset H$ is a closed convex cone, we denote by P_K the projection onto K, then for each $x \in H$, $P_K(x)$ is the unique element in K such that $\|x - P_K(x)\| = \inf_{y \in K} \|x - y\|$. If Y is a topological space and $D \subset Y$ is a subset, we say that D is a *residual set* if it is a countable intersection of open dense subsets of Y.

3 COMPLEMENTARITY PROBLEMS

We will consider in this paper the following kinds of complementarity problems.

Let $\langle E, E^* \rangle$ be a dual system of Banach spaces, $K \subset E$ a closed pointed convex cone and $f : K \to E^*$ a mapping. The *nonlinear complementarity problem* associated with f and K is:

$$NCP(f, K) : \begin{cases} \text{find } x_* \in K \text{ and} \\ f(x_*) \in K^* \text{ and } \langle x_*, f(x_*) \rangle = 0 \end{cases}$$

Let $(H, \langle , \rangle)$ be a Hilbert space, $K \subset H$ a closed pointed convex cone and $f, g : H \to H$ two mappings. The *implicit complementarity problem* associated with f, g and K is:

$$ICP(f, g, K) : \begin{cases} \text{find } x_* \in K \text{ such that ,} \\ g(x_*) \in K, f(x_*) \in K^* \text{ and} \\ \langle g(x_*), f(x_*) \rangle = 0 \end{cases}$$

In a Hilbert space $(H, \langle , \rangle)$ to the problem $NCP(f, K)$ we associate the mapping $\Phi(x) = P_K[x - f(x)]$ and to the problem $ICP(f, g, K)$ the mapping $\Psi : K \to K$ defined by $\Psi(x) = P_K[g(x) - f(x)]$ for each $x \in K$. In Isac (1988), it is proved that the problem $NCP(f, K)$ is solvable if and only if the mapping Φ has a fixed point in K; and the problem $ICP(f, g, K)$ is solvable if and only if, the couple (g, Ψ) has a coincident point in K.

4 CONNECTEDNESS OF SOLUTION SET

Let $(\mathbb{R}^n, \langle , \rangle)$ be a given Euclidean space. If $M \in \mathcal{M}_{n \times n}(\mathbb{R})$ and $q \in \mathbb{R}^n$, *the linear complementarity problem* associated with M, q and the cone $\mathbb{R}^n_+$ is:

$$LCP(M, q, \mathbb{R}^n_+) : \begin{cases} \text{find } x_* \in \mathbb{R}^n_+ \text{ such that} \\ Mx_* + q \in \mathbb{R}^n_+ \text{ and } \langle x_*, Mx_* + q \rangle = 0. \end{cases}$$

In this paper we denote by $SLCP(M, q, \mathbb{R}^n_+)$ the *solution set* of the problem $LCP(M, q, \mathbb{R}^n_+)$.

4.1 The Linear Case

Consider the following classes of matrices. We say that a matrix M is (1): an E_0*-matrix* (semimonotone matrix) if $SLCP(M, q, \mathbb{R}^n_+) = \{0\}$ for all $q > 0$;

(2): a P_0-*matrix* if all principal minors of M are non-negative; (3): an E_0^f-*matrix* (fully semimonotone matrix) if every principal pivotal transform (PPT) of M belongs to E_0 (e.g., see Cottle et al (1992)); and (4): an R_0-*matrix* if $SLCP(M, 0, \mathbb{R}^n_+) = \{0\}$.

The connectedness of the solution set for both linear and nonlinear complementarity problem were considered for the first time by Rapcsak (1994) and he proved that if some special assumptions are satisfied, then some subsets of the solution set are connected. Cao and Ferris (1996) were the first to consider explicitly the class of LCP-connected matrices. We say that a matrix M is LCP-connected if $SLCP(M, q, \mathbb{R}^n_+)$ is topologically connected for each $q \in \mathbb{R}^n$. In Cao and Ferris (1996), such matrices are also called P_c-matrices and the following results are proved:

(i) $P_0 \subseteq P_c \subseteq E_0$, *for* 2×2-*matrices*;

(ii) *for* $M \in P_0$ *and for all* q *except those in a set of measure zero (which depends on* M*), the set* $SLCP(M, q, \mathbb{R}^n_+)$ *is connected.*

In the paper of Cao and Ferris (1996), it is conjectured that $P_c \subseteq E_0$. This conjecture, until now, is still an open problem. Here we have the following result from Jones and Gowda (1996).

Theorem 4.1 *If* $M \in \mathcal{M}_{n\times n}(R)$ *is an LCP-connected matrix, then* $M \in E_0^f$.

Proof: The proof of this result is given by Jones and Gowda (1996). □

In the study of connectedness of solution set for complementarity problems, two mathematical tools have been used until now: the topological degree and Krasnoselskii-Zabreiko's theorem for weekly univalent mappings. The following two results are based on the topological degree due to Jones and Gowda (1996).

Theorem 4.2 *Let* $M \in P_0$ *and* $q \in \mathbb{R}^n$. *If* $SLCP(M, q, \mathbb{R}^n_+)$ *has a bounded connected component, then* $SLCP(M, q, \mathbb{R}^n_+)$ *is connected.*

Theorem 4.3 *If* $M \in R_0$, *then* $M \in P_0$ *if and only if* M *is LCP-connected.*

4.2 The Nonlinear Case

Now, consider the results obtained using the *weakly univalent mappings*.

Let $X \subset \mathbb{R}^n$ be a non-empty subset and $f : X \to \mathbb{R}^n$ a function. We say that f is *univalent* if *it is one-to-one* and continuous. The function f is said to be *weakly univalent* if it is continuous and there exists a sequence of univalent functions $\{f_k\}_{k=0}^{\infty}$ from X into $\mathbb{R}^n$ such that $\{f_k\}_{k=0}^{\infty}$ converges to f uniformly on bounded subsets of X. As examples of weakly univalent functions we cite: (1) univalent functions; (2) affine functions; (3) monotone functions; and (4) P_0-functions.

First we recall the following Krasnoselskii-Zabreiko Theorem.

Theorem 4.4 *Let* $\Omega \subset \mathbb{R}^n$ *be a bounded open set,* $f : \overline{\Omega} \to \mathbb{R}^n$ *be weakly univalent function and* $q \in f(\overline{\Omega}) \backslash f(\partial\Omega)$. *Then* $f^{-1}(q)$ *is connected.*

Proof: It is Theorem 4.8.2 of Krasnoselskii and Zabreiko (1984). □

A variant of Theorem 4.4 is the following result due to Gowda and Sznajder (1997).

Theorem 4.5 *Let $\Omega \subset \mathbb{R}^n$ be a non-empty subset with non-empty interior, $f : \Omega \to \mathbb{R}^n$ a weakly univalent function and $q \in f(\Omega)$. If there exists a non-empty subset $E \subset int(\Omega)$ such that E is compact and clopen in $f^{-1}(q)$, then $f^{-1}(q)$ is connected and $f^{-1}(q) = E$.*

We say that a subset A of Ω is *zero dimensional* if and only if, any open set in A is a union of clopen sets in A. Then we have the following result due to Gowda and Sznajder (1997), too.

Theorem 4.6 *Let Ω be an open set in $\mathbb{R}^n, f : \Omega \to \mathbb{R}^n$ a weakly univalent function and $q \in f(\Omega)$. Then $f^{-1}(q)$ is a singleton under one of the following conditions: (1) $f^{-1}(q)$ is zero dimensional; (2) $f^{-1}(q)$ is countable; and (3) $f^{-1}(q)$ has an isolated point.*

We say that a function $h : \mathbb{R}^n \to \mathbb{R}^n$ is a P_0-function if for every pair x and y with $x \neq y$ in $\mathbb{R}^n$ we have: $\max_{\{i:x_i \neq y_i\}} (h(x) - h(y))_i \cdot (x - y)_i \geq 0$. The following result is also from Gowda and Sznajder (1997).

Theorem 4.7 *Suppose that $f(x) = h(x) + b$, where $h : \mathbb{R}^n \to \mathbb{R}^n$ is a continuous P_0-function and $b \in \mathbb{R}^n$. If $SNCP(f, \mathbb{R}^n_+)$ contains a non-empty bounded clopen subset, then $SNCP(f, \mathbb{R}^n_+)$ is connected. In particular the set $SNCP(f, \mathbb{R}^n_+)$ is connected if it is bounded and $SNCP(f, \mathbb{R}^n_+)$ is a singleton if it has an isolated solution.*

Proof: The solvability of the problem $NCP(f, \mathbb{R}^n_+)$ is equivalent to the solvability of the equation $\Phi(x) = x \bigwedge f(x) = 0$. To the function Φ we apply Theorems 4.5 and 4.6. □

Problem 4.1 *(1): It is important to find new theorems giving that $f^{-1}(q)$ is connected; and (2): Find new necessary and sufficient conditions for LCP-connectedness of matrices.*

5 ESSENTIAL CONNECTED COMPONENTS OF SOLUTION SET FOR COMPLEMENTARITY PROBLEMS

Considering the results presented above, we conclude that the property of solution set to be connected, is a strong property. If $SNCP(f, K)$ is not connected, a natural question is, under what conditions there exists a *connected component* of solution set which is, in some sense, *stable*? This problem can be studied using the notion of *essential* connected component. The notion of *essential component* has as origin the notion of *essential fixed point* introduced by Fort (1950).

Let (X,d) be a compact metric space with the property that every $f \in \mathcal{C}(X,X)$ (the space of all continuous (single-valued) functions from X into X) has a fixed point. The space $\mathcal{C}(X,X)$, with respect to the metric ρ which is defined by $\rho(f,g) = \sup\{d(f(x),g(x)) : x \in X\}$ for each $(f,g) \in \mathcal{C}(X,X)$ is a complete metric space. Now define a set-valued mapping $\mathcal{F} : \mathcal{C}(X,X) \to 2^X$ defined by $\mathcal{F}(f) = \{$the set of fixed point of $f\}$ for each $f \in \mathcal{C}(X,X)$. Then the mapping $\mathcal{F}$ is upper-semicontinuous (e.g., by Wehausen's Theorem). We first recall the following definition due to Fort (1950).

Definition 5.1 *We say that a point $x_* \in \mathcal{F}(f)$ is essential, if and only if, for every neighborhood U of x_*, there exists $\varepsilon > 0$ such that if $g \in \mathcal{C}(X,X)$ and $\rho(f,g) < \epsilon$, then $\mathcal{F}(g)\bigcap U \neq \phi$.*

We mention the following classical result due to Fort (1950).

Theorem 5.1 *Each fixed point of $f \in \mathcal{C}(X,X)$ is essential if and only if, f is a point of continuity of $\mathcal{F}$ (i.e., $\mathcal{F}$ is upper semicontinuous and lower semicontinuous at f).*

The notion of *essential component* of $\mathcal{F}(f)$ was introduced by Kinoshita (1952).

Definition 5.2 *We say that a connected component D of $\mathcal{F}(f)$ is an essential component, if for every open set U such that $D \subset U$, there exists $\delta > 0$ such that, $g \in \mathcal{C}(X,X)$ and $\rho(f,g) < \delta$ imply that $\mathcal{F}(g)\bigcap U \neq \emptyset$.*

In a Hilbert space the explicit complementarity problem $NCP(f,K)$ is equivalent to a fixed point problem and the problem $ICP(f,g,K)$ is equivalent to a coincidence equation. In order to unify the both situations, we first recall the notion of *essential component of coincidence points* as follows.

Definition 5.3 *Let $(E, \|,\|)$ be a Banach space and $X \subset E$ a compact convex set. We say that (1): a set-valued mapping $h : X \to 2^X$ is weakly inward if for each $x \in \partial(X), h(x)\bigcap \overline{I_X(x)} \neq \emptyset$, where $I_X(x) = \bigcup\limits_{\lambda \geq 0} \lambda(X - x)$; and (2): a mapping $h : X \to 2^X$ is weakly outward, if for all $x \in \partial(X)$, $h(x)\bigcap \overline{O_X(x)} \neq \emptyset$, where $O_X(x) = \bigcup\limits_{\lambda \leq 0} \lambda(X - x)$.*

We also need the following notations. Denote by

$$S_X = \{f : X \to 2^X : f \text{ is USC and with non-empty closed convex values}\}$$

and consider on S_X the metric $\rho_*(f,g) = \sup\limits_{x \in X} h(f(x), g(x))$, where h is the Hausdorff metric. Then (S_X, ρ_*) is a complete metric space.

Denoted by $Y_X = \{(f,g) \in S_X \times S_X : f - g$ is weakly inward with respect to $X\}$ and consider on Y_X the metric ρ which is defined by

$$\rho\left((f,g),(f^1,g^1)\right) = \rho_*(f,f^1) + \rho_*(g,g^1)$$

for each $(f,g),(f^1,g^1) \in Y_X$. We can also show that (Y_X,ρ) is a complete metric space.

Finally, we define another set-valued mapping $F : Y_X \to 2^X$ by

$$F(y) = \left\{x \in X : f(x) \bigcap g(x) \neq \emptyset\right\}$$

for each $(f,g) \in Y_X$. Then we have the following result due to Yuan et al. (1998).

Theorem 5.2 *Let X be a non-empty compact convex subset of a Banach space $(E, \|,\|)$ and $f, g : X \to 2^X$ be upper semicontinuous set-valued mappings with non-empty closed convex values. If $f - g$ is weakly inward with respect to X, then there exists $x_* \in X$ such that $f(x_*) \bigcap g(x_*) \neq \emptyset$.*

From Theorem 5.2 we have that the set-valued mapping F has the property that for every $y \in Y_X$, $F(y)$ is non-empty. We note also the following result proved by Isac and Yuan (1999).

Theorem 5.3 *The set-valued mapping $F : Y_X \to 2^X$ is upper semicontinuous with non-empty compact values.*

For each $y \in Y_X$, we can write the set $F(y)$ as the union of its disjoint connected components, i.e., $F(y) = \bigcup_{\alpha \in \Lambda} F_\alpha(y)$.

Definition 5.4 *The component $F_\alpha(y)$, for some $\alpha \in \Lambda$ is said to be an essential component of $F(y)$ if for each open set U containing $F_\alpha(y)$, there exists $\varepsilon > 0$, such that for any $y^1 \in Y_X$ with $\rho(y, y^1) < \varepsilon$, we have $F(y^1) \bigcap U \neq \emptyset$.*

Then we have the following main result on the stability of solutions which was proved by Isac and Yuan (1999).

Theorem 5.4 *Let X be a non-empty compact convex subset of a Banach space $(E, \|,\|)$. For any $y = (f,g) \in Y_X$ there exists at least one essential connected component of $F(y)$.*

6 APPLICATIONS TO COMPLEMENTARITY THEORY

Let $(H, \langle,\rangle)$ be a Hilbert space, $K \subset H$ a closed pointed convex cone and $f : K \to H$ a mapping. Given a mapping $T : K \to H$, we say that T is *bounded* if for any bounded set $D \subset K$, we have that $T(D)$ is bounded. Suppose given a mapping $\varphi : R_+ \to R_+$ such that for some $\gamma > 0$ we have that $\varphi(t) > 0$ whenever $t \geq \gamma$.

Definition 6.1 *We say that $f : K \to H$ is a φ-asymptotically bounded field if and only if $f(x) = x - T(x)$, for all $x \in K$, where $T : K \to H$ has the following properties:*

(i) T is continuous and bounded;

(ii) T is φ-asymptotically bounded, i.e., there exist r, $c \in R_+\backslash\{0\}$, such that for any $x \in K$, with $r \leq \|x\|$, we have $\|T(x)\| \leq c\varphi(\|x\|)$; and

(iii) $\lim_{t\to+\infty} \varphi(t) < +\infty$.

The next result is the first main result of this paper.

Theorem 6.1 *Let $(H, \langle,\rangle)$ be a Hilbert space, $K \subset H$ a locally compact convex cone and $f : K \to H$ a φ-asymptotically bounded field. Consider the problem $NCP(f, K)$ and denote by $SNCP(f, K)$ its solution set. Then, there exists a convex compact set $X \subset K$ such that:*

(1) $SNCP(f, K) \subseteq X$; and

(2) $SNCP(f, K)$ has a connected component F_α with the following property: for any open set U with $F_\alpha \subset U$ there exists $\varepsilon > 0$ such that for any $g : X \to 2^X$ which is upper semicontinuous with $\rho(P_K T, g) < \varepsilon$, then it follows that g has a fixed point in U (we recall that $\rho(P_K T, g) = \sup\{\|P_K(T(x)) - y\| : x \in X, y \in g(x)\}$).

Proof: Consider the mapping $\Phi : K \to K$ defined by $\Phi(x) = P_K[x - (x - T(x))] = P_K[T(x)]$, for all $x \in K$. From the complementarity theory we know (e.g, see Isac (1992)) that the set $SNCP(f, K)$ coincides with the set of fixed points of the mapping Φ. The mapping Φ is φ-asymptotically bounded. Indeed, we have $\|\|\Phi(x)\| = \|P_K[T(x)]\|| = \|P_K[T(x)] - P_K(0)\| \leq \|T(x)\| \leq c\varphi(\|x\|)$ for all $x \in K$ with $\|x\| \geq r$ (where r and c are defined in assumption (ii) of Definition 6.1). Observe that Φ is also a bounded and continuous mapping. We show that $\Phi(K)$ is bounded. Indeed, since Φ is φ-asymptotically bounded, and $\lim_{t\to+\infty} \varphi(t) < +\infty$, there exist $M > 0$ and $\alpha > 0$ such that $\|\Phi(x)\| \leq M$ for all $x \in K$ with $\|x\| > \alpha$. Obviously, $\Phi(K) = \{\Phi(x) : x \in K \text{ and } \|x\| > \alpha\} \cup \{\Phi(x) : x \in K \text{ and } \|x\| \leq \alpha\}$. So, it suffices to observe that, the set $\{\Phi(x) : x \in K \text{ and } \|x\| \leq \alpha\}$ is bounded, since Φ is a bounded mapping. Further, let us define $X := cl\,(\text{conv}(\Phi(K)))$. Since K is closed and convex, and $\Phi(K) \subseteq K$, we have that $X \subseteq K$. Therefore $\Phi(X) \subseteq \Phi(K)$, and by the definition of X we have that $\Phi(X) \subseteq X$. Clearly X is a closed bounded set and because K is locally compact, we deduce that X is compact. By Brouwer's fixed point theorem, the set of fixed points of Φ, with respect to X is non-empty. Observe also that any fixed point of Φ in K must be in X. Indeed, if $x_* \in K$ is a fixed point of Φ, then we have $\Phi(x_*) = x_*$ and by the definition of X we have that $x_* \in X$. We conclude that $SNCP(f, K) \subseteq X$. If we denote by I_X the identity mapping of X, and because $\Phi(X) \subseteq X$, we have that $y = (\Phi, I_X) \in Y_X$ and $F(y)$ is exactly the set of fixed points of Φ in X. The conclusion of theorem is now obtained by applying Theorem 5.4 to the set X and the element $y = (\Phi, I_X) \in Y_X$. □

Remark 6.1 In the case of Theorem 6.1 the solution set $SNCP(f, K)$ of the problem $NCP(f, K)$ has a privileged connected component F_α which is *stable* with respect to the fixed points of USC set-valued mappings $g : X \to 2^X$.

Theorem 6.1 has a variant for implicit complementarity problem $ICP(f, g, K)$ defined in Section 3. In this case the following notion is necessary.

Definition 6.2 *Suppose that $g : K \to H$ is a continuous mapping. We say that $f : K \to H$ is a (g, φ)-asymptotically bounded field if and only if, $f(x) = g(x) - T(x)$, for all $x \in K$, where $T : K \to H$ has the following properties:*
(i) T is continuous and bounded;
(ii) T is φ-asymptotically bounded; and
(iii) $\lim\limits_{t \to +\infty} \varphi(t) < +\infty$.

Denoting by $SICP(f, g, K)$ the solution set of the problem $ICP(f, g, K)$, we have the following result.

Theorem 6.2 *Let $(H, \langle , \rangle)$ be a Hilbert space and $K \subset H$ a locally compact convex cone. If $f : K \to H$ is a (g, φ)-asymptotically bounded field, consider the problem $ICP(f, g, K)$. If $g(K) \subset cl\,(conv(P_K[T(K)]))$ and $P_K T - g$ is weakly inward with respect to $X = cl(conv\,((P_K[T(K)])))$, then the non-empty set $SICP(f, g, K)$ has a connected component with respect to X, F_α such that for every neighborhood U of F_α, with respect to X, there exists $\varepsilon > 0$ such that for every $y' \in Y_X$ with $\rho\,((P_K \circ T, g), y') < \varepsilon$ we have $F(y') \cap U \neq \emptyset$.*

Proof: The proof is similar to the proof of Theorem 5.4. □

7 THE GENERAL CASE

Let $\langle E, E^* \rangle$ be a duality of Banach spaces, $K \subset E$ a closed convex cone and $\mathcal{K}(K)$ the collection of non-empty compact subsets of K. Consider on the set $\mathcal{C}(K, E^*) = \{f : K \to E^*\text{: } f \text{ is continuous }\}$ the generalized metric $\rho(f, f') = \sup\limits_{x \in K} \|f(x) - f'(x)\|$, for all $f, f' \in \mathcal{C}(K, E^*)$. It is known that $(\mathcal{C}(K, E^*), \rho)$ is a *complete metric space.* On the set $Y = \mathcal{C}(K, E^*) \times \mathcal{K}(K)$ we consider the metric d defined by $d(y, y') = \rho(f, f') + h(A, A')$ for all $y = (f, A)$ and $y' = (f', A') \in Y$. We know from Isac and Yuan (1999b) that (Y, d) is also a complete metric space. By defining the set M as $M = \{y = (f, A) : y \in Y$ and the problem $NCP(f, K)$ is such that $SNCP(f, K) \subseteq A\}$.
Then we can show that (M, d) is a metric subspace of (Y, d). Moreover, it is closed and hence it is complete (e.g., see Isac and Yuan (1997b)). We note that $(\mathcal{K}(K), h)$ is a complete metric space. If $y = (f, A) \in M$, this means that $SNCP(f, K) \subset A$.
Now define a set-valued mapping $S : M \to \mathcal{K}(K)$ by $S(y) = SNCP(f, K)$ for each $y \in M$. It is known that the solution mapping S is upper semicontinuous from Isac and Yuan (1999b).

Definition 7.1 *If $y = (f, A) \in M$, then a solution $x_* \in S(y)$ is said to be an essential solution of the problem $NCP(f, K)$ if, for any $\varepsilon > 0$, there exists $\sigma > 0$ such that for any $y' = (f', A') \in M$ with $d(y, y') = \rho(f, f') + h(A, A') < \sigma$, there exists a solution $x'_* \in S(y')$ with $\|x_* - x'_*\| < \varepsilon$.*

Remark 7.1 The notion of *essential solution* is similar to the notion of *stable solution* used by Ha (1987).

For each $y \in M$, we can write down the solution set $S(y)$ of y as the union of its disjoint components $S_\alpha(y)$, where $\alpha \in \Gamma$ and Γ is an index set, i.e., $S(y) = \cup_{\alpha\in\Gamma} S_\alpha(y)$. Now we have the following definition.

Definition 7.2 *For each $y \in M$, the component $S_\alpha(y)$ for some $\alpha \in \Gamma$ is said to be an essential component of y if for each open set U containing $S_\alpha(y)$, there exists $\delta > 0$ such that for all other $y' = (f', A') \in M$ with $\rho(y, y') = \sup_{x\in K} \|f(x) - f'(x)\| + h(A, A') < \delta$, we have that $S(y') \cap U \neq \emptyset$.*

Problem 7.1 *Under what condition, for each $y \in M$, $S(y)$ has an essential connected component ?*

We will show now that this problem has a solution in a particular case. Before we do it, we have the following definition.

Definition 7.3 *Let $M_1 \subset M$ be a non-empty subset. If $y = (f, A) \in M$ and $S(y) = \cup_{\alpha\in\Gamma} S_\alpha(y)$, then we say that a connected component $S_\alpha(y)$ for some $\alpha \in \Gamma$ is an essential connected component of y with respect to M_1 if for each open set U containing $S_\alpha(y)$, there exists $\delta > 0$ such that for any other $y' = (f', A') \in M_1$ with $\rho(y, y') = \sup_{x\in K} \|f(x) - f'(x)\| + h(A, A') < \delta$, we have that $S(y') \cap U \neq \emptyset$.*

Let $D \subset K$ be a non-empty compact set. We also introduce the following definition.

Definition 7.4 *We say that an existence condition (P) for the problem $NCP(f, K)$ is stable with respect to D if for any continuous mapping $f : K \to E^*$ satisfying condition (P) we have that $SNCP(f, K) \subset D$ and the set of all continuous functions from K into E^* satisfying condition (P) is closed with respect to convex combinations.*

Remark 7.2 (1) The following *strong Karamardian's condition* is stable. We say that $f : K \to E^*$ satisfies the strong Karamardian's condition with respect to $D \subset K$ (here D is supposed to be convex and compact) if for all $x \in K \setminus D$, we have that $\langle x - y, f(x)\rangle > 0$ for all $y \in D$; and

(2) The following existence condition is also stable. Consider the Euclidean space $\mathbb{R}^n$ and $K \subset \mathbb{R}^n$ a closed convex cone. Suppose that the continuous function $f : K \to \mathbb{R}^n$ satisfies the following condition: there exists $\rho > 0$ such that for all $x \in K$ with $\|x\| > \rho$, we have $\langle x, f(x)\rangle > 0$ (e.g., see Isac and Carbone (1999)).

The following theorem is the second main result of this paper.

Theorem 7.1 *Let $D \subset K$ be a non-empty and compact subset of K and $y = (f, D) \in M$. Suppose f satisfies with respect to D a stable existence condition for the problem $NCP(f, K)$. Let $M_1 = \{y = (g, D) \in M : g : K \to E^*$ is continuous and satisfies the same condition as f $\}$. Then there exists at least one essential connected component of $S(y)$ with respect to M_1.*

Proof: The proof follows the idea used in the proof of Theorem 3.3 from the paper of Isac and Yuan (1999b). □

Remark 7.3 It is interesting to extend Theorems 6.1 and 7.1 to the non-compact case. To realize this, it is necessary to extend before Theorem 5.4 to non-compact sets, for example, to unbounded sets. This development can be the subject of another paper.

References

Bulavsky, V.A., Isac, G. and Kalashnikov, V.V. (1998), Applications of topological degree theory to complementarity problem, in *Multilevel Optimization: Algorithms and Applications*, edited by Migdalas A. et al., Kluwer Academic Publishers, pp.333-358.

Cao, M. and Ferris, M.C. (1996), P_0-matrices and the linear complementarity problem, *Linear Algebra Appl.*, Vol. 246, pp. 299-312.

Cottle, R.W., Pang, J.S. and Stone, R.E. (1992), *The linear Complementarity Problem*, Academic Press.

Deimling, K. (1985), *Nonlinear Functional Analysis*, Springer-Verlag, Berlin.

Engelking, R. (1980), *General Topology*, Heldermann, Verlag, Berlin.

Fort, M. K. (1950), Essential and nonessential fixed point, *Amer. J. Math.*, Vol. 72, pp. 315-322.

Fort, M. K. (1951), Points of continuity of semi-continuous functions, *Publ. Math. Debrecen*, Vol. 2, pp. 100-102.

Gale, D. and Nikaido, H. (1965), The Jacobian matrix and global univalence of mappings, *Math. Ann.*, Vol. 159, pp. 81-93.

Gowda, M.S. and Sznajder, R. (1997), Weak univalence and connectedness of inverse images of continuous functions, Preprint, Dept. of Math., Univ. of Maryland, 1997.

Ha, C.D. (1987), Applications of degree theory to stability of the complementarity problems, *Math. Oper. Res.*, Vol. 12, pp. 368-376.

Hu, S. and Papageorgiou, N.S. (1997), *Handbook of Multivalued Analysis*, Vol. 1, Kluwer Academic Publishers.

Hyers, D.H., Isac, G. and Rassias, Th.M. (1997), *Topics in Nonlinear Analysis and Applications*, World Scientific, Singapore.

Isac, G. (1992), *Complementarity Problems*, Vol.1528 in Lecture Notes in Mathematics, Springer-Verlag.

Isac, G. (1988), Fixed point theory, coincidence equations on convex cones and complementarity problems, *Contemp. Math.*, Vol. 472, pp. 139-155.

Isac, G. (1986) Complmentarity problem and coincidence equations on convex cones, *Boll. U.M.I.*, Vol. 5-B, pp. 925-943.

Isac, G. (1989) The numerical range theory and boundedness of solutions of the complementarity problem, *J. Math. Anal. Appl.*, Vol. 143, pp. 235-251.

Isac, G. (1998), *Exceptional families of elements for k-set fields in Hilbert spaces and complementarity theory*, in *Optimization Techniques and Applications*, edited by Caccetta, L., Teo, K.L., Siew, P.F., Leung, Y.H., Jennings, L.S., and Rehbock, V., School of Mathematics and Statistics, Curtin University of Technology, Perth, Australia, Vol. 2, pp. 1135-1143.

Isac, G., Bulavsky, V. and Kalashnikov, V. (1997), Exceptional families, topological degree and complementarity problems, *J. Global. Opt.*, Vol. 10, pp. 207-225.

Isac, G. and Carbone, A. (1999), Exceptional families of elements for continuous functions: some applications to complementarity theory, *J. Global Opt.*, Vol. 15, pp. 181-196.

Isac, G. and Yuan, G.X.Z. (1999), The essential components of coincident points for weakly inward and outward set-valued mappings, *Appl. Math. Letters*, Vol. 12, pp. 121-126.

Isac, G. and Yuan, G.X.Z. (1999b), Essential components and connectedness of solutions set for complementarity problems, in *Recent Advance on Variational Inequalities* (in press) edited by F. Giannessi, Kluwer Academic Publisher.

Jiang, J.H. (1962), Essential fixed points of the multivalued mappings, *Scientia Sinica*, Vol. 11, pp. 293-298.

Jones, C. and Gowda, M.S. (1996), On the connectedness of solution set in linear complementarity problems, Preprint, Dept. of Maths, Maryland University

Karamardian, S. (1971), Generalized complementarity problem, *J. Optim. Theory Appl.*, Vol. 8, pp. 161-168.

Kinoshita, S. (1952), On essential component of the set of fixed points, *Osaka J. Math.*, Vol.4, pp. 19-22.

Klein, E. and Thompson, A.C. (1984), *Theory of Correspondences: Including Applications to Mathematical Economics*, John Wiley and Sons.

Krasnoselskii, M.A. and Zabreiko, P.P.(1984), *Geometrical Methods in Nonlinear Analysis*, Springer-Verlag, Berlin.

Luo, Q., Yuan, G. and Sun, T.X. (1997), The existence of essential components for coincident points of set-valued mappings, CADSMAP Research Report, 97-15, The University of Queensland.

More, J.J. and Rheinboldt, W. (1973), On P- and S-functions and related classes of n-dimensional nonlinear mappings, *Linear Algebra*, Vol. 6, pp. 45-68.

Radulescu, M. and Radulescu, S. (1980), Global inversion theorems and applications to differential equations, *Nonlinear Anal., T.M.A.*, Vol. 4, pp. 951-965.

Rapcsak, T. (1994), On the connectedness of the solution set to linear complementarity systems, *J. Optim. Theory Appl.*, Vol. 80, pp. 501-512.

Tan, K. K., Yu, J. and Yuan, X.Z. (1995), The stability of coincident points for multivalued mappings, *Nonlinear Anal. T.M.A.*, Vol. 25, pp. 163-168.

Weber, V. B. (1984), ϕ-asymptotisches spectrum und surektivitatssatze vom Fredholm type fur nichtlinare operatoren mit anwendungen, Math. Nachr., Vol. 117, pp. 7-35.

Yu, J. and Xiang, S.W. (1997), On essential components of Nash equilibrium points, Research Report, Institute of Applied Math., Guizhou University of Technology.

Yuan, G.X.Z. (1999), *KKM Theory with Applications in Nonlinear Analysis*, Marcel Dekker, Inc., USA, New York.

Yuan, X.Z., Smith, B. and Lou, S. (1998), Fixed point and coincidence theorems of set-valued mappings in topological vector spaces with some applications, *Nonlinear Anal., T.M.A.*, Vol. 23, pp. 183-199.

12 ON RELATIONS BETWEEN VECTOR VARIATIONAL INEQUALITY AND VECTOR OPTIMIZATION PROBLEM

Gue Myung Lee

Department of Applied Mathematics,
Pukyong National University,
Pusan 608-737, Korea

Abstract: In this paper, we study equivalent relations between vector variational inequalities for subdifferentials and nondifferentiable convex vector optimization problems. Futhermore, using the equivalent relations, we give existence theorems for solutions of convex vector optimization problems.

Key words: vector variational inequality, vector optimization problem, equivalent relations.

X. Yang et al. (eds.), Progress in Optimization, 167–179.

1 INTRODUCTION

A vector variational inequality in a finite dimensional Euclidean space was first introduced by Giannessi (1980). Since then, existence theorems for solutions of general versions of the vector variational inequality have been studied in Chen and Li (1996), Konnov and Yao (1997), Lee et al (1998), Lee and Kum (1998), Lin et al (1997), and Yu and Yao (1996), and relationships between vector variational inequalities and vector optimization problems have been investigated in Giannessi (1997), Lee et al (1998), and Yang (1993). In particular, using a vector variational-like inequality, Lee et al (1998) gave some existence theorems for solutions of nondifferentiable invex vector optimization problmes, which are generalizations of an existence result of Chen and Craven (1994) for a differentiable convex vector optimization problem. Very recently, Giannessi (1997) showed the equivalence between efficient solutions of a differentiable convex vector optimization problem and solutions of a Minty type vector variational inequality for gradients which is a vector version of the classical Minty variational inequality for gradients in Baiocchi and Capelo (1994). Moreover, he proved the equivalence between solutions of weak Minty type and Stampacchia type vector variational inequalities for gradients and weakly efficient solutions of a differentiable convex vector optimization problem. Yang (1997) gave the equivalnce between solutions of a Stampacchia type vector variational inequality for gradients and efficient solutions of a linear fractional vector optimization problem of which the numerators of the objective functions are linear and the denominators of the objective functions are the same linear functions.

In this paper, following the approaches of Giannessi (1997), we study the equivalence between solutions of a Minty type vector variational inequality for subdifferentials and efficient solutions of a nondifferentiable convex vector optimization problem, and show the equivalence between solutions of weak Minty type and Stampacchia type vector variational inequalities for subdifferentials and weakly efficient solutions of a nondifferentiable convex vector optimization problem.

Moreover, using the equivalent relations, we give existence theorems for weakly efficient solutions of convex vector optimization problems under relaxed compact assumptions.

2 MINTY TYPE VECTOR VARIATIONAL INEQUALITY

Let X be a nonempty subset of $\mathbb{R}^n$ and $f_i : \mathbb{R}^n \to \mathbb{R},\ i = 1, \cdots, p$, functions. Consider the following vector optimization problem **(P)**;

$$\textbf{(P)} \qquad \text{Minimize} \qquad f(x) := (f_1(x), \cdots, f_p(x))$$

$$\text{subject to} \qquad x \in X.$$

Solving **(P)** means to fined the efficient solutions which are defined as follows;

Definition 2.1 (1) *$y \in X$ is said to be an efficient solution of* **(P)** *if for any $x \in X$, $(f_1(x) - f_1(y), \cdots, f_p(x) - f_p(y)) \notin -\mathbb{R}^p_+ \backslash \{0\}$, where $\mathbb{R}^p_+$ is the nonnegative orthant of $\mathbb{R}^p$.*

(2) $y \in X$ is called a properly efficient solution of **(P)** *if $y \in X$ is an efficient solution of* **(P)** *and there exists $M > 0$ such that for each $i = 1, \cdots, p$, we have*

$$\frac{f_i(x) - f_i(y)}{f_j(y) - f_j(x)} \leq M$$

for some j such that $f_j(x) > f_j(y)$ whenever $x \in X$ and $f_i(x) < f_i(y)$.

(3) $y \in X$ is said to be an weakly efficient solution of **(P)** *if for any $x \in X$,*

$$(f_1(x) - f_1(y), \cdots, f_p(x) - f_p(y)) \notin -int\mathbb{R}^p_+,$$

where $int\mathbb{R}^p_+$ is the interior of $\mathbb{R}^p_+$.

Let $f_i : \mathbb{R}^n \to \mathbb{R}, i = 1, \cdots, p$, be convex functions. We formulate the following Minty type vector variational inequality **(MVI)** for subdifferentials:

(MVI) Find $y \in X$ such that for any $x \in X$ and any $\xi_i \in \partial f_i(x)$, $i = 1, \cdots, p$, $(\xi_1^t(x - y), \cdots, \xi_p^t(x - y)) \notin -\mathbb{R}^p_+ \backslash \{0\}$,

where $\partial f_i(x)$ is the subdifferential of f_i at x.

When $f_i : \mathbb{R}^n \to \mathbb{R}$, $i = 1, \cdots, p$, are differentiable convex functions, **(MVI)** becomes the following Minty type vector variational inequality for gradients which was studied by Giannessi (1997);

Find $y \in X$ such that for any $x \in X$,
$(\nabla f_1(x)^t(x - y), \cdots, \nabla f_p(x)^t(x - y)) \notin -\mathbb{R}^p_+ \backslash \{0\}$.

Consider the following Stampacchia type vector variational inequality **(SVI)** for subdifferentials;

(SVI) Find $y \in X$ such that for any $x \in X$, there exist $\xi_i \in \partial f_i(y)$, $i = 1, \cdots, p$, such that

$$(\xi_1^t(x - y), \cdots, \zeta_p^t(x - y)) \notin -\mathbb{R}^p_+ \backslash \{0\}.$$

When $f_i : \mathbb{R}^n \to \mathbb{R}$, $i = 1, \cdots, p$, are differentiable convex functions, (SVI) becomes the following Stampacchia type vector variational inequality for gradients which was studied by Giannessi (1997);

Find $y \in X$ such that for any $x \in X$,
$(\nabla f_1(y)^t(x - y), \cdots, \nabla f_p(y)^t(x - y)) \notin -\mathbb{R}^p_+ \backslash \{0\}$.

Theorem 2.1 *Let X be a nonempty convex subset of $\mathbb{R}^n$ and $f_i, i = 1, \cdots, p$, convex functions. Then $y \in X$ is an efficient solution of* **(P)** *if and only if it is a solution of* **(MVI)**.

Proof: Let $y \in X$ be a solution of **(MVI)**. Suppose to the contrary that $y \in X$ is not an efficient solution of **(P)**. Then there exists $z \in X$ such that

$$(f_1(z) - f_1(y), \cdots, f_p(z) - f_p(y)) \in -\mathbb{R}^p_+ \backslash \{0\}. \tag{1}$$

Since X is convex, we have $z(\alpha) := \alpha y + (1 - \alpha)z \in X$ for any $\alpha \in [0, 1]$. Since f_i is convex, $f_i(z(\alpha)) \leq \alpha f_i(y) + (1 - \alpha) f_i(z)$ for any $\alpha \in [0, 1]$ and hence $f_i(z(\alpha)) - f_i(y) \leq (\alpha - 1)[f_i(y) - f_i(z)]$ for any $\alpha \in [0, 1]$. So we have

$$\frac{f_i(z(\alpha)) - f_i(z(1))}{\alpha - 1} \geq f_i(y) - f_i(z) \text{ for any } \alpha \in (0, 1).$$

By Lebourg's Mean Value Theorem in Clarke (1983), p.41, there exist $\alpha_i \in (0, 1)$ and $\xi_i \in \partial f_i(z(\alpha_i))$ such that

$$\xi_i^t (y - z) \geq f_i(y) - f_i(z), \quad i = 1, \cdots, p. \tag{2}$$

Suppose that $\alpha_1, \cdots, \alpha_p$ are equal. Then it follows from (1) and (2) that $y \in X$ is not a solution of **(MVI)**. This contradicts to the fact that $y \in X$ is a solution of **(MVI)**.

Suppose that $\alpha_1, \cdots, \alpha_p$ are not equal. Let $\alpha_1 \neq \alpha_2$. From (2), we have

$$\xi_1^t (y - z) \geq f_1(y) - f_1(z) \tag{3}$$

$$\text{and} \quad \xi_2^t (y - z) \geq f_2(y) - f_2(z). \tag{4}$$

Since f_1 and f_2 are convex, we have

$$(\xi_1 - \xi_2^*)^t [z(\alpha_1) - z(\alpha_2)] \geq 0 \text{ for any } \xi_2^* \in \partial f_1(z(\alpha_2)) \tag{5}$$

and

$$(\xi_1^* - \xi_2)^t [z(\alpha_1) - z(\alpha_2)] \geq 0 \text{ for any } \xi_1^* \in \partial f_2(z(\alpha_1)) \tag{6}$$

If $\alpha_1 < \alpha_2$, from (5), $(\xi_1 - \xi_2^*)^t (y - z) \leq 0$ and hence from (3), we have

$$\xi_2^{*^t} (y - z) \geq f_1(y) - f_1(z) \text{ for any } \xi_2^* \in \partial f_1(z(\alpha_2)).$$

If $\alpha_2 < \alpha_1$, from (6), $(\xi_1^* - \xi_2)^t (y - z) \geq 0$ and hence from (4), we have

$$\xi_1^{*^t} (y - z) \geq f_2(y) - f_2(z) \text{ for any } \xi_1^* \in \partial f_2(z(\alpha_1)).$$

Therefore, if $\alpha_1 \neq \alpha_2$, letting $\hat{\alpha}^* = \max\{\alpha_1, \alpha_2\}$, we can find $\bar{\xi}_i \in \partial f_i(z(\hat{\alpha}^*))$ such that $\bar{\xi}_i^t (y - z) \geq f_i(y) - f_i(z), i = 1, 2$.

By continuing this process, we can find $\hat{\alpha} \in (0,1)$ and $\bar{\xi}_i \in \partial f_i(z(\hat{\alpha}))$ such that

$$\bar{\xi}_i^t(y-z) \geq f_i(y) - f_i(z), \quad i = 1, \cdots, p. \tag{7}$$

From (1) and (7), $\bar{\xi}_i \in \partial f_i(z(\hat{\alpha})), \quad i = 1, \cdots, p$, and

$$(\bar{\xi}_1^t(y-z), \cdots, \bar{\xi}_p^t(y-z)) \in \mathbb{R}_+^p \backslash \{0\}. \tag{8}$$

Multiplying both sides of (8) by $\hat{\alpha} - 1$, we obtain

$$(\bar{\xi}_1^t(z(\hat{\alpha}) - y), \cdots, \bar{\xi}_p^t(z(\hat{\alpha}) - y)) \in -\mathbb{R}_+^p \backslash \{0\},$$

which contradicts to the fact that $y \in X$ is a solution of **(MVI)**.
Conversely, let $y \in X$ be an efficient solution of **(P)**. Then we have

$$(f_1(y) - f_1(x), \cdots, f_p(y) - f_p(x)) \notin \mathbb{R}_+^p \backslash \{0\}, \quad \text{for any} \quad x \in X. \tag{9}$$

Since f_i is convex, we have

$$f_i(y) - f_i(x) \geq \xi_i^t(y-x) \text{ for any } \quad \xi_i \in \partial f_i(x). \tag{10}$$

Hence, it follows from (9) and (10) that $y \in X$ is a solution of **(MVI)**. □

Theorem 2.2 *Let X be a nonempty convex subset of $\mathbb{R}^n$. If $f_i, \quad i = 1, \cdots, p$, are convex and $y \in X$ is a solution of* **(SVI)**, *then $y \in X$ is an efficient solution of* **(P)** *and $y \in X$ is a solution of* **(MVI)**.

Proof: Since $y \in X$ is a solution of **(SVI)**, for any $x \in X$, there exist $\zeta_i \in \partial f_i(y), \quad i = 1, \cdots, p$, such that,

$$(\zeta_1^t(x-y), \cdots, \zeta_p^t(x-y)) \notin -\mathbb{R}_+^p \backslash \{0\}. \tag{11}$$

Since $f_i, i = 1, \cdots, p$, are convex, we have, for any $x \in X$,

$$f_i(x) - f_i(y) \geq \zeta_i^t(x-y). \tag{12}$$

From (11) and (12), we have

$$(f_1(x) - f_1(y), \cdots, f_p(x) - f_p(y)) \notin -\mathbb{R}_+^p \backslash \{0\}, \quad \text{for any} \quad x \in X.$$

Thus $y \in X$ is an efficient solution of **(P)**. By Theorem 2.1, $y \in X$ is a solution of **(MVI)**. □

From Theorem 2.2, **(SVI)** is a sufficient optimality condition for an efficient solution of **(P)**. However, it was shown in Giannessi (1997) and Yang (1997) that **(SVI)** is, in general, not a necessary optimality condition for an efficient solution of **(P)**. For the completeness, we give this example.

Example 2.1 *Let $f(x) := (x, x^2)$ and $X := [-1, 0]$. Consider the following differentiable convex vector optimization problem* **(P)***;*

$$\textbf{(P)} \qquad \text{Minimize} \qquad f(x)$$
$$\text{subject to} \qquad x \in X.$$

Then $y = 0$ is an efficient solution of **(P)** *and $y = 0$ is a solution of the following Minty type vector variational inequality for gradients: Find $y \in X$ such that for any $x \in X$,*

$$(\nabla f_1(x)^t(x-y), \nabla f_2(x)^t(x-y)) = (x-y, 2x(x-y)) \notin -R^2_+ \backslash \{0\}.$$

However, $y = 0$ is not a solution of the following Stampacchia type vector variational inequality for gradients : Find $y \in X$ such that for any $x \in X$,

$$(\nabla f_1(y)^t(x-y), \nabla f_2(y)^t(x-y)) = (x-y, 2y(x-y)) \notin -R^2_+ \backslash \{0\}.$$

We notice that $y = 0$ is not a properly efficient solution of **(P)** in the above example. However, we will show that if $y \in X$ is an properly efficient solution of **(P)**, then $y \in X$ is a solution of **(SVI)**.

Proposition 2.1 *Let X be a nonempty convex subset of $\mathbb{R}^n$. If $f_i, i = 1, \cdots, p$, are convex, then the following are equivalent;*

(a) *$y \in X$ is a properly efficient solution of* **(P)**.

(b) *there exist $\lambda_i > 0,\ i = 1, \cdots, p$, such that $y \in X$ is a solution of the following scalar variational inequality: Find $y \in X$ such that there exist $\zeta_i \in \partial f_i(y), i = 1, \cdots, p$, such that for any $x \in X$,*

$$(\lambda_1 \zeta_1 + \cdots + \lambda_p \zeta_p)^t (x-y) \geq 0.$$

(c) *there exist $\lambda_i > 0,\ i = 1, \cdots, p$, such that $y \in X$ is a solution of the following scalar variational inequality: Find $y \in X$ such that for any $x \in X$, there exists $\zeta_i \in \partial f_i(y), i = 1, \cdots, p$, such that*

$$(\lambda_1 \zeta_1 + \cdots, + \lambda_p \zeta_p)^t (x-y) \geq 0.$$

Proof: By Theorem 1 and 2 in Geoffrion [8], (a) is equivalent to the fact that there exists $\lambda_i > 0,\ i = 1, \cdots, p$, such that $y \in X$ is an optimal solution of the following scalar optimization problem **(SP)**;

$$\textbf{(SP)} \qquad \text{Minimize} \quad \lambda_1 f_1(x) + \cdots + \lambda_p f_p(x)$$
$$\text{subject to} \quad x \in X.$$

Moreover, it is well known that the fact that $y \in X$ is an optimal solution of **(SP)** is equivalent to the fact that there exist $\zeta_i \in \partial f_i(y),\ i = 1, \cdots, p$, such that for any $x \in X$,

$$(\lambda_1\zeta_1 + \cdots + \lambda_p\zeta_p)^t(x - y) \geq 0.$$

Hence (a) is equivalent to (b).

On the other hand, (b) is equivalent to the fact that there exist $\lambda_i > 0, i = 1, \cdots, p$, such that

$$\max_{\zeta_i \in \partial f_i(y)} \inf_{x \in X} (\lambda_1\zeta_1 + \cdots + \lambda_p\zeta_p)^t(x - y) \geq 0.$$

Thus, by Extended Minimax Theorem of Jeyakumar [11], (b) is equivalent to the fact that there exists $\lambda_i > 0,\ i = 1, \cdots, p$, such that

$$\inf_{x \in X} \max_{\zeta_i \in \partial f_i(y)} (\lambda_1\zeta_1 + \cdots + \lambda_p\zeta_p)^t(x - y) \geq 0.$$

Hence (b) is equivalent to (c). □

From Proposition 2.1, we can easily obtain the following theorem;

Theorem 2.3 *Let X be a nonempty convex subset of $\mathbb{R}^n$. If $f_i,\ i = 1, \cdots, p$, are convex and $y \in X$ is a properly efficient solution of* **(P)**, *then $y \in X$ is a solution of* **(SVI)**.

3 WEAK VECTOR VARIATIONAL INEQUALITY

We consider the following weak Minty vector variational inequality **(WMVI)** and weak Stampacchia type vector variational inequality **(WSVI)**:

(WMVI) Find $y \in X$ such that for any $x \in X$ and any

$$\xi_i \in \partial f_i(x),\ i = 1, \cdots, p,\ (\xi_1^t(x - y), \cdots, \xi_p^t(x - y)) \notin -int\mathbb{R}^p_+.$$

(WSVI) Find $y \in X$ such that for any $x \in X$, there exists $\zeta_i \in \partial f_i(y)$, $i = 1, \cdots, p$, such that

$$(\zeta_1^t(x - y), \cdots, \zeta_p^t(x - y)) \notin -int\mathbb{R}^p_+.$$

Theorem 3.1 *Let X be a nonempty convex subset of $\mathbb{R}^n$ and $f_i,\ i = 1, \cdots, p$, convex functions. Then $y \in X$ is a solution of* **(WSVI)** *if and only if $y \in X$ is a solution of* **(WMVI)**.

Proof: Let $y \in X$ be a solution of **(WSVI)**. Then for any $x \in X$, there exist $\zeta_i \in \partial f_i(y),\ \ i = 1, \cdots, p$, such that

$$(\zeta_1^t(x-y), \cdots, \zeta_p^t(x-y)) \notin -\mathrm{int}\mathbb{R}_+^p. \tag{13}$$

Since $\partial f_i, i = 1, \cdots, p$, are monotone, we have, for any $x \in X$ and any $\xi_i \in \partial f_i(x)$,

$$(\zeta_i - \xi_i)^t(x-y) \leq 0,\ \ i = 1, \cdots, p. \tag{14}$$

From (13) and (14), we have for any $x \in X$ and any $\xi_i \in \partial f_i(x),\ \ i = 1, \cdots, p$,

$$(\xi_1^t(x-y), \cdots, \xi_p^t(x-y)) \notin -\mathrm{int}\mathbb{R}_+^p.$$

Hence $y \in X$ is a solution of **(WMVI)**.

Conversely let $y \in X$ be a solution of **(WMVI)**. Consider any $z \in X$ and any sequence $\{\alpha_n\} \downarrow 0$ with $\alpha_n \in (0,1]$. By the convexity of X, $z_n := (1-\alpha_n)y + \alpha_n z \in X$. Since $y \in X$ is a solution of **(WMVI)**, there exist $\xi_i^n \in \partial f_i(z_n),\ \ i = 1, \cdots, p$, such that

$$(\xi_1^{n^t}(z-y), \cdots, \xi_p^{n^t}(z-y)) \notin -\mathrm{int}\mathbb{R}_+^p.$$

Since f_i is convex, f_i is locally Lipschitz, see Aubin (1993), p. 25 and hence there exists $l > 0$ such that for sufficiently large n and all i, $\|\xi_i^n\| \leq l$. So we can assume that the sequence $\{\xi_i^n\}$ converges to ζ_i for each $i = 1, \cdots, p$. Since the multifunction $x \mapsto \partial f_i(x)$ is closed (see Clarke (1983), p. 29) and $\xi_i^n \in \partial f_i(z_n)$, $\zeta_i \in \partial f_i(y)$ for each $i = 1, \cdots, p$. Thus, for any $z \in X$, there exist $\zeta_i \in \partial f_i(y),\ \ i = 1, \cdots, p$, such that

$$(\zeta_1^t(z-y), \cdots, \zeta_p^t(z-y)) \notin -\mathrm{int}\mathbb{R}_+^p.$$

Hence $y \in X$ is a solution of **(WSVI)**. □

Theorem 3.2 *Let X be a nonempty convex subset of $\mathbb{R}^n$ and $f_i,\ \ i = 1, \cdots, p$, convex functions. Then $y \in X$ is a weakly efficient solution of* **(P)** *if and only if $y \in X$ is a solution of* **(WSVI)**.

Proof: Suppose that $y \in X$ is not a weakly efficient solution of **(P)**. Then there exists $z \in X$ such that

$$f_i(y) > f_i(z),\ \ i = 1, \cdots, p. \tag{15}$$

Since $f_i,\ \ i = 1, \cdots, p$, are convex, we have, for any $\zeta_i \in \partial f_i(y)$,

$$f_i(z) - f_i(y) \geq \zeta_i^t(z-y). \tag{16}$$

From (15) and (16), we have, for any $\zeta_i \in \partial f_i(y)$,

$$(\zeta_1^t(z-y), \cdots, \zeta_p^t(z-y)) \in -\mathrm{int}\mathbb{R}_+^p.$$

Hence $y \in X$ is not a solution of **(WSVI)**.

Conversely suppose that $y \in X$ is not a solution of **(WSVI)**. By Theorem 3.1, $y \in X$ is not a solution of **(WMVI)**. So, there exist $\bar{x} \in X$ and $\xi_i \in \partial f_i(\bar{x}),\ i = 1, \cdots, p$, such that

$$(\xi_1^t(\bar{x} - y), \cdots, \xi_p^t(\bar{x} - y)) \in -\text{int}\mathbb{R}_+^p. \tag{17}$$

By the convexity of f_i, we have,

$$f_i(y) - f_i(\bar{x}) \geq \xi_i^t(y - \bar{x}) \tag{18}$$

From (17) and (18), we have

$$(f_1(\bar{x}) - f_1(y), \cdots, f_p(\bar{x}) - f_p(y)) \in -\text{int}\mathbb{R}_+^p.$$

Thus $y \in X$ is not a weakly efficient solution of **(P)**. □

Proposition 3.1 *Let X be a nonempty convex subset of $\mathbb{R}^n$. If $f_i, i = 1, \cdots, p$, are convex, then the followings are equivalent;*

(a) *$y \in X$ is a weakly efficient solution of* **(P)**.

(b) *there exist $\lambda_i \geq 0,\ i = 1, \cdots, p$, not all zero, such that $y \in X$ is a solution of the following scalar variational inequality: Find $y \in X$ such that there exist $\zeta_i \in \partial f_i(y),\ i = 1, \cdots, p$, such that for any $x \in X$,*

$$(\lambda_1\zeta_1 + \cdots + \lambda_p\zeta_p)^t(x - y) \geq 0.$$

(c) *there exist $\lambda_i \geq 0, i = 1, \cdots, p$, not all zero, such that $y \in X$ is a solution of the following scalar variational inequality: Find $y \in X$ such that for any $x \in X$, there exists $\zeta_i \in \partial f_i(y),\ i = 1, \cdots, p$, such that*

$$(\lambda_1\zeta_1 + \cdots + \lambda_p\zeta_p)^t(x - y) \geq 0.$$

Proof: By Proposition 1 and Theorem 1 in Aubin (1979) pp. 46-47, (a) is equivalent to the fact that there exists $\lambda_i \geq 0, i = 1, \cdots, p$, not all zero, such that $y \in X$ is an optimal solution of the following scalar optimization problem **(SP)**;

$$\textbf{(SP)} \qquad \text{Minimize} \quad \lambda_1 f_1(x) + \cdots + \lambda_p f_p(x)$$
$$\text{subject to} \quad x \in X.$$

So, by the method similar to the proof in Proposition 2.1, we can obtain the conclusions of the above proposition. □

By Theorems 3.1 and 3.2, and Proposition 3.1, we can obtain the following theorem;

Theorem 3.3 *Let X be a nonempty convex subset of* $\mathbb{R}^n$ *and* f_i, $i = 1, \cdots, p$, *convex functions. Then the following are equivalent;*

(a) $y \in X$ *is a weakly efficient solution of* **(P)**.

(b) $y \in X$ *is a solution of* **(WMVI)**.

(c) $y \in X$ *is a solution of* **(WSVI)**.

(d) *there exist* $\lambda_i \geq 0$, $i = 1, \cdots, p$, *not all zero, such that* $y \in X$ *is a solution of the following scalar variational inequality: Find* $y \in X$ *such that there exist* $\zeta_i \in \partial f_i(y)$, $i = 1, \cdots, p$, *such that for any* $x \in X$,

$$(\lambda_1 \zeta_1 + \cdots + \lambda_p \zeta_p)^t (x - y) \geq 0.$$

(e) *there exist* $\lambda_i \geq 0$, $i = 1, \cdots, p$, *not all zero, such that* $y \in X$ *is a solution of the following scalar variational inequality: Find* $y \in X$ *such that for any* $x \in X$, *there exists* $\zeta_i \in \partial f_i(y)$, $i = 1, \cdots, p$, *such that*

$$(\lambda_1 \zeta_1 + \cdots + \lambda_p \zeta_p)^t (x - y) \geq 0.$$

4 EXISTENCE THEOREMS FOR VECTOR OPTIMIZATION PROBLEM

Now we introduce a particular form of Theorem 1 in Park (1992); this is modified in order to achieve our existence theorems for **(P)**. This theorem is a generalization of the well-known fixed point theorem of Fan-Browder (see Theorem 1 in Browder (1968)).

Theorem 4.1 *Let X be a nonempty convex subset of a Hausdorff topological vector space E, and K a nonempty compact subset of X. Let* $A, B : X \to 2^X$ *be two multifunctions. Suppose that*

(1) *for any* $x \in X, Ax \subset Bx$;

(2) *for any* $x \in X, Bx$ *is convex;*

(3) *for any* $x \in K, Ax \neq \emptyset$;

(4) *for any* $y \in X, A^{-1}y$ *is open; and*

(5) *for each finite subset N of X, there exists a nonempty compact convex subset L_N of X containing N such that for each* $x \in L_N \backslash K, Ax \cap L_N \neq \emptyset$.

Then B has a fixed point $\bar{x}$, that is, $\bar{x} \in B\bar{x}$.

Now we give existence theorems for weakly efficient solutions of convex vector optimization problems under relaxed compact assumptions.

Theorem 4.2 *Let X be a nonempty convex subset of* $\mathbb{R}^n$ *and* f_i, $i = 1, \cdots, p$, *convex functions. Assume that there exists a nonempty compact subset K of X such that for each finite subset N of X, there exists a nonempty compact convex subset L_N of X containing N such that for each* $x \in L_N \backslash K$, *there exists* $z \in L_N$ *such that there exist* $\xi_i \in \partial f_i(z)$, $i = 1, \cdots, p$, *satisfying*

$$(\xi_1^t (z - x), \cdots, \xi_p^t (z - x)) \in -\text{int}\mathbb{R}_+^p.$$

Then there is a weakly efficient solution of **(P)**.

Proof: Denote $\partial f_1(x) \times \cdots \times \partial f_p(x)$ by $\partial f(x)$ and $(s_1^t(x-y), \cdots, s_p^t(x-y))$ by $s^t(x-y)$, where $s = (s_1, \cdots, s_p) \in \mathbb{R}^p$.

Define two multifunctions $A, B : X \to 2^X$ to be

$$Ax := \{z \in X | \ \exists t \in \partial f(z) \text{ s.t. } t^t(z-x) \in -\text{int}\mathbb{R}_+^p\},$$

$$Bx := \{z \in X | \ \forall s \in \partial f(x), \ \ s^t(z-x) \in -\text{int}\mathbb{R}_+^p\}.$$

The proof is organized in the following parts.

(i) Since ∂f_i is monotone, for any $x \in X, Ax \subset Bx$.

(ii) It is clear that for each $x \in X, Bx$ is convex.

(iii) For each $z \in X, A^{-1}z$ is open. In fact, let $\{x_n\}$ be a sequence in $(A^{-1}z)^c$ convergent to $x \in X$. Then $z \notin Ax_n$ and hence for any $t \in \partial f(z), t^t(z-x_n) \notin -\text{int}\mathbb{R}_+^p$. Hence for any $t \in \partial f(z), t^t(z-x) \notin -\text{int}\mathbb{R}_+^p$, and hence $z \notin Ax$, namely, $x \in (A^-1z)^c$. Therefore $(A^{-1}z)^c$ is closed, whence $A^{-1}y$ is open.

(iv) By assumption, for each finite subset N of X, there exists a nonempty compact convex subset L_N of X containing N such that for each $x \in L_N \backslash K$, there exists $z \in L_N$ such that there exist $\xi_i \in \partial f_i(z), \ \ i = 1, \cdots, p$, such that

$$(\xi_1^t(z-x), \cdots, \xi_p^t(z-x)) \in -\text{int}\mathbb{R}_+^p.$$

Thus for each $x \in L_N \backslash K$, there exists $z \in L_N$ such that $z \in Ax$ and hence $L_N \cap Ax \neq \emptyset$.

(v) B has no fixed point. If not, there exists $x \in X$ such that for any $s \in \partial f(x), s^t(x-x) \in -\text{int}\mathbb{R}_+^p$. This is a contradiction.

From (i)-(v), we see, by Theorem 4.1, that there must be $y \in K$ such that $Ay = \emptyset$, namely, for any $z \in X, z \notin Ay$, that is, for any $t \in \partial f(z), t^t(z-y) \notin -\text{int}\mathbb{R}_+^p$. Hence $y \in X$ is a solution of **(WMVI)**. By Theorem 3.3, $y \in X$ is a weakly efficient solution of **(P)**. So, there is a weakly efficient solution of **(P)**. □

Corollary 4.1 *(cf. Chen and Craven (1994)) Let X be a nonempty convex subset of $\mathbb{R}^n$ and f_i, $i = 1, \cdots, p$, differentiable convex functions. Assume that there exists $z_0 \in X$ such that the set*

$$\{x \in X : (\nabla f_1(z_0)^t(z_0 - x), \cdots, \nabla f_p(z_0)^t(z_0 - x)) \notin -\text{int}\mathbb{R}_+^p\}$$

is compact. Then there is a weakly efficient solution of **(P)**.

Proof: We let

$$K := \{x \in X : (\nabla f_1(z_0)^t(z_0 - x), \cdots, \nabla f_p(z_0)^T(z_0 - x)) \notin -\text{int}\mathbb{R}_+^p\}$$

and for each finite subset N of X, let $L_N := co(N \cup K)$, where coA is the convexhull of a set A. Then for each $x \in L_N \backslash K \subset X \backslash K$, we have

$$(\nabla f_1(z_0)^t(z_0 - x), \cdots, \nabla f_p(z_0)^t(z_0 - x)) \in -\mathrm{int}\mathbb{R}^p_+.$$

Since L_N is a compact convex subset of X and $z_0 \in L_N$, it follows from Theorem 4.1 that the conclusion of Corollary 4.1 holds. □

Acknowledgments

This research was supported in part by the KOSEF Grant 961-0102-009-2, and in part by BSRI-97-1440.

References

Aubin, J. P. (1979), *Applied Functional Analysis*, John Wiley & Sons, Inc..

Aubin, J. P. (1993), *Optima and Equilibria*, Springer-Verlag, Berlin Hidelberg New York.

Baiocchi, C. and Capelo, A. (1994), *Variational and Quasi-Variational Inequalities Applications to Free-boundary Problems*, John Wiley.

Browder, F. E. (1968), The fixed point theory of multivalued mappings in topological vector space, *Math. Ann.*, Vol. 177, pp. 284-301.

Chen, G. Y. and Craven, B. D. (1994), Existence and continuity of solutions for vector optimization, *J. Optim. Theory Appl.*, Vol. 81, pp. 459-468.

Chen, G. Y. and Li, S. J. (1996), Existence of solutions for generalized vector quasivariational inequality, *J. Optim. Theory Appl.*, Vol. 90, pp. 331-334.

Clarke, F. H. (1983), *Optimization and Nonsmooth Analysis*, Wiley-Interscience, New York.

Geoffrion, A. M. (1968), Properly efficiency and the theory of vector maximization, *J. Math. Anal. Appl.*, Vol. 22, pp. 618-630.

Giannessi, F. (1980), Theorems of alternative, quadratic programs and complementarity problems, in *Variational Inequalities and Complementarity Problems*, edited by Cottle, R. W., Giannessi, F. and Lions, J. L., pp. 151-186, John Wiley and Sons, Chichester, England.

Giannessi, F. (1997), On Minty variational principle, in *New Trends in Mathematical programming*, Kluwer.

Jeyakumar, V. (1986), A generalization of a minimax theorem of Fan via a theorem of the alternative, *J. Optim. Theory Appl.*, Vol. 48, pp. 525-533.

Konnov, I. V. and Yao, J. C. (1997), On the generalized vector variational inequality problem, *J. Math. Anal. Appl.*, Vol. 206, pp. 42-58.

Lee, G. M., Lee, B. S. and Chang, S. S. (1996), On vector quasivariational inequalities, *J. Math. Anal. Appl.*, Vol. 203, pp. 626-638.

Lee, G. M., Kim, D. S., Lee, B. S. and Yen, N. D. (1998), Vector variational inequality as a tool for studying vector optimization problems, *Nonlinear Anal., Th. Meth. Appl.*, Vol. 34, pp. 745-765.

Lee, G. M., Kim, D. S. and Kuk, H. (1998), Existence of solutions for vector optimization problems, *J. Math. Anal. Appl.*, Vol. 220, pp. 90-98.

Lee, G. M. and Kum, S. (1998), On implicit vector variational inequality, *J. Optim. Theory Appl.*, (accepted).

Lin, K. L., Yang, D. P. and Yao, J. C. (1997), Generalized vector variational inequalities, *J. Optim. Theory Appl.*, Vol. 92, pp. 117-125.

Park, S. (1992), Some coincidence theorems on acyclic multifunctions and applications to KKM theory, in *Fixed point Theory and Applications*, pp. 248-277.

Yang, X. Q. (1993), Generalized convex functions and vector variational inequalities, *J. Optim. Theory Appl.*, Vol. 79, pp. 563-580.

Yang, X. Q. (1997), Vector variational inequality and multiobjective pseudolinear programming, *J. Optim. Theory Appl.*, Vol. 95, pp. 729-734.

Yu, S. J. and Yao, J. C. (1996), On vector variational inequalities, *J. Optim. Theory Appl.*, Vol. 89, pp. 749-746.

III OPTIMIZATION METHODS

13 PARAMETER ESTIMATION IN DYNAMIC SYSTEMS

Klaus Schittkowski

Department of Mathmatics,
The University of Bayreuth,
95440 Bayreuth, Germany

Abstract: The intention of the paper is to present a review on numerical methods to identify parameters in systems of ordinary differential equations, differential algebraic equations, or systems of one-dimensional time-dependent partial differential equations with or without algebraic equations. Proceeding from given experimental data, i.e. observation times and measurements, the minimum least squares distance of measured data from a fitting criterion is computed, which depends on the solution of the dynamic system. We present a typical *black box* approach that is easily implemented proceeding from some standard numerical analysis tools. A list of possible applications is included.

Key words: parameter estimation, data fitting, least squares optimization, ordinary differential equations, partial differential equations, algebraic equations, method of lines

X. Yang et al. (eds.), Progress in Optimization, 183–204.

1 INTRODUCTION

Data fitting plays an important role in many natural science, engineering and other disciplines. The key idea is to estimate unknown parameters in a mathematical model that describes a real life situation, by minimizing the distance of some known experimental data from the theoretical model function values. Thus model parameters which cannot be measured directly can be identified by a least squares fit and analysed subsequently in a quantitative way. Parameter estimation or data fitting, respectively, is extremely important in all practical situations, where a mathematical model and corresponding experimental data are available to describe the behaviour of a dynamic system.

The mathematical model that must be provided by the system analyst, has to belong to one of the following categories:

- ordinary differential equations with initial values
- differential algebraic equations
- one-dimensional time-dependent partial differential equations
- one-dimensional partial differential algebraic equations

Often the general mathematical model contains additional features with the goal to apply it to a large variety of different applications, e.g.:

1. More than one fitting criterion is to be defined, i.e. more than one experimental data set can be fitted within a model formulation.

2. The fitting criteria are arbitrary functions depending on the parameters to be estimated, the solution of the underlying dynamic system, and the time variable.

3. An additional independent model parameter is introduced called *concentration* for simplicity.

4. The model possesses arbitrary equality or inequality constraints and upper and lower bounds for parameters to be estimated. Constraints may depend also on the solution of the dynamic equation at given time and concentration or spatial values, respectively.

5. Dynamic models contain so-called switching points where the integration is restarted, e.g. for non-continuous input functions or model changes.

6. Differential algebraic equations are formulated up to index 3. Consistent initial values for index-1-formulations are to be computed internally.

7. In case of partial differential equations, also coupled ordinary differential equations and non-continuous transitions between different areas are taken into account.

8. Partial differential equations may possess algebraic equations e.g. to solve higher order systems. Consistent initial values are computed internally.

9. Alternatively partial differential equations are defined by their flux functions to facilitate the input. In some situations hyperbolic equations must be solved by special discretization techniques, for example the upwind formula.

10. Data are be fitted w.r.t. the L_2 (sum odf squares), L_1 (sum of absolute values), or L_∞ (maximum of absolute values) norm.

For illustration purposes only we denote the first independent model variable the time variable of the system, the second one the concentration variable and the dependend data as *measurement* values of an *experiment*. By this the most frequent application is probably described. On the other hand, these terms may have another meaning within a model depending on the underlying application problem.

The data fitting model, alternative phrases are parameter estimation or system identification, is outlined in Section 2. We give a brief review of typical numerical optimization algorithms based on the Gauss-Newton method. In Sections 3 to 6 is is shown, how the considered dymamic systems can be adapted to fit into the least squares formulation required for starting an optimization algorithm. Only some basic features of the underlying ideas are presented. More details are found in the references.

Some *real life* problems are listed in Section 7, i.e. problems with a realistic practical background. Application areas are pharmacy, biochemistry, chemical engineering, physics, and mechanical engineering.

2 DATA FITTING METHODS

Besides of some very special cases, available optimization algorithms are unable to exploit the special structure of a data fitting formulation, i.e., that the norm of certain differences of a model function from given measurements is to be minimized. The most frequently used L_2-norm leads to a least squares problem of the form

$$\begin{array}{ll} \min & \sum_{i=1}^{l} f_i(p)^2 \\ \text{s.t.} & p \in \mathbb{R}^n \\ & g_j(p) = 0, \ j = 1, ..., m_e, \\ & g_j(p) \geq 0, \ j - m_e + 1, ..., m, \\ & p_l \leq p \leq p_u . \end{array} \tag{2.1}$$

Here we assume that the parameter vector p is n-dimensional and that all nonlinear functions are continuously differentiable with respect to p. Upper and lower bounds are treated independently from the remaining constraints.

The assumption that all problem functions must be smooth is essential. The efficient numerical algorithms under consideration are based more or less on the Gauss-Newton method, which requires first derivatives. To understand their basic structure, we omit constraints and introduce the notation

$$F(p) := (f_1(p), \ldots, f_l(p))^T$$

for the objective function vector, also called the residual, and let

$$f(p) := \frac{1}{2} \sum_{i=1}^{l} f_i(p)^2 \ .$$

Then

$$\nabla f(p) = \nabla F(p) F(p) \tag{2.2}$$

defines the gradient of the objective function with

$$\nabla F(p) = (\nabla f_1(p), \ldots, \nabla f_l(p)) \ .$$

If we assume now that the problem functions $f_1, \ldots, f_l$ are twice continuously differentiable, we get the Hessian matrix of f by

$$\nabla^2 f(p) = \nabla F(p) \nabla F(p)^T + B(p) \ , \tag{2.3}$$

where

$$B(p) := \sum_{i=1}^{l} f_i(p) \nabla^2 f_i(p) \ . \tag{2.4}$$

Proceeding from a given iterate p_k, Newton's method can be applied to (2.1) to get a search direction $d_k \in \mathbb{R}^n$ by solving the linear system

$$\nabla^2 f(p_k) d_k + \nabla f(p_k) = 0$$

or, alternatively,

$$\nabla F(p_k) \nabla F(p_k)^T d + B(p_k) d_k + \nabla F(p_k) F(p_k) = 0 \ . \tag{2.5}$$

Let us assume for a moment that

$$F(p^\star) = (f_1(p^\star), \ldots, f_l(p^\star))^T = 0$$

at an optimal solution $p^\star$, i.e. that the residual $F(p)$ vanishes at $p^\star$. Then we neglect matrix $B(p_k)$ in (2.5), cf. (2.4) for justification, and (2.5) defines the normal equations of the linear least squares problem

$$\min_{d \in \mathbb{R}^n} \|\nabla F(p_k)^T d + F(p_k)\| \ . \tag{2.6}$$

A new iterate is obtained by $p_{k+1} := p_k + \alpha_k d_k$, where d_k is a solution of (2.6) and where α_k denotes a suitable steplength parameter. It is obvious that a quadratic convergence rate is achieved when starting sufficiently close to an optimal solution. The above calculation of a search direction is known as the

Gauss-Newton method and represents the traditional way to solve nonlinear least squares problems, see Björck (1990) for more details.

In general the Gauss-Newton method possesses the attractive feature that it converges quadratically, although we do not provide any second order information. A typical theorem that is found in any textbook about numerical analysis can be formulated in the following way:

Theorem 2.1 *Assume that the unconstrained least squares problem (2.1) has an optimal solution $p^\star$ with $F(p^\star) = 0$, and that the Jacobian matrix of F possesses full rank and is Lipschitz continuous in a neighbourhood of $p^\star$. If the starting point p_0 of the Gauss-Newton method with steplength one is sufficiently close to $p^\star$, then the iterates p_k converge quadratically to $p^\star$, i.e. there is a positive constant γ with*

$$\|p_{k+1} - p^\star\| \leq \gamma \|p_k - p^\star\|^2$$

for all k.

However the assumptions of the convergence theorem for the Gauss-Newton method are very strong and cannot be satisfied in real situations. We have to expect problems with non-zero residuals, with rank-deficient Jacobian matrices, with non-continuous derivatives, and with starting points far away from a solution.

In particular difficulties arise when we want to solve problems with a large residual, i.e. if $F(p^\star)^T F(p^\star)$ is not sufficiently small relative e.g. to $\|\nabla F(p^\star)\|$. Numerous proposals have been made in the past to deal with this situation, and it is outside the scope of this chapter to give a review on all possible attempts developed in the last 20 years. Only a few remarks are presented to illustrate basic features of the main approaches, for further reviews see Gill, Murray, and Wright (1981), Ramsin and Wedin (1977) or Dennis (1977).

A very popular method is known as Levenberg-Marquardt algorithm, see Levenberg (1944) and Marquardt (1963). The key idea is to replace the Hessian in (2.5) by a multiple of the identity matrix, say $\lambda_k I$, with a suitable nonnegative factor λ_k. Obviously we get a regular system of linear equations of the form

$$\nabla F(p_k) \nabla F(p_k)^T d + \lambda_k d + \nabla F(p_k) F(p_k) = 0 \ .$$

For the choice of λ_k and the relationship to the so-called *trust region methods*, see e.g. Moré (1977). A more sophisticated idea is to replace $B(p_k)$ in (2.5) by any quasi-Newton matrix B_k, cf. e.g. Dennis (1973). But some additional safeguards are necessary to deal with indefinite matrices $\nabla F(p_k) \nabla F(p_k)^T + B_k$ in order to get a descent direction. A modified algorithm was proposed by Gill and Murray (1978), where B_k is either equal to $B(p_k)$, a second order approximation of $B(p_k)$, or a quasi-Newton matrix. In this case a diagonal matrix is added to $\nabla F(p_k) \nabla F(p_k)^T + B_k$ to obtain a positive definite matrix. Lindström (1982) proposed a combination of a Gauss-Newton and a Newton method by using a certain subspace minimization.

If, however, the residual is too large then it is not possible to exploit the special structure and a general unconstrained minimization algorithm, e.g. a quasi-Newton method, can be applied as well. Some Fortran codes are now available which are also able to handle constraints.

DFNLP: By transforming the original problem into a general nonlinear programming problem in a special way, typical features of a Gauss-Newton and quasi-Newton least squares method are retained, see Schittkowski (1988) for details. In case of minimizing a sum of absolute function values or the maximum of absolute function values, the problem is transformed into a smooth nonlinear programming problem by introducing additional variables and constraints. The resulting optimization problem is solved by a standard sequential quadratic programming code called NLPQL, cf. Schittkowski (1985/86).

NLSNIP: The code is a special purpose implementation for solving constrained nonlinear least squares problems by a combination of Gauss-Newton, Newton and quasi-Newton techniques, cf. Lindström (1982).

DN2GB: The subroutine is a frequently used unconstrained least squares algorithm and was developed by Dennis, Gay, and Welsch (1981). The mathematical method is also based on a combined Gauss-Newton and quasi-Newton approach.

DSLMDF: First successive line searches are performed along the unit vectors by comparing function values only. The one-dimensional minimization is based on a quadratic interpolation. After the search cycle the Gauss-Newton-type method DFNLP is executed with a given number of iterations. If a solution is not obtained with sufficient accuracy, the search cycle is repeated, see Nickel (1995) for details.

Note that DFNLP and NLSNIP are able to take linear or nonlinear constraints into account. However the other algorithms allow upper and lower bounds of the parameters to be estimated. All algorithms are capable of solving problems with large residuals.

Alternatively the L_2-norm may be changed to the L_1-norm, i.e.,

$$\begin{array}{ll} \min & \sum_{i=1}^{l} |f_i(p)| \\ \text{s.t.} & p \in \mathbb{R}^n \\ & g_j(p) = 0, \ j = 1, ..., m_e, \\ & g_j(p) \geq 0, \ j = m_e + 1, ..., m, \\ & p_l \leq p \leq p_u \end{array} \tag{2.7}$$

or the L_∞-norm leading to

$$\begin{array}{ll} \min & \max_{i=1,\dots,l} |f_i(p)| \\ \text{s.t.} & p \in \mathbb{R}^n \\ & g_j(p) = 0, \ j = 1, ..., m_e, \\ & g_j(p) \geq 0, \ j = m_e + 1, ..., m, \\ & p_l \leq p \leq p_u. \end{array} \tag{2.8}$$

By introducing additional variables and constraints, both problems can be transformed into general, smooth nonlinear programming problems, see Schittkowski (1988), for which standard SQP codes are available, e.g. the code NLPQL of Schittkowski (1985/86).

3 SYSTEMS OF ORDINARY DIFFERENTIAL EQUATIONS

We proceed from r measurement sets, given in the form

$$(t_i, c_j, y_{ij}^k) \ , \ i = 1, \ldots, n_t, \ j = 1, \ldots, n_c, \ k = 1, \ldots, r \ , \tag{3.1}$$

where n_t time values, n_c concentration values and $n_t n_c r$ corresponding measurement values are defined. Again any nonlinear constraints for the parameters to be estimated, are permitted. The vector-valued model function

$$h(p,t,c) = (h_1(p,t,c), \ldots, h_r(p,t,c))^T$$

depends on a concentration parameter c and in addition on the solution $y(p,t,c)$ of a system of ordinary differential equations, i.e.

$$h(p,t,c) = \overline{h}(p, y(p,t,c), t, c) \ , \tag{3.2}$$

where $y(p,t,c)$ denotes the solution of a system of s ordinary differential equations

$$\begin{aligned} \dot{y}_1 = & \ F_1(p,y,t,c) \ , \ y_1(0) = y_1^0(p,c) \ , \\ & \ldots \\ \dot{y}_s = & \ F_s(p,y,t,c) \ , \ y_s(0) = y_s^0(p,c) \ . \end{aligned} \tag{3.3}$$

Without loss of generality we assume that the initial time is zero. The initial values of the differential equation system $y_1^0(p,c)$, ..., $y_s^0(p,c)$ may depend on one or more of the system parameters to be estimated, and on the concentration parameter c. In this case, we have to assume in addition that the observation times are strictly increasing.

There are many practical situations where a model changes during integration over the time variable, and where initial values must be adopted. A typical example is a pharmacokinetic application with an initial infusion and subsequent applications of drug doses by injection. Particularly in case of short peaks, the integration routine may not recognize the peak at all because of a big time step. Moreover the numerical approximation of gradients could become unstable in case of discontinuities. In a formal way we describe the model changes as follows:

$$\begin{array}{lll} \dot{y}_1^0 = & F_1^0(p,y^0,t,c) & , \; y_1^0(0) = y_1^0(p,c) \;, \\ & \ldots & \\ \dot{y}_s^0 = & F_s^0(p,y^0,t,c) & , \; y_s^0(0) = y_s^0(p,c) \;, \\ \dot{y}_1^i = & F_1^i(p,y^i,t,c) & , \; y_1^i(\tau_i) = y_1^i(p,c,y_1^{i-1}(p,\tau_i,c)) \;, \\ & \ldots & \\ \dot{y}_s^i = & F_s^i(p,y^i,t,c) & , \; y_s^i(\tau_i) = y_s^i(p,c,y_s^{i-1}(p,\tau_i,c)) \;, \end{array} \tag{3.4}$$

where $i = 1, \ldots, n_b$ and n_b is the number of break or switching points τ_i with $0 < \tau_1 < \ldots < \tau_{n_b}$. The initial values of each subsystem are given by functions $y_j^i(p, c, y)$ depending on the parameters to be estimated, the actual concentration value, and the solution of the previous interval at the switching point τ_i. Internally the integration of the differential equation is restarted at a break point. It is possible that these points become variables to be adapted during the optimization process, if they are not known in advance.

Constraints of the form

$$\begin{array}{lll} g_j(p) & = 0 \;, & j = 1, ..., m_e \;, \\ g_j(p) & \geq 0 \;, & j = m_e + 1, ..., m \end{array} \tag{3.5}$$

are allowed, where the restriction functions may depend on the solution of the dynamic system at predetermined time and concentration values, i.e.

$$g_j(p) = \overline{g}_j(p, y(p, t_j, c_j), t_j, c_j) \tag{3.6}$$

for $j = 1, ..., m$. The m predetermined time and, if available, concentration values must coincide with some of the given measurement values. If not, the given data are rounded to the nearest experimental value. If constraints are to be defined independently from given measurement data, it is recommended that dummy experimental values with zero weights are inserted at the desired time and concentration points t_j and c_j, respectively.

There are two different methodologies to estimate parameters in systems of ordinary differential equations, either the *black box* approach or full discretization. In the latter case the differential equation is discretized by a collocation method, difference formula, Tschebyscheff polynomials, or any other type of approximation with the intention to replace the set of differential equations by algebraic ones. Then artificial optimization parameters, for example coefficients of polynomials, and the nonlinear algebraic equations are added to the existing data fitting problem. One advantage of the discretization approach is that analytical derivatives of the state functions w.r.t. the parameters to be optimized are much easier to obtain than in the other case.

In the *black box* approach we apply an available ODE solver for the numerical integration of the system, e.g. the explicit codes DOPRI5, DOP853, or ODEX of Hairer, Nørsett, and Wanner (1993), and the implicit codes RADAU5, SDIRK4, or SEULEX of Hairer and Wanner (1991) and DASSL of Petzold (1982), Ascher and Petzold (1998). Most of them are based on Runge-Kutta fomulae of higher order with adaptive stepsizes, see also Prince and Dormand (1981) or Shampine and Watts (1979). The list is by no means complete and any other ODE solver can be used as well.

Implicit methods are also able to solve stiff equations. Gradients of the right-hand side of the differential equation w.r.t. state variables are evaluated by numerical approximation or analytically using either user-provided derivatives or automatic differentiation.

Some of the codes use dense output, i.e. the integration is performed over the whole interval given by first and last time value. Intermediate solution values needed for computing the fitting criterion, are interpolated. In this case it is possible to save a significant number of function evaluations of the right-hand side of the ODE.

When using a standard ODE solver, gradients w.r.t. parameters to be estimated are usually obtained by external numerical differentiation. Other approaches are the inclusion of sensitivity equations or internal numerical differentiation, see Bock (1983). In the latter case the Runge-Kutta scheme is differentiated and gradient values are computed within the integration routine, see e.g. the code IND-DIR of Benecke (1993).

In some cases it is possible that multiple shooting reduces the number of iterations of the least squares algorithm drastically, in particuler if measurements are available for all state variables. So-called shooting points are introduced which may coincide with the experimental time values. Then the integration is performed successively from one shooting point to the subsequent one, cf. Ascher and Petzold (1998). To guarantee that a continuous numerical solution of the ODE is obtained, we have to add initial values for each shooting interval to the parameters to be optimized, and nonlinear equality constraints obtained by the requirement that a solution variable at the end of a shooting interval must coincide with the initial value of the subsequent one. A special advantage is that starting values needed for the optimization algorithm are easily obtained from the measurement values, see Bock (1983).

4 SYSTEMS OF DIFFERENTIAL ALGEBRAIC EQUATIONS

Parameter estimation problems based on differential algebraic equations are very similar to those based on ordinary differential equations. The only difference is that we allow additional algebraic equations in the system, i.e. we get additional system variables.

Thus we have to replace (3.3) by the extended system

$$\begin{array}{rll} \dot{y}_1 = & F_1(p,y,z,t,c) & , \ y_1(0) = y_1^0(p,c) \ , \\ & \cdots & \\ \dot{y}_{s_1} = & F_{s_1}(p,y,z,t,c) & , \ y_{s_1}(0) = y_{s_1}^0(p,c) \ , \\ 0 = & G_1(p,y,z,t,c) & , \ z_1(0) = z_1^0(p,c) \ , \\ & \cdots & \\ 0 = & G_{s_2}(p,y,z,t,c) & , \ z_{s_2}(0) = z_{s_2}^0(p,c) \ . \end{array} \tag{4.1}$$

Now $y(p,t,c)$ and $z(p,t,c)$ are solution vectors of a joint system of $s_1 + s_2$ differential and algebraic equations (DAE). The system is called an index-1-problem or an index-1-DAE, if the algebraic equations can be solved w.r.t. z, i.e. if the matrix

$$\nabla_z G(p, y, z, t, c) \tag{4.2}$$

possesses full rank. In all other cases we get DAE's with a higher index, see e.g. Hairer and Wanner (1991) or any other textbook for a suitable definition and more details.

Similar to systems of ordinary differential equations, it is possible to define switching or break points. Note also that break points can be treated as optimization variables, i.e. may vary from one iteration step to the other.

For simplicity we consider now only problems of index 1, although the numerical algorithms are capable to treat also higher index models. Moreover problems with higher index can be transformed to problems of index 1 by successive differentiation of the algebraic equations. However we must be careful to avoid drift effects, see Ascher and Petzold (1998).

Initial values of the model must satisfy the consistency equations

$$\begin{aligned} G_1(p, y^0(p,c), z^0(p,c), t, c) &= 0 \ , \\ \ldots & \\ G_{s_2}(p, y^0(p,c), z^0(p,c), t, c) &= 0 \ . \end{aligned} \tag{4.3}$$

Here the initial values $y_1^0(p,c), \ldots, y_{s_1}^0(p,c)$ and $z_1^0(p,c), \ldots, z_{s_2}^0(p,c)$ are functions depending on the parameters to be estimated, and the concentration variable. If the consistency condition is not satisfied, consistent initial values are computed by solving the above nonlinear system of equations w.r.t. z after insertion of the initial values for the differential variables. Also if switching points exist, consistent initial values must be evaluated before restarting the integration procedure at such a point.

Again we have to assume that the observation times are strictly increasing, and get a fitting criterion of the form

$$h(p, t, c) = \overline{h}(p, y(p, t, c), z(p, t, c), t, c) \ .$$

As before, constraints of the form

$$\begin{aligned} g_j(p) &= 0, \quad j = 1, ..., m_e \ , \\ g_j(p) &\geq 0, \quad j = m_e + 1, ..., m \end{aligned} \tag{4.4}$$

are allowed, where the restriction functions may depend on the solution of the dynamic system at predetermined time and concentration values, i.e.

$$g_j(p) = \overline{g}_j(p, y(p, t_j, c_j), z(p, t_j, c_j), t_j, c_j) \tag{4.5}$$

for $j = 1, ..., m$.

In the *black box* approach, the algebraic differential equation is to be solved by any of the available implicit solution methods, e.g., RADAU5, SDIRK4, or SEULEX of Hairer and Wanner (1991) and DASSL of Petzold (1982), Ascher and Petzold (1998). DAE's with an index higher than 1 can be solved by RADAU5. If consistent initial values cannot be provided by the user, the corresponding nonlinear system of equations is treated as general nonlinear programming problem with equality constraints, and solved e.g. by NLPQL of Schittkowski (1985/86).

As in case of ODE's, the shooting technique and internal numerical evaluation of gradients can also be applied.

5 SYSTEMS OF ONE-DIMENSIONAL TIME-DEPENDENT PARTIAL DIFFERENTIAL EQUATIONS

Partial differential equations are extensions of ordinary differential equations, where an additional spatial variable is introduced together with its first and optionally also some higher partial derivatives. Again we assume without loss of generality that the initial time is zero. This assumption facilitates the description of the mathematical model and is easily satisfied in practice by a suitable linear transformation of the time variable.

The model we want to investigate now is defined by a system of n_p one-dimensional partial differential equations in one or more spatial intervals, see also Schittkowski (1997a). These intervals which could describe e.g. certain areas with different diffusion coefficients, are given by the outer boundary values x_L and x_R that define the total integration interval w.r.t. the space variable x, and optionally some additional internal transition points $x_1^a, \ldots, x_{n_a-1}^a$. Thus we get a sequence of $n_a + 1$ boundary and transition points

$$x_0^a := x_L < x_1^a < \ldots < x_{n_a-1}^a < x_{n_a}^a := x_R \ . \tag{5.1}$$

For each integration interval, we define a one-dimensional, time-dependent partial differential equation of the form

$$u_t^i = F^i(x, t, f^i(x, t, v, u^i, u_x^i), f_x^i(x, t, v, u^i, u_x^i), v, u^i, u_x^i, u_{xx}^i, p) \ , \tag{5.2}$$

where $x \in R$ is the spatial variable with $x_{i-1}^a \leq x \leq x_i^a$ for $i = 1, \ldots, n_a$, $t \in R$ the time variable with $0 < t \leq T$, $v \in \mathbb{R}^{n_o}$ the solution vector of coupled system of ordinary differential equations defined below (5.7), $u^i \in \mathbb{R}^{n_p}$ the system variable we want to compute, and $p \in \mathbb{R}^n$ the parameter vector to be identified by the data fitting algorithm.

Optionally the right-hand side may depend also on a so-called flux function $f^i(x, t, v, u^i, u_x^i)$ which is introduced either to facilitate the declaration of the function side or for being able to apply special upwind formulae in case of hyperbolic equations. Standard approximation schemes can break down e.g. if shocks propagate over the integration interval. In this case we need special model formulations to apply the discretization technique, and the underlying equation is usually of the form

$$u_t^i = (f^i(x, t, u^i, p))_x \ , \quad i = 1, \ldots n_a \ . \tag{5.3}$$

A solution of the coupled system depends on the spatial variable x, the time variable t, the parameter vector p, and is therefore written in the form $v(t, p)$ and $u^i(x, t, p)$ for $i = 1, \ldots, n_a$.

For both end points x_L and x_R we allow boundary conditions of the form

$$\begin{aligned} u^1(x_L,t,p) &= u^L(t,p) \ , \\ u^{n_a}(x_R,t,p) &= u^R(t,p) \ , \\ u^1_x(x_L,t,p) &= \hat{u}^L(t,p) \ , \\ u^{n_a}_x(x_R,t,p) &= \hat{u}^R(t,p) \end{aligned} \tag{5.4}$$

for $0 < t \leq t_s$. It is essential to understand that we do not require the evaluation of all boundary functions. Instead a user may omit some of them depending on the structure of the PDE model. Note that boundary information is also contained in coupled ordinary differential equations.

Transition conditions between the different areas may be defined in addition. They are allowed at transition points and have the form

$$\begin{aligned} u^i(x^a_i,t,p) &= c^R_i(t,u^{i+1}(x^a_i,t,p),p) \ , \\ u^{i+1}(x^a_i,t,p) &= c^L_i(t,u^i(x^a_i,t,p),p) \ , \\ u^i_x(x^a_i,t,p) &= \hat{c}^R_i(t,u^{i+1}(x^a_i,t,p),u^{i+1}_x(x^a_i,t,p),p) \ , \\ u^{i+1}_x(x^a_i,t,p) &= \hat{c}^L_i(t,u^i(x^a_i,t,p),u^i_x(x^a_i,t,p),p) \end{aligned} \tag{5.5}$$

with $0 < t \leq t_s$, $i = 1,\ldots,n_a - 1$. Again, the user may omit any of these functions if a transition condition does not exist at a given x^a_i-value. To avoid internal access conflicts and to get a well-defined problem, either u^i or u^{i+1} may be defined, also either u^i_x or u^{i+1}_x, respectively, but not both of them.

Since the starting time is assumed to be zero, initial conditions have the form

$$u^i(x,0,p) = u^i_0(x,p) \ , \quad i = 1,\ldots,n_a \tag{5.6}$$

and are defined for all $x \in [x^a_{i-1}, x^a_i]$, $i = 1,\ldots,n_a$.

If the partial differential equations are to be coupled to ordinary differential equations, we proceed from an additional ODE-system of the form

$$\dot{v}_j = G_j(t,v,u^{i_j}(x_j,t,p),u^{i_j}_x(x_j,t,p),u^{i_j}_{xx}(x_j,t,p),p) \tag{5.7}$$

for $j = 1,\ldots,n_o$ with initial values

$$v(0,p) = v_0(p) \ , \tag{5.8}$$

that may depend again on the parameters to be estimated. The system has n_o components, i.e. $v = (v_1,\ldots,v_{n_o})^T$. Coupling of ordinary differential equations is allowed at arbitrary points within the integration interval and the corresponding area is denoted by the index i_j. The spatial variable value x_j belongs to the j-th area, i.e. $x_j \in [x^a_{i_j-1}, x^a_{i_j})$ or $x_j \in [x^a_{n_a-1}, x^a_{n_a}]$, respectively, $j = 1,\ldots,n_o$, and is called coupling point.

Coupling points are rounded to their nearest line when discretizing the system. The right-hand side of the coupling equation may depend on the corresponding solution of the partial equation and its first and second derivative w.r.t. the space variable at the coupling point under consideration.

Similar to systems of ordinary differential equations, it is possible to define break points and modified initial values at these points, that may depend on the parameters to be estimated, the spatial variable and the solution w.r.t. the previous interval.

To indicate that the k-th fitting criterion $h_k(p,t)$ depends also on the solution of the differential equation at the corresponding fitting point, where k denotes the index of a measurement set, we use the notation

$$h_k(p,t) = \overline{h}_k(p,t,v(t,p),u^{i_k}(x_k,t,p),u_x^{i_k}(x_k,t,p),u_{xx}^{i_k}(x_k,t,p)) \tag{5.9}$$

and insert $\overline{h}_k$ instead of h_k into the data fitting function. Again the fitting criteria may depend on solution values at a given spatial variable value w.r.t. to an integration interval defined by the index i_k. The spatial variable x_k belongs to the i_k-th integration area, i.e. $x_k \in \left[x_{i_k-1}^a, x_{i_k}^a\right)$ or $x_k \in \left[x_{n_a-1}^a, x_{n_a}^a\right]$, respectively, $k = 1,\ldots,r$, where r denotes the total number of measurement sets. The fitting criterion may depend on the solution of the partial equation and its first and second derivative w.r.t. the space variable at the fitting point. Fitting points are rounded to their nearest line when discretizing the system.

As for ordinary differential equations, dynamic constraints of the form

$$\begin{aligned} g_j(p) &= 0\,, \quad j = 1,\ldots,m_e\,, \\ g_j(p) &\geq 0\,, \quad j = m_e+1,\ldots,m \end{aligned} \tag{5.10}$$

are allowed, where the restriction functions may depend on the solution of the partial differential equation and its first and second spatial derivatives at predetermined time and spatial values, and the solution of the coupled ordinary differential equation at predetermined time values, i.e.

$$g_j(p) = \overline{g}_j(x_j,t_j,p,v(t_j,p),u^{j_k}(x_j,t_j,p),u_x^{j_k}(x_j,t_j,p),u_{xx}^{j_k}(x_j,t_j,p)) \tag{5.11}$$

for $j = 1,\ldots,m$. Here u^{j_k} denotes the corresponding integration area that contains the spatial parameter x_j. The m predetermined time values must coincide with some of the given measurement values. If not, the given data are rounded to the nearest experimental data. The corresponding spatial values must coincide with a line obtained by the equidistant discretization of the underlying integration interval. If constraints are to be defined independently from given measurement data, it is recommended that dummy experimental values with zero weights are inserted at the desired time and space coordinates t_j and x_j, respectively.

In order to achieve smooth fitting criteria and constraints, we assume that all model functions are continuously differentiable w.r.t. the parameter vector p. Moreover we assume that the discretized system of differential equations is uniquely solvable for all p with $p_l \leq p \leq p_u$. A collection of 20 examples of partial differential equations that can be solved by the presented approach and comparative numerical results are found in Schittkowski (1997a).

A frequently used idea is to transform the partial differential into a system of ordinary differential equations by discretizing the model functions w.r.t. the spatial variable x. This approach is known as the method of lines, see e.g. Schiesser (1991).

For the i-th integration interval of the spatial variable, we denote the number of discretization points by n_i, $i = 1, \ldots, n_a$. We proceed from uniform grid points within each interval and get a discretization of the whole space interval from x_L to x_R. To approximate the first and second partial derivatives of $u(x, t, p)$ w.r.t. the spatial variable at a given point x, different alternatives are available, see also Schittkowski (1998b):

a) Polynomial approximation: We compute an interpolating polynomial subject to some neighbouring values. The number of interpolation points depends on the polynomial degree selected. Polynomials of order 3, 5, 7 or 9 are computed by Newton's formula. By differentiating the interpolation formulae, first and second order approximations are obtained. In case of Neumann boundary conditions, Hermite interpolation is used for being able to exploit known derivative values.

b) Difference Formulae: First and second derivatives can be approximated by difference formulae, see e.g. Schiesser (1991). Difference formulae with 3 and 5 points for first derivatives are available. These can also be applied recursively to get the second derivatives. Alternatively a 5-point difference formula for second derivatives is implemented as well. The difference formulae are adapted at the boundary to accept given function and gradient values.

c) Upwind Formulae for Hyperbolic Equations: In case of a hyperbolic equation, i.e.

$$u_t^i = (f^i(x, t, u^i, p))_x \tag{5.12}$$

with a so-called flux function f, approximation by polynomials or difference formulae might become unstable, especially if non-continuous boundary conditions are supplied to describe e.g. the propagation of shocks, see again Schiesser (1991) for some numerical examples. Thus, special difference formulae have been derived in the past to overcome the instabilities, based e.g. on the so-called upwind formulae simply given by forward differences, or more advanced higher order TVD formulae. For more information, see the original literature, e.g. Yee (1985), Chakravarthy and Osher (1984a), Chakravarthy and Osher (1984b), Chakravarthy and Osher (1985), Sweby (1984), Wang and Richards (1991), and Yang and Przekwas (1992). TVD stands for total variation diminishing and the correponding one parameter family of upwind formulae was proposed by Chakravarthy and Osher (1984a). In this case a certain stability criterion requires that the internal time stepsizes of the ODE solver do not become too small compared to the spatial discretization accuracy. Because of the black box approach used, the stepsizes however cannot be modified and we have to assume that the criterion remains satisfied.

Whenever a boundary condition in Dirichlet form is given for $1 \leq k \leq n_p$, we know the value of the boundary function and use it to interpolate or approximate the function $u(x, t, p)$ as decribed above. In other words, the corresponding function value in the right-hand side of the discretized system is replaced by the value given. Alternatively, a boundary condition may appear in Neumann form. In this case, the derivative values at the boundary are replaced by the given ones before evaluating the second order spatial derivative approximations.

Ordinary differential equations are added to the discretized system without any further modification. Since arbitrary coupling points are allowed, they are rounded to the nearest line of the discretized system. In the same way fitting criteria can be defined at arbitrary values of the spatial variable.

When defining the transition function, it is important to have the underlying flux direction in mind. If, for example, the flux is in the direction of the spatial variable and we want to define a continuous transition at x_i^a, then we have to formulate the corresponding transition function in the form $u_k^{i+1}(x_i^a, t, p) = u_k^i(x_i^a, t, p)$ in order to guarantee that the boundary values w.r.t. x_L are spread over the interval.

For the same reasons outlined in the previous sections it is possible that the right-hand side of a PDE becomes non-continuous w.r.t. integration time. Thus, it is possible to supply time values, where the integration of the ODE is restarted with initial tolerances, e.g. with the initially given stepsize. The integration in the preceeding interval is stopped at the time value given minus a relative error in the order of the machine precision. Break or switching points are either constant or optimization variables to be adapted by the optimization code.

It can be shown that the resulting large system of ordinary differential equations becomes stiff in some situations, when discretization accuracy increases. Thus the usage of implicit ODE solvers is recommended. Since the Jacobian of the discretized right-hand side has a band structure, it is essential that the selected implicit ODE solver is capable to exploit it efficiently.

6 SYSTEMS OF ONE-DIMENSIONAL PARTIAL DIFFERENTIAL ALGEBRAIC EQUATIONS

Partial differential algebraic equations are based on the same model structure as one-dimensional, time-dependent PDE's. The only difference is that additional algebraic equations are permitted as in case of DAE's. Typical examples are higher order partial differential equations, e.g.,

$$u_t = f(x, t, u, u_{xxxx}, p) \ ,$$

transformed into a second order equation by introducing additional variables, e.g.,

$$\begin{aligned} u_t &= f(x, t, u, w_{xx}, p) \ , \\ 0 &= w - u_{xx} \ , \end{aligned}$$

or distributed systems of the form

$$\begin{aligned} u_t^1 &= f_1(x, t, u^1, u^2, p) \ , \\ u_x^2 &= f_2(x, t, u^1, u^2, p) \ , \end{aligned}$$

with initial values $u^1(x,0,p) = u_1(x,p)$, $u^2(0,t,p) = u_2(t,p)$, transformed into the PDAE

$$\begin{array}{rcl} u_t^1 & = & f_1(x,t,u^1,u^2,p) \ , \\ 0 & = & u_x^2 - f_2(x,t,u^1,u^2,p) \end{array}$$

or

$$\begin{array}{rcl} u_x^2 & = & f_2(x,t,u^1,u^2,p) \ , \\ 0 & = & u_x^1 - f_1(x,t,u^1,u^2,p) \ , \end{array}$$

respectively.

Thus we proceed again from a sequence of $n_a + 1$ boundary and transition points

$$x_0^a := x_L < x_1^a < \ldots < x_{n_a-1}^a < x_{n_a}^a := x_R \tag{6.1}$$

and consider the PDAE-system

$$\begin{array}{rcl} u_t^i & = & F_1^i(x,t,f^i(x,t,v,u^i,u_x^i),f_x^i(x,t,v,u^i,u_x^i),v,u^i,u_x^i,u_{xx}^i,p) \ , \\ 0 & = & F_2^i(x,t,f^i(x,t,v,u^i,u_x^i),f_x^i(x,t,v,u^i,u_x^i),v,u^i,u_x^i,u_{xx}^i,p) \ , \end{array} \tag{6.2}$$

$i = 1, \ldots n_a$, where $x \in R$ is the spatial variable with $x_{i-1}^a \leq x \leq x_i^a$ for $i = 1, \ldots, n_a$, $t \in R$ the time variable with $0 < t \leq T$, $v \in \mathbb{R}^{n_o}$ the solution vector of the coupled system of ordinary differential equations, $u^i \in \mathbb{R}^{n_p}$ the system variable we want to compute, and $p \in \mathbb{R}^n$ the parameter vector to be identified by the data fitting algorithm.

But now the state variables u^i are divided into so-called differential variables and algebraic variables, i.e. $u^i = (u_1^i, u_2^i)^T$, where the number of algebraic variables is identical to the number of algebraic equations summarized by the vector F_2. Also, we allow flux functions, switching points, constraints for parameters and state functions, and coupled ordinary differential equations. All these extensions are treated in the same way as for partial differential equations without algebraic equations.

However we must handle at least the initial conditions with more care. We have to guarantee that they satisfy the algebraic equations

$$0 = F_2^i(x,0,f^i(x,0,v,u^i,u_x^i),f_x^i(x,0,v,u^i,u_x^i),v,u^i,u_x^i,u_{xx}^i,p) \tag{6.3}$$

for $i = 1, \ldots n_a$. If initial conditions for discretized algebraic equations are violated, then the corresponding system of nonlinear equations is solved internally proceeding from initial values given. In other words, consistent initial values are computed automatically, where user-provided data serve as starting parameters for the nonlinear programming algorithm. Consequently we allow only index-1-systems unless it is guaranteed that consistent initial values for the discretized DAE are available. Jumps or discontinuities at initial values of algebraic equations do not make sense.

A straightforward idea is to now transform the partial differential into a system of differential algebraic equations by discretizing the model functions w.r.t. the spatial variable x. It is assumed that always the last n_{ae} equations of the n_p given ones are algebraic.

Again we denote the number of discretization points by n_i, $i = 1, \ldots, n_a$, for the i-th integration interval of the spatial variable. We proceed from uniform grid points within each interval and get a discretization of the whole space interval from x_L to x_R. To approximate the first and second partial derivatives of $u(x,t,p)$ w.r.t. the spatial variable at a given point x, we may apply polynomial approximation or difference formulae as outlined in the previous section. Thus we get a system of differential algebraic equations which can then be solved by any of the available integration routines.

7 APPLICATIONS

Parameter estimation in dynamic systems is a widely used technique and it is out of the scope of this paper to present a review. Only a few examples are to be listed, mostly obtained by the author with a software system called EASY-FIT, see Schittkowski (1997b), that realizes the numerical concepts discussed before. EASY-FIT consists of a database containing models, data and results, and of underlying numerical algorithms for solving the parameter estimation problem depending on the mathematical structure, i.e.

MODFIT	parameter estimation in explicit functions, Laplace transforms, ordinary differential and differential algebraic equations, see Schittkowski (1995)
SYSFIT	parameter estimation in steady-state systems, see Schittkowski (1994)
PDEFIT	parameter estimation in one-dimensional, time-dependent partial differential algebraic equations, see Schittkowski (1998b)

Model functions are either interpreted and evaluated symbolically by a program called PCOMP, see Dobmann, Liepelt, and Schittkowski (1995), that allows in addition automatic differentiation of nonlinear model functions, or by user-provided FORTRAN subroutines.

The author's database of EASY-FIT contains more than 550 *real life* and academic data fitting problems based on about 400 different mathematical models. In some cases it is also possible to transform an optimal control problem to a parameter estimation problem. A few practical applications are listed:

1. Substrate diffusion in a metabolically active cutaneous tissue, see Schittkowski (1998a), Boderke, Schittkowski, Wolf and Merkle (1998), Steinsträsser (1994), or Blatt and Schittkowski (1998) for an optimal control model

2. Acetylene reactor, see Wansbrough (1985) or Birk, Liepelt, Schittkowski, and Vogel (1998) for an optimal control approach

3. Optimal design of corrugated satellite horns, see Kühn and Hombach (1983) or Hartwanger, Schittkowski, and Wolf (1998)

4. Drying of maltodextrin in a convection oven, see Frias, Oliveira, and Schittkowski (1998)

5. Receptor-ligand binding study, see Rominger and Albert (1985) or Schittkowski (1994)

6. Multibody system of a truck, see Simeon, Grupp, Führer, and Rentrop (1994), Simeon and Rentrop (1993), or Führer and Leimkuhler (1991)
7. Manutec r3 robot, see Otter and Türk (1988)
8. Mass transfer in sorbing porous media, see Hoch (1995), Van Genuchten and Wierenga (1976), or Andersson and Olsson (1985)
9. Ammonium and nitrate fertilization in forest soils, see Fischer (1996)
10. Distillation column, see Jourdan (1997), or Kuhn and Schmidt (1987)
11. Cooling during rolling-mill operations, see Hedrich (1996) or Groch (1990)
12. Signal evaluation in periodic mechanical systems, see Zschieschang and Rockhausen (1996)
13. Isomarization in protein unfolding and refolding, see Odefey, Mayr, and Schmid (1995), or Mayr, Odefey, Schutkowski, and Schmid (1996)
14. Thermodynamic coupling for GroEL-mediated unfolding, see Walter, Lorimer, and Schmid (1996)

8 CONCLUSIONS

We show an approach to compute unknown parameters in a dynamical model consisting of ordinary or partial differential equations, by a least squares data fit. One-dimensional partial differential equations are discretized by the method of lines leading to large systems of ODE's or DAE's, respectively. Ordinary differential equations are integrated by a standard solvers, and the least squares optimization problem is solved iteratively by a Gauss-Newton method.

The intention of the paper is to present a review on some techniques that are routinely used to estimate data in dynamical systems. A list of applications is included.

References

Andersson, F., and Olsson, B. (eds.) (1985), *Lake Gådsjön: An Acid Forest Lake and its Catchment, Ecological Bulletins*, Vol. 37, Stockholm.

Ascher, U.M., and Petzold, L.R. (1998), *Computer Methods for Ordinary Differential Equations and Differential-Algebraic Equations,* SIAM, Philadelphia.

Benecke, C. (1993), *Interne Numerische Differentiation von Gewöhnlichen Differentialgleichungen,* Diploma Thesis, Department of Mathematics, University of Bayreuth.

Birk, J., Liepelt, M., Schittkowski, K., and Vogel, F. (1998), Computation of optimal feed rates and operation intervals for tubular reactors, *Journal of Process Control,* (to appear).

Blatt, M., and Schittkowski, K. (1998), Optimal control of one-dimensional partial differential equations applied to transdermal diffusion of substrates, in *Optimization Techniques and Applications*, edited by Caccetta, L., Teo,

K.L., Siew, P.F., Leung, Y.H., Jennings, L.S., and Rehbock, V., School of Mathematics and Statistics, Curtin University of Technology, Perth, Australia, Vol. 1, pp. 81-93.

Bock, H.G. (1983), Recent advantages in parameter identification techniques for ODE, in Proceedings of the International Workshop on *Numerical Treatment of Inverse Problems in Differential and Integral Equations*, Birkhäuser, pp. 95-121.

Boderke, P., Schittkowski, K., Wolf, M. and Merkle, H.P. (1998), A mathematical model for diffusion and concurrent metabolism in metabolically active tissue, (submitted for publication).

Björck, A. (1990), *Least Squares Methods,* Elsevier.

Chakravarthy, S.R., and Osher, S.(1984), High resolution schemes and the entropy condition, *SIAM Journal on Numerical Analysis*, Vol. 21, No. 5, pp. 955-984

Chakravarthy, S.R., and Osher, S. (1984), Very high order accurate TVD schemes, *ICASE Report* No. 84-44.

Chakravarthy, S.R., and Osher, S. (1985), Computing with high resolution upwind schemes for hyperbolic equations, *Lectures in Applied Mathematics,* Vol. 22, pp. 57-86.

Dennis, J.E.jr. (1977), Nonlinear least squares, in *The State of the Art in Numerical Analysis*, edited by Jacobs, D., Academic Press.

Dennis, J.E.jr. (1973), Some computational technique for the nonlinear least squares problem, in *Numerical Solution of Systems of Nonlinear Algebraic Equations.* edited by Byrne, G.D., and Hall, C.A., Academic Press.

Dennis, J.E.jr., Gay, D.M., and Welsch R.E. (1981), Algorithm 573: NL2SOL-an adaptive nonlinear least-squares algorithm, *ACM Transactions on Mathematical Software*, Vol. 7, No. 3, pp. 369-383.

Dobmann, M., Liepelt, M., Schittkowski, K. (1995), Algorithm 746: PCOMP: A FORTRAN code for automatic differentiation, *ACM Transactions on Mathematical Software*, Vol. 21, No. 3, pp. 233-266.

Fischer, P. (1996), *Modellierung und Simulation der Ammonium- und Nitrat-Dynamik in strukturierten Waldböden under besonderer Berücksichtigung eines dynamischen, hierarchischen Wurzelsystems,* Diploma Thesis, Department of Mathematics, University of Bayreuth.

Frias, J.M., Oliveira, J.C, Schittkowski, K. (1998), Modelling of maltodextrin DE12 drying process in a convection oven, (submitted for publication).

Führer, C., Leimkuhler, B. (1991), Numerical solution of differential-algebraic equations for constrained mechanical motion, *Numerische Mathematik*, Vol. 59, pp. 55-69.

Gill, P.E., and Murray, W. (1978), Algorithms for the solution of the nonlinear least-squares problem, *SIAM Journal on Numerical Analysis*, Vol. 15, pp. 977–992.

Gill, P.E., Murray, W. and Wright, M.H. (1981), *Practical Optimization,* Academic Press.

Groch, A.G. (1990), Automatic control of laminar flow cooling in continuous and reversing hot strip mills, *Iron and Steel Engineer*, pp. 16-20.

Hairer, E., and Wanner, G. (1991), *Solving Ordinary Differential Equations II. Stiff and Differential-Algebraic Problems,* Springer Series Computational Mathematics, Vol. 14, Springer.

Hairer, E., Nørsett, S.P., and Wanner, G. (1993), *Solving Ordinary Differential Equations I: Nonstiff Problems,* Springer Series Computational Mathematics, Vol. 8, Springer.

Hartwanger, C., Schittkowski, K., and Wolf, R. (1998), Optimal design of corrugated satellite horns, Report, Department of Mathematics, University of Bayreuth.

Hedrich, C. (1996), *Modellierung, Simulation und Parameterschätzung von Kühlprozessen in Walzstraßen,* Diploma Thesis, Department of Mathematics, University of Bayreuth.

Hoch, R. (1995), *Modellierung von Fließwegen und Verweilzeiten in einem Einzugsgebiet unter stationären Fließbedingungen,* Diplomarbeit, Fakultät für Biologie, Chemie und Geowissenschaften, Universität Bayreuth.

Jourdan, M. (1997), *Simulation und Parameteridentifikation von Destillationskolonnen,* Diploma Thesis, Department of Mathematics, University of Bayreuth.

Kuhn, U. and Schmidt, G. (1987), Fresh look into the design and computation of optimal output feedback controls for linear multivariable systems, *International Journal on Control,* Vol. 46, No. 1, pp. 75-95.

Kühn, E., and Hombach, V. (1983), Computer-aided analysis of corrugated horns with axial or ring-loaded radial slots, Report, Research Center of the Deutsche Bundespost

Levenberg, K. (1944), A method for the solution of certain problems in least squares, *Quarterly Applied Mathematics,* Vol. 2, pp. 164-168.

Lindström, P. (1982), A stabilized Gauss-Newton algorithm for unconstrained least squares problems, Report UMINF-102.82, Institute of Information Processing, University of Umea, Umea, Sweden.

Lindström, P, (1983), A general purpose algorithm for nonlinear least squares problems with nonlinear constraints, Report UMINF-103.83, Institute of Information Processing, University of Umea, Umea, Sweden.

Marquardt, D. (1963), An algorithm for least-squares estimation of nonlinear parameters, *SIAM Journal of Applied Mathematics,* Vol. 11, pp. 431-441.

Mayr, L.M., Odefey, C., Schutkowski, M. and Schmid, F.X. (1996), Kinetic analysis of the unfolding and refolding of ribonuclease T1 by a stopped-flow double-mixing technique, *Biochemistry,* Vol. 35, No. 17, pp. 5550-5561.

Moré, J.J. (1977), The Levenberg-Marquardt algorithm: implementation and theory, in *Numerical Analysis,* edited by Watson, G., Lecture Notes in Mathematics, Vol. 630, Springer.

Nelder, J.A. and Mead, R. (1965), A simplex method for function minimization, *The Computer Journal,* Vol. 7, pp. 308.

Nickel, B. (1995), *Parameterschätzung basierend auf der Levenberg-Marquardt-Methode in Kombination mit direkter Suche,* Diploma Thesis, Department of Mathematics, University of Bayreuth.

Odefey, C., Mayr, L.M., and Schmid, F.X. (1995), Non-prolyl cis-trans peptide bond isomerization as a rate-determining step in protein unfolding and refolding, *Journal of Molecular Biology*, Vol. 245, pp. 69-78

Otter, M. and Türk, S. (1988), The DFVLR models 1 and 2 of the Manutec r3 robot, DFVLR-Mitt. 88-3, DFVLR, Oberpfaffenhofen.

Petzold, L.R. (1982), A description of DASSL: A differential/algebraic system solver, in *Proceedings of the 10th IMACS World Congress*, Montreal, Canada.

Prince, P.J. and Dormand, J.R. (1981), High order embedded Runge-Kutta formulae, *Journal on Computational Applied Mathematics*, Vol. 7, pp. 67-75.

Ramsin, H. and Wedin, P.A. (1977), A comparison of some algorithms for the nonlinear least squares problem, *Nordisk Tidstr. Informationsbehandlung (BIT)*, Vol. 17, pp. 72–90.

Rominger K.L. and Albert, H.J. (1985), Radioimmunological determination of Fenoterol. Part I: Theoretical fundamentals, *Arzneimittel-Forschung/Drug Research*, Vol.35, No.1a, pp. 415-420.

Schiesser, W.E. (1991), *The Numerical Method of Lines,* Academic Press, San Diego.

Schittkowski, K. (1985/86), NLPQL: A FORTRAN subroutine solving constrained nonlinear programming problems, *Annals of Operations Research*, Vol. 5, pp. 485–500.

Schittkowski, K. (1988), Solving nonlinear least squares problems by a general purpose SQP-method, in *Trends in Mathematical Optimization*, edited by Hoffmann, K.-H., Hiriart-Urruty, J.-B., Lemarechal, C. and Zowe, J., *International Series of Numerical Mathematics*, Vol. 84, Birkhäuser, pp. 295-309.

Schittkowski, K. (1994), Parameter estimation in systems of nonlinear equations, *Numerische Mathematik*, Vol. 68, pp. 129-142.

Schittkowski, K. (1995), *Parameter estimation in differential equations,* in *Rc cent Trends in Optimization Theory and Applications*, edited by Agarwal, R.P., WSSIAA, World Scientific Publishing Co., 2, pp. 353-370.

Schittkowski, K. (1997a), Parameter estimation in partial differential equations, *Optimization Methods and Software*, Vol. 7, No. 3-4, pp. 165-210.

Schittkowski, K. (1997b), EASY-FIT: Parameter estimation in dynamical systems - User's guide, Report, Mathematisches Institut, Universität Bayreuth.

Schittkowski, K. (1998), Parameter estimation in a mathematical model for substrate diffusion in a metabolically active cutaneous tissue, *Progress in Optimization II: Contributions from Australiasia*, edited by Yang, X.Q., Mees, A.I., Fisher, M. and Jennings, L.S., Kluwer Academic Publisher, pp. ??.

Schittkowski, K. (1998), PDEFIT: A FORTRAN code for parameter estimation in partial differential equations, *Optimization Methods and Software* (to appear).

Shampine, L.F. and Watts, H.A. (1979), *The art of writing a Runge-Kutta code,* Applied Mathematics and Computations, Vol. 5, pp. 93-121.

Simeon, B. and Rentrop, P. (1993), *An extended descriptor form for the simulation of constrained mechanical systems,* in *Advanced Multibody System*

Dynamics, edited by Schiehlen, W., Kluwer Academic Publishers, pp. 469-474.

Simeon, B., Grupp, F., Führer, C. and Rentrop, P. (1994), A nonlinear truck model and its treatment as a multibody system, *Journal of Computational and Applied Mathematics*, Vol. 50, pp. 523-532.

Steinsträsser, I. (1994), *The Organized HaCaT Cell Culture Sheet: A Model Approach to Study Epidermal Peptide Drug Metabolism,* Dissertation, Pharmaceutical Institute, ETH Zürich.

Sweby, P.K. (1984), High resolution schemes using flux limiters for hyperbolic conservation laws, *SIAM Journal on Numerical Analysis*, Vol. 21, No. 5, pp. 995-1011.

Van Genuchten, M.T. and Wierenga, P.J. (1976), Mass transfer studies in sorbing porous media. 1. Analytical solutions, *Soil Sci. Soc. Am. Journal*, Vol. 44, pp. 892-898.

Walter, S., Lorimer, G.H. and Schmid, F.X. (1996), A thermodynamic coupling mechanism for GroEl-mediated unfolding, *Biochemistry*, Vol. 93, pp. 9425-9430.

Wang, Z. and Richards, B.E. (1991), High resolution schemes for steady flow computation, *Journal of Computational Physics*, Vol. 97, pp. 53-72.

Wansbrough, R.W. (1985), Modelling chemical reactors, *Chemical Engineering*, Vol. 5.

Yang, H.Q. and Przekwas, A.J. (1992), A comparative study of advanced shock-capturing schemes applied to Burgers' equation, *Journal of Computational Physics*, Vol. 102, pp. 139-159.

Yee, H.C. (1985), Construction of a class of symmetric TVD schemes, *Lectures in Applied Mathematics*, Vol. 22, pp. 381-395.

Zschieschang, T. and Rockhausen, L. (1996), Zur Signalauswertung und Modellierung bei periodischen Vorgängen mit transienten Anteilen, Report, Institute of Mechanics, Technical University of Chemnitz-Zwickau, Chemnitz, Germany.

14 METHODS OF FEASIBLE DIRECTIONS: A REVIEW

Xibin Chen and Michael M. Kostreva

Department of Mathematical Sciences,
Clemson University,
Clemson, SC 29634-1907,
USA

Abstract: Since the theoretical basis for the method of feasible directions (MFD) was originally developed by Zoutendijk in 1960's, several basic variations and modifications of MFD were proposed and investigated. Even though faster algorithms for solving nonlinear programming problems exist, MFD has never been abandoned because of several important advantages such as the descent property, feasibility of all iterates, conceptual simplicity and overall computational efficiency. In fact, MFD is especially popular in the engineering community where it is heavily used in structural optimization and other engineering design optimization. In engineering design, it is extremely important to end up with a design which satisfies the hard specifications expressed by a set of inequalities. This paper reviews several typical methods of feasible directions and their variants: Zoutendijk's MFD, Topkis - Veinott's MFD, Pironneau - Polak's MFD, Wiest - Polak's MFD, Cawood - Kostreva's MFD, Panier - Tits' FSQP and a modified MFD. Then, the performances of three MFD: Pironneau - Polak's MFD, Cawood - Kostreva's MFD and the modified MFD are compared numerically and graphically. The results suggest that the modified MFD is more efficient than the other two MFD.

Key words: nonlinear programming, method of feasible directions (MFD), sequential quadratic programming (SQP), direction finding subproblem (DFS), engineering design optimization, structural optimization.

X. Yang et al. (eds.), Progress in Optimization, 205–219.

1 INTRODUCTION

The design of engineering systems such as mechanical systems, structures and electronic circuits involves the selection of a configuration and of design parameters in this configuration in such a way that a set of specifications are met and a design objective is optimized. In engineering design, it is extremely important to end up with a design which satisfies the hard specifications expressed by a set of inequalities. The engineering design optimization problem is to minimize or maximize an arbitrary objective function subject to nonlinear equality and inequality constraints. We consider the following formulation with inequality constraints:

$$\min\{f^0(x) \mid f^j(x) \leq 0;\ j = 1, 2, \cdots, m\} \tag{1.1}$$

where $f^j : \mathbb{R}^n \to \mathbb{R}^1$ $(j = 0, 1, 2, \cdots, m)$ are smooth functions, and x is a vector of design variables. Let $\Gamma = \{x \in \mathbb{R}^n \mid f^j(x) \leq 0,\ j = 1, 2, \cdots, m\}$ be the set of all feasible designs x for the problem (1.1).

The method of feasible directions is a very important method for solving the nonlinear programming problem (1.1). The fundamental concept of MFD is that of the feasible direction of descent. If $x \in \Gamma$, then $h \neq 0$ is called a feasible direction of descent for x if there exists $\bar{\alpha}$ such that $\forall \alpha \in (0, \bar{\alpha}]$ the following two properties hold: (i) $x + \alpha h \in \Gamma$. (ii) $f^0(x + \alpha h) < f^0(x)$. Although methods of feasible directions are considered by some to be less interesting than the superlinearly convergent algorithms, they have important advantages in engineering design optimization. They always produce a feasible design. If an initial design is feasible, all subsequent designs are feasible. They improve the design at each iteration, so that an acceptable solution may be achieved in early steps. In addition, MFD involves only first-order derivatives of objective and constraint functions. Due to these properties, MFD is especially popular in the engineering community where it is heavily used in structural optimization (Belegundu et al (1995), Korycki et al (1996b), Vanderplaats (1984a) and Vanderplaats (1984b)) and other engineering design problems (Tits et al (1986)). There are successful commercial codes which use MFD to support the engineering design process (Vanderplaats (1993)). The family of MFD includes Zoutendijk's MFD, Topkis - Veinott's MFD, Pironneau - Polak's MFD, Wiest - Polak's MFD, Cawood - Kostreva's MFD, Panier - Tits' FSQP and a modified MFD presented here.

In Section 2, several typical MFD are reviewed. After giving the general algorithm of MFD, this section describes the direction finding subproblem (DFS) of Zoutendijk's MFD, Topkis - Veinott's MFD, Pironneau - Polak's MFD, Wiest - Polak's MFD, Cawood - Kostreva's MFD, Panier - Tits' FSQP and the modified MFD. Section 3 discusses the numerical comparison and graphical interpretation of three MFD: Pironneau - Polak's MFD, Cawood - Kostreva's MFD and the modified MFD. These three MFD use quadratic programming problems as their DFS. The number of iterations for twelve test problems are listed and the paths of iterations are plotted. Section 4 gives the concluding remarks.

2 TYPICAL METHODS OF FEASIBLE DIRECTIONS

After the theoretical basis for the method of feasible directions was originally developed by Zoutendijk (1960), several basic variations and modifications of MFD were proposed and investigated. The basic steps in methods of feasible directions involve solving a linear or nonlinear programming subproblem to find the direction vector and then finding the step-length along this direction by performing a constrained one-dimensional search. After updating the current point, the above steps are repeated until the termination criterion is satisfied. The general algorithm of MFD can be described as follows:

- **Step 1:** Find or give $x_0 \in \Gamma$. Set $i = 0$.
- **Step 2: Direction Finding Subproblem (DFS)):** Given an iterate $x_i \in \Gamma$, set up the DFS to find h_i, a feasible direction of descent at x_i. If none exists, then stop the algorithm and x_i is the solution to the problem (1.1).
- **Step 3: One-Dimensional Search:** Find α_i such that (i) $x_i + \alpha_i h_i \in \Gamma$, (ii) $f^0(x_i + \alpha_i h_i) < f^0(x_i)$.
- **Step 4:** Set $x_{i+1} = x_i + \alpha_i h_i, i = i + 1$, go to Step 2.

The main differences among various algorithms of MFD are the differences in the direction finding subproblem (DFS). Most of recent research on MFD has been focused on the improvements and the modifications of the direction finding subproblem. The convenient way to capture most of variants of Zoutendijk's family of MFD is to follow the style of Polak (1971), and introduce a special set S bounding the length of the direction h. The set is compact and contains the origin in its interior. The direction finding subproblem at x_i can be then written as:

$$\min_{h \in S} \{ \max \{\nabla f^0(x_i)^T h;\ f^j(x_i) + \nabla f^j(x_i)^T h,\ j = 1, 2, \cdots, m\}\} \tag{2.1}$$

Equivalently,

$$\min \quad \{h^0\} \tag{2.2}$$
$$s.t. \quad \nabla f^0(x_i)^T h \leq h^0 \tag{2.3}$$
$$f^j(x_i) + \nabla f^j(x_i)^T h \leq h^0, \quad j = 1, 2, \cdots, m \tag{2.4}$$
$$h \in S \tag{2.5}$$

where h^0 is an auxiliary scalar variable. It is easy to demonstrate that this type of DFS produces a feasible direction of descent $h \neq 0$ if x_i is not a Fritz John point or its solution is $(h^0, h) = (0, 0)$. A constraint on the length of the direction h is necessary, otherwise, the DFS could be unbounded.

2.1 *Zoutendijk's MFD*

Notice that DFS (2.2) - (2.5) includes linearized versions of all constraints of the problem (1.1). One may be tempted to include only active constraints at x_i. In

1960, this idea was originally used by Zoutendijk (1960) in one of variants of the DFS he proposed. He also chose the simple box $\{h, |-1 \leq h_k \leq 1,\ k = 1, \cdots, n\}$ as the constraining set S and obtained the following LP as a DFS:

$$\begin{aligned} \min \quad & \{h^0\} \\ s.t. \quad & \nabla f^0(x_i)^T h \leq h^0 \\ & \nabla f^j(x_i)^T h \leq h^0, \quad j \in J(x_i) \\ & -1 \leq h_k \leq 1, \quad k = 1, 2, \cdots, n \end{aligned}$$

where $J(x_i)$ is the set of indices of all active constraints. Unfortunately, in Zoutendijk's simplified method, neither the direction-finding map nor the line-search map are closed. This algorithm was thus not protected against jamming, or zigzagging, a term which is used to denote the phenomenon of producing a sequence convergent to a point which is not a Fritz John point. The counterexample, given in Wolfe (1967), shows that jamming effects may take place if inactive constraints are removed from the DFS, as in Zoutendijk's simplified DFS. Due to the fact that inactive constraints are not included in the DFS, the difficulty is that the distance moved along the generated directions may tend toward zero, causing jamming at a non-optimal point.

2.2 *Topkis - Veinott's MFD*

A modification of Zoutendijk's MFD, developed by Topkis et al (1967), assures convergence of the algorithm to a Fritz John point. Topkis - Veinott's MFD uses the following problem as the DFS:

$$\begin{aligned} \min \quad & \{h^0\} \\ s.t. \quad & \nabla f^0(x_i)^T h \leq h^0 \\ & f^j(x_i) + \nabla f^j(x_i)^T h \leq h^0, \quad j = 1, 2, \cdots, m \\ & -1 \leq h_k \leq 1, \quad k = 1, 2, \cdots, n \end{aligned}$$

Here, both active and inactive constraints play a role in determining the direction of movement. As opposed to the Zoutendijk's DFS, no sudden change in the direction is encountered when approaching the boundary of a currently inactive constraint.

Further, a variant of their DFS is as follows:

$$\begin{aligned} \min \quad & \{h^0\} \\ s.t. \quad & \nabla f^0(x_i)^T h \leq h^0 \\ & f^j(x_i) + \nabla f^j(x_i)^T h \leq h^0, \quad j = 1, 2, \cdots, m \\ & h^T h \leq 1 \end{aligned}$$

This DFS also includes all the constraints and uses the constraining set $S = \{h \mid h^T h \leq 1\}$ to limit the length of feasible directions. Even though it is not a QP problem, a numerical procedure for solving this type of problem can be easily developed on the basis of the KKT conditions of optimality (Vanderplaats

(1984a)). But, a counterexample in Pironneau et al (1972) shows that Topkis - Veinott's MFD with the constraining set $S = \{h \mid -1 \leq h_k \leq 1,\ k = 1, 2, \cdots, n\}$ does not converge linearly even under convexity assumptions such as requiring that $f^j(x)$ $(j = 0, 1, 2, \cdots, m)$ be convex functions. For Topkis - Veinott's MFD with the constraining set $S = \{h \mid h^T h \leq 1\}$, no better results could be obtained for this same example.

2.3 Pironneau - Polak's MFD

The Pironneau - Polak's MFD (Pironneau et al (1973)) was derived from the Topkis - Veinott's MFD by replacing the normalization set $S = \{h \mid h^T h \leq 1\}$ by adding the term $\frac{1}{2}||h||^2$ to the objective function of the DFS. The DFS is as follows:

$$\begin{array}{ll} \min & \{h^0 + \frac{1}{2}||h||^2\} \\ s.t. & \nabla f^0(x_i)^T h \leq h^0 \\ & f^j(x_i) + \nabla f^j(x_i)^T h \leq h^0, \quad j = 1, 2, \cdots, m \end{array}$$

The MFD of Pironneau and Polak was extensively studied and proved to be globally convergent. This modification results in a linearly convergent algorithm. Pironneau and Polak derived a bound on a constant of linear convergence under the assumptions convexity (Pironneau et al (1972) and Pironneau et al (1973)). At the same time, this algorithm is insensitive to the number of variables in the problem (1.1) and its DFS is usually easy to solve. Furthermore, the assumptions of convexity were relaxed in Chaney (1976a) and Chaney (1976b), who derived a similar bound under the assumptions relative to those in the second order sufficiency conditions for local minimum.

2.4 Wiest - Polak's MFD

Polak proposed a MFD, called generalized quadratic programming (GQP) (Polak (1971)). By adding the second-derivative terms into the objective function and constraint functions of the DFS of Topkis - Veinott's MFD with $S = \{h \mid -1 \leq h \leq 1\}$, a new DFS with a Newton-Raphson-like appearance, given by Polak (1971), can be obtained. The DFS is as follows:

$$\begin{array}{ll} \min & \{h^0\} \\ s.t. & \nabla f^0(x_i)^T h + \frac{1}{2} h^T H_0 h \leq h^0 \\ & f^j(x_i) + \nabla f^j(x_i)^T h + \frac{1}{2} h^T H_j h \leq h^0, \quad j = 1, 2, \cdots, m \\ & -1 \leq h_k \leq 1, \quad k = 1, 2, \cdots, n \end{array}$$

where H_j $(j = 0, 1, 2, \cdots, m)$ are some matrices, possibly approximations of Hessian matrices of $f^j(x)$ at x_i. This DFS requires us to minimize a linear form on a constraint set defined by linear and quadratic inequalities. There are no known methods for guaranteed solution of such problems in a finite number of iterations, according to the statements in Polak (1971). So this method remains conceptual but not implementable.

Wiest and Polak modified this DFS in Wiest et al (1992). They simplified all matrices H_j to a multiple of the identity, γI, with the same scalar $\gamma > 0$ for all j to obtain a finitely solvable procedure for the DFS. The DFS is as follows:

$$\begin{array}{ll} \min & \{h^0\} \\ s.t. & \nabla f^0(x_i)^T h + \frac{\gamma}{2} h^T h \leq h^0 \\ & f^j(x_i) + \nabla f^j(x_i)^T h + \frac{\gamma}{2} h^T h \leq h^0, \quad j = 1, 2, \cdots, m \\ & -1 \leq h_k \leq 1, \quad k = 1, 2, \cdots, n \end{array}$$

where the scalar $\gamma > 0$. With such simplification, the resulting GQP can be to solved in a finite number of iterations by means of one QP plus one projection operation. As a result, they obtain a MFD with a DFS parameterized by a single scalar. They showed that GQP - based algorithm converges linearly with a smaller bound on the cost convergence ratio than that obtained for the Polak - Mayne - Trahan algorithm in Pironneau et al (1972) and Polak et al (1979). Numerical experiments also show this algorithm to be generally superior to the Polak - Mayne - Trahan algorithm, and competitive with the feasible descent algorithm in Herskovits (1986).

2.5 Cawood - Kostreva's MFD

Another method, first proposed by Cawood and Kostreva (Cawood et al (1994)) and called the norm-relaxed MFD or Cawood - Kostreva's MFD, generalizes Pironneau - Polak's MFD. Instead of the Euclidean norm in the DFS of Pironneau - Polak's MFD, the DFS of this MFD uses an arbitrary elliptical norm generated by a scaling symmetric positive definite matrix. The DFS of Cawood - Kostreva MFD is as follows:

$$\begin{array}{ll} \min & \{h^0 + \frac{\gamma}{2} h^T Q h\} \\ s.t. & \nabla f^0(x_i)^T h \leq h^0 \\ & f^j(x_i) + \nabla f^j(x_i)^T h \leq h^0, \quad j = 1, 2, \cdots, m \end{array}$$

where $\gamma > 0$ is a scalar and Q is a symmetric, positive-definite matrix.

In fact, Cawood - Kostreva's MFD and Wiest - Polak's MFD are related to each other by the idea of GQP. In Wiest - Polak's MFD, all functions are quadratically approximated but all matrices generating quadratic terms are of the same form γI. In Cawood - Kostreva's MFD, the quadratic approximations of all functions are abandoned and an arbitrary symmetric, positive definite matrix is introduced to generate an elliptical norm to bound the length of the direction. The resulting DFS is a classical QP subproblem which can be conveniently solved in its dual form. Since it is a strictly convex problem, it can be numerically solved without any difficulty. In terms of the number of degrees of freedom in controlling the speed of convergence of the algorithm, Cawood -

Kostreva's MFD offers broader scope of possibilities than Pironneau - Polak's MFD (Pironneau et al (1973)) or Wiest - Polak's MFD (Wiest et al (1992)).

Cawood et al (1994) showed that this method is globally convergent under a rather mild and standard assumption that the set $\{x \in \Gamma \mid f^0(x) \leq f^0(x_0)\}$ is compact and has an interior, where x_0 is the initial feasible point, given as input to the algorithm. On the basis of this algorithm, Cawood - Kostreva's MFD was further investigated by Korycki, Kostreva and Chen in Korycki (1995), Korycki et al (1996a), Korycki et al (1996b), Korycki et al (1997) and Kostreva et al (1996). Korycki and Kostreva showed that Cawood - Kostreva's MFD is linearly convergent and derived a bound of convergence (Korycki et al (1996a)), presented three update techniques: BFGS technique, the multi-layer grid technique and the trust region technique (Korycki et al (1997)). Kostreva, Chen and Korycki further demonstrated its numerical performance on some larger randomly generated problems than the classical test problems (Kostreva et al (1996)). The numerical experiments of Cawood - Kostreva's MFD have been considered in Korycki (1995), Korycki et al (1996b), Korycki et al (1997) and Kostreva et al (1996). Three groups of test problems were used in the numerical experiments of Cawood - Kostreva's MFD. The first group contained fifteen classical test problems from Hock et al (1981) and Schittkowski (1987). The second group consisted of five classical structural optimization problems. The third group contains nine comparably larger problems generated by a randomized test problem generator. All numerical and theoretical results showed that they provided an improvement of Pironneau – Polak's MFD.

2.6 Panier - Tits' FSQP

Successive quadratic programming(SQP) methods, also known as sequential, or recursive, quadratic programming methods, employ Newton's method or Quasi-Newton's method to directly solve the KKT conditions for the original problem. As a result, the accompanying subproblem turns out to be the minimization of a quadratic approximation of the Lagrangian function optimized over a linear approximation to the constraint. Hence, this type of process is also known as a projected Lagrangian, or the Lagrange Newton, approach. At each iteration, the following DFS is solved and then the line-search procedure determines the step-length along this direction. DFS is as follows:

$$\min \quad \{\nabla f^0(x_i)^T h + \frac{1}{2} h^T H h\} \tag{2.6}$$

$$s.t. \quad f^j(x_i) + \nabla f^j(x_i)^T h \leq 0, \quad j = 1, 2, \cdots, m \tag{2.7}$$

where H represents the Hessian of the Lagrangian at x_i, either given explicitly or approximated by finite differences or variable metric techniques such as BFGS.

The methods based on the DFS (2.6) - (2.7) nowadays are probably the most popular class of algorithms for the solution of problem (1.1): successive quadratic programming. This class of methods is superlinearly convergent. Numerical experiments have showed this class of methods often dramatically outperforms algorithms of other classes (Hock et al (1981) and Powell (1977)). However, the major disadvantage of this method is that the iterates produced by it may not be feasible. This excludes SQP from some applications in engineering design, where each iterate is supposed to represent a realistic design, which frequently has no meaning when the feasibility is not maintained. Moreover, from the computational point of view, infeasibility of an iterate x_i implies that the DFS (2.6) - (2.7) at x_i may have no solution. Because of these factors, special precautions have to be taken in practical implementations of SQP. Some of the techniques to remedy the infeasibility of the iterates in the SQP are referred as feasible sequential quadratic programming (FSQP) methods and the variants of FSQP methods credited to Panier, Tits and Herskovits (Panier et al (1987), Panier et al (1988), Panier et al (1993)).

Following an idea used by Herskovits in a different context (Herskovits (1986) and Panier et al (1987)) proposed an algorithm of the successive quadratic programming type, an ancestor of FSQP method, which is shown to converge locally superlinearly and globally under mild assumptions. This method is a method of feasible direction and numerical tests have proven it to be competitive with most successful available nonlinear programming algorithms. Later, Panier, Tits and Herskovits (Panier et al (1988)) proposed a QP-free, globally convergent, locally superlinearly convergent algorithm for solving the problem (1.1). Unlike successive quadratic programming type methods such as SQP and the previous algorithm of Panier and Tits, this algorithm does not involve the solution of quadratic programming problems, but merely that of linear systems of equations. Locally the iteration can be viewed as a perturbation of a quasi-Newton iteration on both the primal and dual variables for the solution of the equations in the Kuhn-Tucker first order condition of optimality. It is observed that, provided the current iterate is feasible and the current multiplier estimates are strictly positive, the primal component of the quasi-Newton direction is a direction of descent for f^0. This fact is used to induce global convergence, without the need of a surrogate merit function. A "bending" of the search direction, similar to that in the previous algorithm of Panier and Tits, prevents any 'Maratos-effect' and local superlinear convergence is proven. While the algorithm requires that an initial feasible point be available, it constructs the successive iterates which are feasible as well. The FSQP method developed by Panier and Tits in Panier et al (1993) generalized the the previous algorithm proposed by Panier and Tits (Panier et al (1987)).

The process of the generalized algorithm credited by Panier and Tits (Panier et al (1993)) is as follows:

- Compute a descent direction d^0 by solving the DFS of SQP method (2.6) - (2.7). If $d^0 = 0$, then stop.

- Compute a feasible direction d^1, for example, by solving the DFS of Pironneau - Polak's MFD.

- Determine the search direction by computing $\rho(\cdot) : \mathbb{R}^n \rightarrow [0,1]$ such that $\rho(d^0) = 1$ when $||d^0||$ is larger than some given threshold, $\rho(d^0)$ is bounded away from zero outside every neighborhood of zero and, for $||d^0||$ small, $\rho(d^0) = O(||d^0||^2)$. Compute the direction $d = (1-\rho)d^0 + \rho d^1$.

- Compute a correction $\hat{d}$ by solving the following quadratic programming problem:

$$\min_{\hat{d}} \quad \frac{1}{2}(d+\hat{d})^T H(d+\hat{d}) + \nabla f^0(x)^T(d+\hat{d})$$
$$f^j(x+d) + \nabla f^j(x)^T\hat{d} \leq ||d||^\tau, \quad j = 1, 2, \cdots, m.$$

where $\tau \in (2,3)$ is preselected.

- Compute $\bar{t}$, the first number of the sequence $\{1, \beta, \beta^2, \cdots\}$ satisfying

$$\begin{aligned} f^0(x+td+t^2\hat{d}) &\leq f^0(x) + \alpha t \nabla f^0(x)^T d \\ f^j(x+td+t^2\hat{d}) &\leq 0, \quad j = 1, 2, \cdots, m. \end{aligned}$$

where $\alpha \in (0, \frac{1}{2})$ and $\beta \in (0,1)$ are preselected parameters. Update $x := x + \bar{t}d + \bar{t}^2 d$.

FSQP updates H according to the BFGS formula with Powell's modification in Powell (1977) and the initial H_0 is the identity matrix.

The FSQP approaches are globally convergent, locally superlinear convergent and feasible for solving inequality constrained optimization problem (1.1) under mild assumptions. However, each iteration in these algorithms requires the solution of two QP problems and a linear least squares problem, or two linear systems of equations and a linear least squares problem, or three QP problems. In each case, the three subproblems must be solved in sequence. Even though they converge locally superlinearly, their computing cost at a single iteration is rather high as compared with classical MFD.

2.7 A Modified MFD

So far, the price of all superlinearly convergent algorithms for achieving superlinearity is either the loss of feasibility of iterates, as in the case of SQP methods, or rather high computational cost (solving three QP or LP subproblems) at a single iteration, as in the case of the FSQP method. The loss of feasibility of iterates in SQP methods can cause catastrophic failure in the computation, especially when an interface to a finite element package is required. In addition, the classical MFD only needs to solve one QP subproblem or one LP subproblem and the cost is low at each iterate as compared with FSQP algorithm. The classical MFD with linear convergence rate are still advantageous relative to the existing superlinearly convergent methods such as SQP and FSQP for solving the engineering design optimization problems.

This section describes a modified MFD proposed by Chen and Kostreva (1999). This MFD generalizes Cawood - Kostreva's MFD. This modified MFD introduces more parameters which can be adjusted to speed up the convergence.

The general DFS of MFD was described in (2.1). Similarly, by introducing an auxiliary variable h^0 and weighting scalars $\gamma_j > 0$ $(j = 0, 1, 2, \cdots, m)$ and using the similar idea of Pironneau - Polak's MFD and Cawood - Kostreva's MFD, we add the second order term into the objective function of the DFS to limit the length of the directions. The new DFS is as follows:

$$\min \quad \{h^0 + \frac{1}{2}h^T Q h\} \tag{2.8}$$

$$s.t. \quad \nabla f^0(x)^T h \leq \gamma_0 h^0 \tag{2.9}$$

$$f^j(x) + \nabla f^j(x)^T h \leq \gamma_j h^0, \quad j = 1, 2, \cdots, m \tag{2.10}$$

where Q a positive definite, symmetric matrix and $\gamma_j > 0$ $(j = 0, 1, 2, \cdots, m)$. For example,, we can select Q to represent the Hessian of the Lagrangian at x, either given explicitly or approximated by finite differences or variable metric techniques such as BFGS.

Observe that the above is a generalization of Cawood - Kostreva's MFD and Pironneau - Polak's MFD. In this new DFS, if $\gamma_j = \gamma$ $(j = 0, 1, 2, \cdots, m)$ and $\bar{h^0} = \gamma h^0$, the new DFS (2.8) - (2.10) reduces to the following:

$$\min \quad \{\frac{1}{\gamma}(\bar{h^0} + \frac{\gamma}{2}h^T Q h)\}$$

$$s.t. \quad \nabla f^0(x)^T h \leq \bar{h^0}$$

$$f^j(x) + \nabla f^j(x)^T h \leq \bar{h^0}, \quad j = 1, 2, \cdots, m$$

where γ is a preselected scalar. So, this DFS generates the same directions as those generated by Cawood - Kostreva's MFD when they have the same weight γ and matrix Q. Furthermore, it is also equivalent to Pironneau - Polak's MFD by selecting $Q = I$ and $\gamma = 1$.

2.8 Features of Typical MFD

Table 2.1 lists the features of typical methods of feasible directions: Zoutendijk's MFD, Topkis - Veinott's MFD, Pironneau - Polak's MFD, Wiest - Polak's MFD, Cawood - Kostreva's MFD, Panier - Tits' FSQP and the modified MFD.

3 THE COMPARISONS OF THREE MFD

The DFS of Pironneau - Polak's MFD and Cawood - Kostreva's MFD may be induced from the DFS of the modified MFD. The DFS of the modified MFD generates the same directions generated by that of Cawood - Kostreva's MFD when $\gamma_j = \gamma$ $(j = 0, 1, 2, \cdots, m)$ while the DFS of the modified MFD is the same as that of Pironneau - Polak's MFD when $\gamma_j = 1$ $(j = 0, 1, 2, \cdots, m)$ and $Q = I$. After the implementation of these three MFD, their performances are compared numerically and graphically. During the numerical experiments, the initial step-length $\alpha_{max} = 10.0$.

Table 2.1 The Features of Typical MFD

MFD	Convergence	Convergence Rate	Parameter	Subprob.
Modified	Global	At least Linear	Multiple	One
Panier-Tits'	Global	Local Superlinear	No	Three
Cawood - Kostreva's	Global	Linear with a Smaller Bound	Multiple	One
Wiest - Polak's	Global	Linear with a Smaller Bound	One	One
Pironneau - Polak's	Global	Linear	No	One
Topkis - Veinott's	Global	Sublinear	No	One
Zoutendijk's	Not Global	N/A	No	One

3.1 Numerical Comparison

Table 3.1 The Number of Iterations for Three Methods

Problem No.	modified MFD			Cawood-Kostreva MFD	Pironneau-Polak MFD
	1/8	1/4	1/2		
29	18	12	9	14	38
31	22	25	31	40	115
34	75	27	29	17	33
35	15	11	21	21	18
43	13	16	22	34	53
76	14	13	16	22	20
86	60	40	66	100	351
100	21	19	23	29	105
113	29	48	75	82	79
284	55	39	37	43	108
315	13	12	13	12	12
323	7	9	11	11	23

We solve twelve test problems selected from Hock et al (1981) and Schittkowski (1987) by using Pironneau - Polak's MFD, Cawood - Kostreva's MFD and our modified MFD, respectively. When using Cawood - Kostreva's MFD, we selected $\gamma = 1.0$ and used BFGS to update the matrix. In our algorithm, $\gamma_0 = 1.0$ is used and three cases: $\gamma_j = \frac{1}{2}, \frac{1}{4}, \frac{1}{8}$ $(j = 1, \cdots, m)$ are considered. The number of iterations solving each problem for three methods are listed in Table 3.1.

Even though there is no best constant γ_j for all problems in our algorithm, the numerical results generally show that the numbers of iterations when using this modified MFD are smaller than those when using Pironneau - Polak's

MFD or Cawood - Kostreva's MFD for most test problems. It is likely that the algorithm with a set of suitable weights will converge faster than Pironneau - Polak's MFD and Cawood - Kostreva's MFD.

3.2 Graphical Interpretation

We consider the following simple problem with only one constraint and two variables to conveniently represent the iterating process graphically:

$$\begin{array}{ll} \min & (x_1 - 10)^2 + (x_2 - 10)^2 \\ s.t. & x_1^2 + \dfrac{x_2^2}{4} \leq 25 \end{array}$$

We use $(0, 0)$ as the initial feasible point. In the modified MFD, $\gamma_0 = 1.0$ and $\gamma_1 = 0.1$ are used. The first three iterates and the optimal solutions for three methods when $\varepsilon = 10^{-3}$ and $\varepsilon = 10^{-6}$ are listed in Table 3.2. In Table 3.2, $\bar{x}$ and $f(\bar{x})$ are optimal solution and optimal value for $\varepsilon = 10^{-3}$ while $\hat{x}$ and $f(\hat{x})$ are optimal solution and optimal value for $\varepsilon = 10^{-6}$. $\overline{NI}$ and $\widehat{NI}$ stand for the numbers of iterations for $\varepsilon = 10^{-3}$ and $\varepsilon = 10^{-6}$, respectively. The iterating paths of Pironneau - Polak's MFD, Cawood - Kostreva's MFD and the modified MFD for $\varepsilon = 10^{-3}$ are shown in Figure 3.1. The results in

Table 3.2 The Iterating Results for Three Different Methods

Items	Pironneau - Polak's	Cawood - Kostreva's	modified MFD
$x^{(0)}$	$(0.0, 0.0)$	$(0.0, 0.0)$	$(0.0, 0.0)$
$f(x^{(0)})$	200.0	200.0	200.0
$x^{(1)}$	$(4.447038, 4.447038)$	$(4.447038, 4.447038)$	$(4.447038, 4.447038)$
$f(x^{(1)})$	61.670770	61.670770	61.670770
$x^{(2)}$	$(3.581151, 5.756942)$	$(3.581151, 5.756942)$	$(3.690019, 6.747559)$
$f(x^{(2)})$	59.205163	59.205163	50.394233
$x^{(3)}$	$(2.973699, 7.396035)$	$(3.652881, 6.770030)$	$(3.623103, 6.891452)$
$f(x^{(3)})$	56.149537	50.718622	50.327894
$\bar{x}$	$(3.605759, 6.927462)$	$(3.601581, 6.936337)$	$(3.612598, 6.913489)$
$f(\bar{x})$	50.326815	50.325797	50.325453
$\overline{NI}$	16	7	5
$\hat{x}$	$(3.605409, 6.928499)$	$(3.605548, 6.928209)$	$(3.605488, 6.928334)$
$f(\hat{x})$	50.324911	50.324910	50.324910
$\widehat{NI}$	23	19	10

Table 3.2 show that the modified MFD converges much faster than Pironneau - Polak's MFD and faster than Cawood - Kostreva's MFD. It also can be seen that Cawood - Kostreva's MFD is faster than Pironneau - Polak's MFD. From Table 3.2 and Figures 3.1, we can see that three methods have different iterating paths, i.e., converging processes. Pironneau - Polak's MFD takes zigzag approach to the optimal solution and Cawood - Kostreva's MFD obtusely

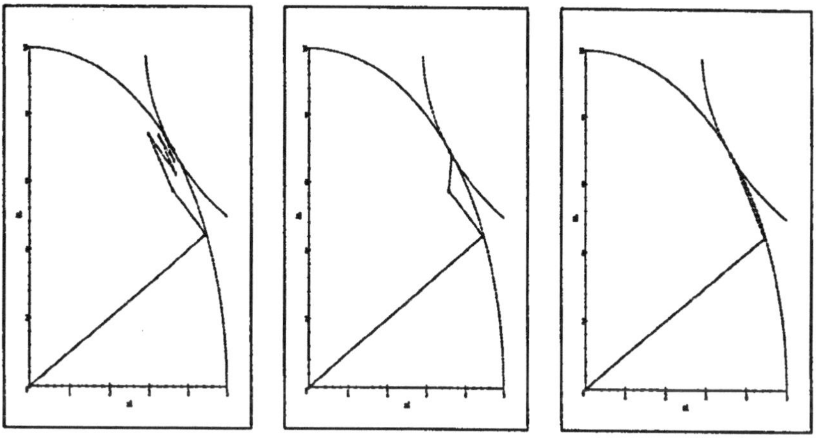

Figure 3.1 Iterating Paths for Three Methods: Pironneau - Polak's MFD (left), Cawood - Kostreva's MFD (middle) and the modified MFD (right)

approaches the optimal solution while the modified MFD with $\gamma_0 = 1.0$ and $\gamma_1 = 0.1$ almost directly points to the optimal solution from the feasible solution $(4.447038, 4.447038)$. These observations mean that the DFS of the modified MFD found more favorable feasible descent directions for three methods.

4 CONCLUDING REMARKS

MFD has several important advantages such as the descent property, feasibility of all iterates, conceptual simplicity and overall computational efficiency and satisfies the requirements of engineering design problems. MFD has been successfully and heavily used to solve engineering design optimization problems such as structural optimization problems and control system design problems. This paper reviews typical MFD: Zoutendijk's MFD, Topkis - Veinott's MFD, Pironneau - Polak's MFD, Wiest - Polak's MFD, Cawood - Kostreva's MFD, Panier - Tits' FSQP and the modified MFD. Then, the performance of three MFD: Pironneau - Polak's MFD, Cawood - Kostreva's MFD and the modified MFD are compared numerically and graphically. The results show that the modified MFD is more efficient than Pironneau - Polak's MFD and Cawood - Kostreva's MFD for most test problems.

However, how to choose the parameters of the modified MFD has not been addressed here. In addition, the modified MFD is linearly convergent. But, for some special cases, this modified MFD is superlinearly convergent. Also, a variant of the modified MFD combining with suitable parameters may yield a locally superlinearly convergent algorithm. This algorithm should be comparable with Panier - Tits' FSQP. These issues are under investigation and will be discussed in futhure work.

Acknowledgments

The authors would like to thank the referees for their helpful comments.

References

Belegundu, A. D., Berke, L. and Patnaik, S. N. (1995), An optimization algorithm based on the method of feasible directions, *Structural Optimization*, Vol. 9, pp. 83-88.

Cawood, M. E. and Kostreva, M. M. (1994), Norm-relaxed method of feasible directions for solving nonlinear programming problems, *Journal of Optimization Theory and Applications*, Vol. 83, pp. 311-320.

Chaney, R. W. (1976a), On the Pironneau-Polak method of centers, *Journal of Optimization Theory and Applications*, Vol. 20, pp. 269-295.

Chaney, R. W. (1976b), On the rate of convergence of some feasible direction algorithms, *Journal of Optimization Theory and Applications*, Vol. 20, pp. 297-313.

Chen, X. and Kostreva, M. M. (1999), A generalization of the norm-relaxed method of feasible directions, *Applied Mathematics and Computation*, Vol. 102, pp. 257-272.

Herskovits, J. N. (1986), A two-stage feasible directions algorithm for nonlinear constrained optimization, *Mathematical Programming*, Vol. 36, pp. 19-38.

Hock, W. and Schittkowski, K. (1981), *Test Examples for Nonlinear Programming Codes*, Lecture Notes in Economics and Mathematical Systems, Springer Verlag, Berlin, Germany.

Korycki, J. A. (1995), *Norm-relaxed Method of Feasible Directions*, Ph. D. Dissertation, Department of Mathematical Sciences, Clemson University, August, 1995.

Korycki, J. A. and Kostreva, M. M. (1996a), Convergence analysis of norm-relaxed method of feasible directions, *Journal of Optimization Theory and Applications*, Vol. 91, pp. 389-418.

Korycki, J. A. and Kostreva, M. M. (1996b), Norm-relaxed method of feasible directions: application in structural optimization, *Structural Optimization*, Vol. 11, pp. 187-194.

Korycki, J. A. and Kostreva, M. M. (1997), Self-tuning norm-relaxed method of feasible directions, *Journal of Optimization Theory and Applications*, Vol. 92, pp. 311-330.

Kostreva, M. M., Chen, X. B. and Korycki, J. A. (1996), A norm-relaxed method of feasible directions for solving nonlinear programming problems, *Proceedings of 6th AIAA/USAF/NASA/ISSMO Symposium on Multidisciplinary Analysis and Optimization*, Bellevue, WA, September 1996.

Panier, E. R. and Tits, A. L. (1987), A superlinearly convergent feasible method for the solution of inequality constrained optimization problems, *SIAM Journal on Control and Optimization*, Vol. 25, pp. 934-950.

Panier, E. R., Tits, A. L. and Herskovits, J. N. (1988), A QP-free, globally convergent, locally superlinearly convergent algorithm for inequality constrained optimization, *SIAM Journal on Control and Optimization*, Vol. 26, pp. 788-811.

Panier, E. R. and Tits, A. L. (1993), On combining feasibility, descent and superlinear convergence in inequality constrained optimization, *Mathematical Programming*, Vol. 59, pp. 261-276.

Pironneau, O. and Polak, E. (1972), On the rate of convergence of certain methods of centers, *Mathematical Programming*, Vol. 2, pp. 230-257.

Pironneau, O. and Polak, E. (1973), Rate of convergence of a class of methods of feasible directions, *SIAM Journal on Numerical Analysis*, Vol. 10, pp. 161-173.

Polak, E. (1971), *Computational Methods in Optimization*, Academic Press, New York.

Polak, E., Trahan, R. and Mayne, D. Q. (1979), Combined phase I - phase II methods of feasible directions, *Mathematical Programming*, Vol. 17, pp. 32-61.

Powell, M. J. D. (1977), A fast algorithm for nonlinearly constrained optimization calculations, in *Lecture Notes in Mathematics 630*, edited by Watson, G. A., Springer-Verlag, Berlin, New York, pp. 144-157.

Schittkowski, K. (1987), *More Test Examples for Nonlinear Programming Codes*, Lecture Notes in Economics and Mathematical Systems, Springer Verlag, Berlin, Germany.

Tits, A. L., Nye, W. T. and Sangiovanni-Vincentelli, A. L. (1986), Enhanced methods of feasible directions for engineering design problems, *Journal of Optimization Theory and Applications*, Vol. 5, pp. 475-504.

Topkis, D. M. and Veinott, A. F. (1967), On the convergence of some feasible direction algorithms for nonlinear programming, *SIAM Journal on Control*, Vol. 5, pp. 268-279.

Vanderplaats, G. N. (1984a), *Numerical Optimization Techniques for Engineering Design*, McGraw Hill, New York.

Vanderplaats, G. N. (1984b), Efficient feasible directions algorithm for design synthesis, *AIAA Journal*, Vol. 22, pp. 1633-1640.

Vanderplaats, G. N. (1993), *DOT/DOC Users Manual*, Vanderplaats, Miura and Associates.

Wiest, E. J., Polak, E. (1992), A generalized quadratic programming - based phase I - phase II method for inequality constrained optimization, *Journal of Applied Mathematics and Optimization*, Vol. 26, pp. 223-252.

Wolfe, P. (1967), On the convergence of gradient methods under constraints, IBM Research Rept. RC 1752, Yorktown Heights, New York.

Zoutendijk, G. (1960), *Methods of Feasible Directions*, Elsevier Publishing Company, Amsterdam, Netherlands.

15 COMPUTATIONAL METHOD FOR A CLASS OF OPTIMAL SWITCHING CONTROL PROBLEMS

Y. Liu

Department of Applied Mathematics,
The Hong Kong Polytechnic University,
Kowloon, Hong Kong

and K.L. Teo

Department of Applied Mathematics,
The Hong Kong Polytechnic University,
Kowloon, Hong Kong

Abstract: In this paper, we consider a class of optimal control problems in which the system is to be determined in an optimal way. This class of problems involves the choice of a fixed number of switching time points which divide the system's time horizon into a number of time periods. For each of these time periods, a subsystem is selected, from a finite number of given candidate subsystems, to run during that time periods. The choice of the switching points and the selection of the subsystems are carried out in such a way that a given cost functional is minimized. We consider only problems involving ordinary differential equations over a finite time horizon. A computational method is developed for solving these problems. In our method, the candidate subsystems are combined into a single system by considering their 'linear combination'. By introducing suitable constraints on the coefficients of the linear combination and using a time rescaling technique, the original problem is transformed into an equivalent optimal control problem with system parameters. An algorithm is proposed for solving this transformed problem and the required gradient formulae are derived. To show the effectiveness of the method, a numerical example is solved.

Key words: optimal control, switching time, switching control, continuous and discrete parameters.

X. Yang et al. (eds.), Progress in Optimization, 221–237.

1 INTRODUCTION

In this paper, we consider a class of optimal control problems in which the system is to be determined in an optimal way. This class of problems involves the choice of a fixed number of switching time points t_1, t_2, ..., t_{K-1} which divide the system's time horizon into K time periods, namely

$$[0, t_1), [t_1, t_2), \ldots, [t_{K-2}, t_{K-1}), [t_{k-1}, T]$$

and the selection of a subsystem S_i for each subinterval $[t_{i-1}, t_i)$. At $t = 0$ the first subsystem S_1 starts to operate and continues until the first switching time $t = t_1$ is reached. At time $t = t_1$, S_1 is switched off, and the second subsystem S_2 is switched on, with a starting system state continued from that of S_1 at $t = t_1$. S_2 continues to operate until the second switching time t_2 is reached. In general, at the i-th switching time t_i, the i-th subsystem S_i is switched off and the $(i+1)$-th subsystem is switched on. The starting state of S_{i+1} continues that of S_i. That is, the state of S_i and that of S_{i+1} are equal at $t = t_i$. The $(i+1)$-th system continues until the next switching time $t = t_{i+1}$ is reached. This process repeats until the final time T is reached. All the K subsystems and the $K-1$ switching times will be taken to be decision variables. Each of the subsystems will be referred to as a switching system. For any sequence $S = (S_1, S_2, \ldots, S_K)$ of switching subsystems and any sequence $\xi = (t_1, t_2, \ldots, t_{K-1})^T$ of switching times satisfying $t_i - t_{i-1} \geq 0$, $i = 1, 2, \ldots, K$ with $t_0 = 0$ and $t_K = T$, the pair (ξ, S) will be referred to as a switching sequence. Our optimal control problem is to minimize a cost functional subject to the above described system over all possible switching sequences.

Many control problems arising in economics, management, production, engineering and industry follow the model described above, as factors directly related to the system dynamics may be subject to sudden changes. General optimal switching control problems have been studied in recent years (Evans and Friedman (1979), Li and Yong (1995), Yong (1989), Yong (1991)). The problems are usually formulated as optimal control problems, in which the feasible controls are appropriate switching functions. Problems about the optimality principle, the HJB equation, and the existence of optimal controls are investigated. For more details, see Evans and Friedman (1979), Li and Yong (1995) and the relevant references cited therein.

In this paper, our aim is to develop an efficient computational method for this class of optimal control problem. It is clear that the problem under consideration involves both discrete and continuous decision variables represented, respectively, by the subsystems and the switching times. To deal with this mixed discrete and continuous optimal control problem, a transformation technique is introduced. This technique is then used to transform the mixed continuous-discrete optimal control problem into an optimal parameter selection problem (see Teo et al (1991)), which involves only continuous decision variables. The transformed problem, however, involves variably located switching times. Numerical solution for such an optimal control problem is still difficult. Therefore, a switching time location method is introduced. This method is to further transform the problem into an optimal control problem with system parameters. The control function in the optimal control problem obtained at this stage is a piecewise constant function, of which the values are the lengths of the switching intervals. This optimal control problem can alternatively be treated as an optimal parameter selection problem by taking the lengths of the switching intervals as parameters (hence there is no control). The original problem and the transformed problems are proven to be equivalent. Gradient formulae for the function in the dynamics, and the cost and constraint functionals of the final optimal control problem are derived. On this basis, the problem is readily solvable by using existing optimal control software packages such as MISER3.2 Jennings et al (1991). Comments on the numerical details of the transformed problems are given. For illustration, a numerical example is included.

2 PROBLEM FORMULATION

Let $M > 1$ be an integer, and let $\Omega_1, \Omega_2, \ldots, \Omega_M$ denote M given dynamic systems over $[0,T]$. For $i = 1, 2, \ldots, M$, let the i-th system Ω_i be described by the following system of differential equations:

$$\dot{x}(t) = f_i(t, x(t)), \; 0 \le t \le T, \tag{2.1}$$

where $x = (x_1, x_2, \ldots, x_n)^T \in \mathbb{R}^n$ is the system state, and $f_i = (f_1^i, f_2^i, \ldots, f_n^i)^T$ is a continuously differentiable function from $\mathbb{R}^1 \times \mathbb{R}^n$ to $\mathbb{R}^n$. The M systems will be denoted collectively by

$$\Omega = \{\Omega_1, \Omega_2, \ldots, \Omega_M\}.$$

Let $K > 1$ be a given integer. Define the following sets $\mathcal{T}$, $\mathcal{S}$ and $\mathcal{V}$:

$$\begin{aligned} \mathcal{T} &= \{(t_1, t_2, \ldots, t_{K-1})^T \in R^{K-1} \mid 0 \le t_1 \le t_2 \le \ldots \le t_{K-1} \le T\} && (2.2)\\ \mathcal{S} &= \{(S_1, S_2, \ldots, S_K) \mid S_i \in \Omega, \; i = 1, 2, \ldots, K\} && (2.3)\\ \mathcal{V} &= \{(v_1, v_2, \ldots, v_K) \mid v_i \in \{1, 2, \ldots, M\}, \; i = 1, 2, \ldots, K\}. && (2.4) \end{aligned}$$

The systems in Ω are to be used to construct a new system, hence, each Ω_i is referred to as a subsystem. An element in $\mathcal{T}$ will be called a switching time sequence, and an element in $\mathcal{S}$ will be referred to as a switching subsystem sequence. The set $\mathcal{T} \times \mathcal{S}$ is called the set of feasible switching sequences. Since each subsystem Ω_i is determined by its subscript, $\mathcal{T} \times \mathcal{V}$ and $\mathcal{T} \times \mathcal{S}$ are used interchangeably, and elements in $\mathcal{T} \times \mathcal{V}$ are also called feasible switching sequences.

Corresponding to each feasible switching sequence $(\xi, v) \in \mathcal{T} \times \mathcal{V}$, where $\xi = (t_1, t_2, \ldots, t_{K-1}) \in \mathcal{T}$ and $v = (v_1, v_2, \ldots, v_K) \in \mathcal{V}$, we can construct a system over the time horizon $[0,T]$ as follows:

$$\dot{x}(t) = f_{v_i}(t, x(t)), \;\; t_{i-1} < t < t_i, \;\; 1 \le i \le K, \tag{2.5}$$

$$x(0) = x_0, \tag{2.6}$$

$$x(t_i) = x(t_i - 0), \;\; i = 1, 2, \ldots, K-1. \tag{2.7}$$

This system contains up to K different subsystems, and is to be understood as described in Section 1. Since, for $i \neq j$, we may have $v_i = v_j$, the actual number of different subsystems used to construct the overall system (2.5)-(2.7) can be less than K. The solution of system (2.5)-(2.7) corresponding to $(\xi, v) \in \mathcal{T} \times \mathcal{V}$ is denoted by $x(t|\xi, v)$.

Consider system (2.5)-(2.7). Then for given $(\xi, v) \in \mathcal{T} \times \mathcal{V}$, define the following cost functional:

$$J(\xi, v) = \sum_{i=1}^{K-1} \Phi_i(v_i, v_{i+1}) + \Phi_K(x(T|\xi, v)) + \sum_{i=1}^{K} \int_{t_{i-1}}^{t_i} g(x(t|\xi, v), t, v_i)dt, \quad (2.8)$$

where Φ_i, $i = 1, 2, \ldots, K-1$, represents the cost incurred by switching the system from Ω_{v_i} to $\Omega_{v_{i+1}}$, Φ_K represents the cost on the final state, and the i-th integral term represents the running cost of the subsystem Ω_i during the time interval (t_{i-1}, t_i).

Our optimal switching control problem can now be stated as:

Problem (P). Minimize $J(\xi, v)$ subject to system dynamics (2.5)-(2.7) and constraint $(\xi, v) \in \mathcal{T} \times \mathcal{V}$.

3 PROBLEM TRANSFORMATION

Consider Problem (P) formulated in the previous section. We note that the functions involved in both the dynamical system and the integrand of the cost functional are discontinuous. The problem also involves variably located switching times. All these make the problem very hard for numerical solution. In fact, no optimal control or optimization techniques are available in the literature for solving such problems. Our aim in this paper is to develop an efficient numerical method for this optimal control problem. By introducing a transformation, we transform Problem (P) into an equivalent optimal control problem. This equivalent optimal control problem still contains variably located switching times which cause numerical difficulties. Another transformation technique is then introduced to further transform the problem into a second equivalent optimal control problem solvable by standard optimal control techniques.

Define the following function:

$$h(t, x, \xi, \alpha_1, \alpha_2, \ldots, \alpha_M) = \sum_{i=1}^{M} \rho(t, \xi, \alpha_i) \cdot f_i(t, x) \quad (3.1)$$

where $\xi = (t_1, t_2, \ldots, t_{K-1})^T \in R^{K-1}$, $\alpha_i = (\alpha_i^1, \alpha_i^2, \ldots, \alpha_i^K)^T \in R^K$, $i = 1, 2, \ldots, M$, and

$$\rho(t, \xi, \alpha_i) = \sum_{j=1}^{K} \alpha_i^j \chi_{[t_{j-1}, t_j)}(t). \quad (3.2)$$

In order to understand how the function h is related to Problem (P), we consider the following function $q(\sigma)$:

$$q(\sigma) = \sum_{i=1}^{M} i^2\sigma_i - \left(\sum_{i=1}^{M} i\sigma_i\right)^2 \tag{3.3}$$

where $\sigma = (\sigma_1, \sigma_2, \ldots, \sigma_M)^T \in R^M$. We have the following result (see Wang et al (1996)).

Lemma 3.1 *Let ε_i be the i-th column of the $M \times M$ unit matrix in $R^{M\times M}$ and $\mathcal{B} = \{\varepsilon_1, \varepsilon_2, \ldots, \varepsilon_M\}$. For any $\sigma = (\sigma_1, \sigma_2, \ldots, \sigma_M)^T \in R^M$, $\sigma \in \mathcal{B}$ holds, if and only if*

$$\sigma_i \geq 0, \; i = 1, 2, \ldots, M, \tag{3.4}$$

$$\sum_{i=1}^{M} \sigma_i = 1, \tag{3.5}$$

$$q(\sigma) = 0. \tag{3.6}$$

The following theorem relates the function h to the Problem (P).

Theorem 3.1 *Let $\xi = (t_1, t_2, \ldots, t_{K-1})^T \in \mathcal{T}$. For*

$$\alpha_i = (\alpha_i^1, \alpha_i^2, \ldots, \alpha_i^K)^T \in R^K, \; i = 1, 2, \ldots, M,$$

let $\alpha^j = (\alpha_1^j, \alpha_2^j, \ldots, \alpha_M^j)^T$, and let the function $h(t, x, \xi, \alpha_1, \alpha_2, \ldots, \alpha_M)$ be defined by (3.1). Suppose that the following conditions are satisfied,

$$\alpha_i^j \geq 0, \; i = 1, 2, \ldots, M; \; j = 1, 2, \ldots, K, \tag{3.7}$$

$$\sum_{i=1}^{M} \alpha_i^j - 1 = 0, \quad j = 1, 2, \ldots, K, \tag{3.8}$$

$$q(\alpha^j) = 0, \; j = 1, 2, \ldots, K. \tag{3.9}$$

Then, there exists a sequence $v = (v_1, v_2, \ldots, v_K) \in \mathcal{V}$ such that the function h defined by (3.1) satisfies

$$h(t, x, \xi, \alpha_1, \alpha_2, \ldots, \alpha_M) = f_{v_i}(t, x), \; \text{for } t \in [t_{i-1}, t_i), \; i = 1, 2, \ldots, K. \tag{3.10}$$

On the other hand, for any $v = (v_1, v_2, \ldots, v_K) \in \mathcal{V}$, there exist vectors $\alpha_i \in R^K$, $i = 1, 2, \ldots, M$, satisfying (3.7)–(3.9) such that the function h given by (3.10) satisfies (3.1).

Proof: Suppose α_i, $i = 1, 2, \ldots, M$, satisfy (3.7)- (3.9), then it follows from Lemma 3.1 that $\alpha^j \in \mathcal{B}$, $j = 1, 2, \ldots, K$. That is, each $\alpha^j = (\alpha_1^j, \alpha_2^j, \ldots, \alpha_M^j)$ has one element, denoted by $\alpha_{v_j}^j$, equals to 1, while all the remaining elements are 0. Thus, from (3.1) and (3.2), we obtain

$$h(t, x, \xi, \alpha_1, \alpha_2, \ldots, \alpha_M)$$

$$= \sum_{i=1}^{M} \sum_{j=1}^{K} \left[\alpha_i^j f_i(t,x) \chi_{[t_{j-1},t_j]}(t) \right]$$
$$= \sum_{j=1}^{K} \sum_{i=1}^{M} \left[\alpha_i^j f_i(t,x) \chi_{[t_{j-1},t_j]}(t) \right]$$
$$= \sum_{j=1}^{K} \left[\chi_{[t_{j-1},t_j]}(t) f_{v_j}(t,x) \right]. \tag{3.11}$$

Equality (3.11) clearly implies (3.10), and the first part of the proof is complete.

Now we consider the second part of the corollary. For $v = (v_1, v_2, \ldots, v_K)$ in $\mathcal{V}$, define $\alpha_{v_j}^j = 1$, and $\alpha_i^j = 0$, if $i \neq v_j$, $j = 1, 2, \ldots, K.$, It is clear that α_i given by

$$\alpha_i = (\alpha_i^1, \alpha_i^2, \ldots, \alpha_i^K)^T \in R^K, \ i = 1, 2, \ldots, M$$

satisfy (3.7)–(3.9) and (3.10) holds. The proof is complete. □

Consider the following system dynamics:

$$\dot{x}(t) = h(t, x(t), \xi, \alpha_1, \alpha_2, \ldots, \alpha_M), \tag{3.12}$$
$$x(0) = x_0, \tag{3.13}$$

We denote by

$$x(t|\xi, \alpha_1, \alpha_2, \ldots, \alpha_M)$$

the solution of the above initial value problem, and let

$$\hat{J}(\xi, \alpha_1, \alpha_2, \ldots, \alpha_M)$$
$$= \sum_{j=1}^{K-1} \Phi_j\left(\sum_{i=1}^{M} i\alpha_i^j, \sum_{i=1}^{M} i\alpha_i^{j+1}\right) + \Phi_K(x(T|\xi, \alpha_1, \alpha_2, \ldots, \alpha_M))$$
$$+ \sum_{j=1}^{K} \int_{t_{j-1}}^{t_j} g(t, x(t|\xi, \alpha_1, \alpha_2, \ldots, \alpha_M), \sum_{i=1}^{M} i\alpha_i^j) dt. \tag{3.14}$$

Now we define the following problem:
Problem (P_1). Minimize $\hat{J}(\xi, \alpha_1, \alpha_2, \ldots, \alpha_M)$ subject to system dynamics (3.12)-(3.13), and constraints

$$\xi \in \mathcal{T}, \tag{3.15}$$
$$\alpha_i^j \geq 0, \ 1 \leq j \leq K, \ 1 \leq i \leq M, \tag{3.16}$$
$$\sum_{i=1}^{M} \alpha_i^j - 1 = 0, \ j = 1, 2, \ldots, K, \text{ and} \tag{3.17}$$
$$q(\alpha^j) = 0, \text{ for } \alpha^j = (\alpha_1^j, \alpha_2^j, \ldots, \alpha_M^j)^T, \ j = 1, 2, \ldots, K. \tag{3.18}$$

The relation between Problem (P) and Problem (P_1) are established in the following theorem.

Theorem 3.2 *Problem (P) and Problem (P_1) are equivalent in the sense that $(\xi^*, v^*) \in \mathcal{T} \times V$ is a solution of Problem (P) if and only if $(\xi^*, \alpha_1^*, \alpha_2^*, \ldots, \alpha_M^*)$ is a solution of Problem (P_1), where $v^* = (v_1^*, v_2^*, \ldots, v_K^*)$ and*

$$\alpha_i^* = (\alpha_i^{*1}, \alpha_i^{*2}, \ldots, \alpha_i^{*K})^T, \ i = 1, 2, \ldots, M \tag{3.19}$$

are related by

$$v_j^* = \sum_{i=1}^{M} \alpha_i^{*j} \cdot i, \ j = 1, 2, \ldots, K. \tag{3.20}$$

Furthermore,

$$J(\xi^*, v^*) = \hat{J}(\xi^*, \alpha_1^*, \alpha_2^*, \ldots, \alpha_M^*). \tag{3.21}$$

Proof: Let $(\xi, v) \in \mathcal{T} \times \mathcal{V}$ be any feasible switching sequence for Problem (P), where $\xi = (t_1, t_2, \ldots, t_{K-1})$, and $v = (v_1, v_2, \ldots, v_K)$. For each $1 \leq j \leq K$, define

$$\alpha_i^j = \begin{cases} 0, & \text{if } 1 \leq i \leq M \ \text{ and } i \neq v_j \\ 1, & \text{if } i = v_j. \end{cases} \tag{3.22}$$

Then, $(\xi, \alpha_1, \alpha_2, \ldots, \alpha_M)$ satisfies (3.15)-(3.18) and

$$v_j = \sum_{i=1}^{M} \alpha_i^j \cdot i, \ j = 1, 2, \ldots, K. \tag{3.23}$$

It is easy to see that

$$x(t|\xi, v) = x(t|\xi, \alpha_1, \alpha_2, \ldots, \alpha_M), \ 0 \leq t \leq T, \tag{3.24}$$

where $x(t|\xi, v)$ and $x(t|\xi, \alpha_1, \alpha_2, \ldots, \alpha_M)$ are, respectively, the states of systems (2.5)-(2.7) and (3.12)-(3.13). From (3.23), (3.24), and the definitions of the functions $J(\xi, v)$ and $\hat{J}(\xi, \alpha_1, \alpha_2, \ldots, \alpha_M)$, it follows that

$$J(\xi, v) = \hat{J}(\xi, \alpha_1, \alpha_2, \ldots, \alpha_M). \tag{3.25}$$

On the other hand, for any $(\xi, \alpha_1, \alpha_2, \ldots, \alpha_M)$ satisfying (3.15)-(3.18), where $\alpha_i = (\alpha_i^1, \alpha_i^2, \ldots, \alpha_i^K)$, $i = 1, 2, \ldots, M$, let $v = (v_1, v_2, \ldots, v_K)$ be given by (3.20). From Lemma 3.23, $\alpha^i = (\alpha_1^i, \alpha_2^i, \ldots, \alpha_M^i) \in \mathcal{B}$ and it is easily verified that $(\xi, v) \in \mathcal{T} \times \mathcal{V}$ and $(\xi, \ v)$ satisfies (3.25).

If $(\xi^*, v^*) \in \mathcal{T} \times \mathcal{V}$ is a solution of Problem (P), then $(\xi^*, \alpha_1^*, , \ldots, \alpha_M^*)$ with the α_i^*'s obtained from (3.22) satisfies (3.15)-(3.20). Moreover, $(\xi^*, \alpha_1^*, \ldots, \alpha_M^*)$ is a solution of Problem (P_1). In fact, if $(\xi^*, \alpha_1^*, \alpha_2^*, \ldots, \alpha_M^*)$ is not a solution of Problem (P_1), then there exists $(\hat{\xi}, \hat{\alpha}_1, \hat{\alpha}_2, \ldots, \hat{\alpha}_M)$ satisfying (3.15)-(3.18) such that

$$J(\hat{\xi}, \hat{\alpha}_1, \hat{\alpha}_2, \ldots, \hat{\alpha}_M) < \hat{J}(\xi^*, \alpha_1^*, \alpha_2^*, \ldots, \alpha_M^*). \tag{3.26}$$

Let $\hat{v}$ be obtained from $\hat{\alpha}_1, \hat{\alpha}_2, \ldots, \hat{\alpha}_M$ according to (3.20). Then $(\hat{\xi}, \hat{v}) \in \mathcal{T} \times \mathcal{V}$ and

$$J(\hat{\xi}, \hat{v}) = J(\hat{\xi}, \hat{\alpha}_1, \hat{\alpha}_2, \ldots, \hat{\alpha}_M) < \hat{J}(\xi^*, \alpha_1^*, \alpha_2^*, \ldots, \alpha_M^*) = J(\xi^*, v^*)$$

which contradicts the fact that (ξ^*, v^*) is a solution of Problem (P). Similarly, we can show that if $(\xi^*, \alpha_1^*, \alpha_2^*, \dots, \alpha_M^*)$ is a solution of Problem (P_1), then (ξ^*, v^*) obtained from (3.20) is a solution of Problem (P).

Finally, the equality (3.21) is obtained directly from (3.25). □

In view of Theorem 3.2, it is clear that, in order to solve Problem (P), we need only to consider Problem (P_1) which is an optimal parameter selection problem involving only continuous decision parameters. However, the corresponding system (3.12)-(3.13) contains switching times as decision parameters. This causes numerical difficulties in actual computations. In the following, a time transformation is introduced to further transform Problem (P_1) into a second equivalent problem solvable by existing optimal control software packages, such as MISER3.2.

For $\xi = (t_1, t_2, \dots, t_{K-1})^T \in \mathcal{T}$, let

$$\tau_j = t_j - t_{j-1}, \quad j = 1, 2, \dots, K, \tag{3.27}$$

and let $\tau = (\tau_1, \tau_2, \dots, \tau_K)^T$, where $t_0 = 0$, and $t_K = T$. It is clear that $\xi \in \mathcal{T}$ if and only if τ satisfies

$$\tau_j \geq 0, \; j = 1, 2, \dots, K \;\text{ and }\; \sum_{j=1}^{K} \tau_j - T = 0. \tag{3.28}$$

From (3.27), we see that τ is uniquely determined by ξ, and vice versa.

For any $\tau = (\tau_1, \tau_2, \dots, \tau_K)^T$, define a function $u(\cdot|\tau) : [0, K] \to R$ as follows:

$$u(s|\tau) = \sum_{j=1}^{K} \tau_j \chi_{[j-1,j)}(s), \tag{3.29}$$

where $\chi_{[j-1,j)}(s)$ is the characteristic function of the interval $[j-1, j)$ on $[0, K)$. Let

$$\mathcal{U} = \{u(\cdot|\tau) : [0, K) \to \mathbb{R}^1 | \tau = (\tau_1, \tau_2, \dots, \tau_K)^T \in R^K \text{ satisfies (3.28)}\}. \tag{3.30}$$

Define a mapping $U : \mathcal{T} \to \mathcal{U}$ by

$$U : \xi \longmapsto u(\cdot|\tau(\xi)) \tag{3.31}$$

where $\tau(\xi)$ is determined by (3.27). It is easily seen that U is invertible.

For $u \in \mathcal{U}$, let $t(\cdot|u) : [0, K) \to [0, T)$ be the solution of the following differential equation

$$\frac{dt(s|u)}{ds} = u(s), \quad 0 \leq s < K, \tag{3.32}$$

$$t(0|u) = 0. \tag{3.33}$$

Under the transformation $t = t(s|u)$ defined by (3.32)-(3.33), we have

$$\rho(t(s|U(\xi)), \xi, \alpha_i) = \sum_{j=1}^{K} \alpha_i^j \chi_{[j-1,j)}(s), \; 0 \leq s < K. \tag{3.34}$$

We note that the right hand side of (3.34) does not depend on ξ. Let

$$\eta(s,\alpha_i) = \sum_{j=1}^{K} \alpha_i^j \chi_{[j-1,j)}(s), \ 0 \le s < K, \tag{3.35}$$

then we have

$$\rho(t(s|U(\xi)),\xi,\alpha_i) = \eta(s,\alpha_i). \tag{3.36}$$

Theorem 3.3 *Let $u = U(\xi)$, and let*

$$x(t|\xi,\alpha_1,\alpha_2,\ldots,\alpha_M)$$

be the solution of system (3.12)-(3.13). Then, the function

$$y(s|u,\alpha_1,\alpha_2,\ldots,\alpha_M)$$

with components $y_1, y_2, \ldots, y_{n+1}$ defined by

$$y_i(s|u,\alpha_1,\alpha_2,\ldots,\alpha_M) = x_i(t(s|u)|\xi,\alpha_1,\alpha_2,\ldots,\alpha_M), \ 1 \le i \le n, \tag{3.37}$$
$$y_{n+1}(s|u,\alpha_1,\alpha_2,\ldots,\alpha_M) = t(s|u) \tag{3.38}$$

is the solution of the system

$$\dot{y}(s) = \phi(s,u(s),y(s),\alpha_1,\alpha_2,\ldots,\alpha_M), \tag{3.39}$$
$$y(0) = (x_0^T,0)^T, \tag{3.40}$$

where

$$\phi(s,u,y(s),\alpha_1,\alpha_2,\ldots,\alpha_M) = \begin{bmatrix} u\sum_{i=1}^{M} \eta(s,\alpha_i)p_i(y(s)) \\ u \end{bmatrix} \tag{3.41}$$

and

$$p_i(y(s)) = f_i(y_{n+1}(s),(y_1(s),y_2(s),\ldots,y_n(s))^T), \ i = 1,2,\ldots,M. \tag{3.42}$$

Proof: From (3.1), (3.36) and (3.42), we have

$$\begin{aligned} & h(t(s|u),x(t(s|u)|\xi,\alpha_1,\alpha_2,\ldots,\alpha_M)),\alpha_1,\alpha_2,\ldots,\alpha_M) \\ &= \sum_{i=1}^{M} [\rho(t(s|u),\xi,\alpha_i)\cdot f_i(t(s|u),x(t(s|u)|\xi,\alpha_1,\alpha_2,\ldots,\alpha_M))] \\ &= \sum_{i=1}^{M} [\eta(s,\alpha_i)\cdot f_i(t(s|u),x(t(s|u)|\xi,\alpha_1,\alpha_2,\ldots,\alpha_M))] \\ &= \sum_{i=1}^{M} [\eta(s,\alpha_i)\cdot p_i(y(s|u,\alpha_1,\alpha_2,\ldots,\alpha_M)]. \end{aligned} \tag{3.43}$$

Let $y = (y_1, y_2, \ldots, y_{n+1})^T$ be given by (3.37)-(3.38). Since $x(t|\xi, \alpha_1, \ldots, \alpha_M)$ is the solution of (3.12) and (3.13), it follows from (3.32), (3.43) and (3.41) that

$$
\begin{aligned}
&\frac{d}{ds} y(s|u, \alpha_1, \alpha_2, \ldots, \alpha_M) \\
&= \frac{d}{ds} \begin{bmatrix} x(t(s|u)|\xi, \alpha_1, \alpha_2, \ldots, \alpha_M) \\ t(s|u) \end{bmatrix} \\
&= \begin{bmatrix} u(s) h(t(s|u), x(t(s|u)|\xi, \alpha_1, \alpha_2, \ldots, \alpha_M), \xi, \alpha_1, \alpha_2, \ldots, \alpha_M) \\ u(s) \end{bmatrix} \\
&= \begin{bmatrix} u(s) \sum_{i=1}^{M} [\eta(s, \alpha_i) \cdot p_i(y(s|u, \alpha_1, \alpha_2, \ldots, \alpha_M)] \\ u(s) \end{bmatrix} \\
&= \phi(s, u(s), y(s), \alpha_1, \alpha_2, \ldots, \alpha_M).
\end{aligned}
$$

The proof is complete. □

Define a set $\mathcal{H}$ as follows

$$\mathcal{H} = \{(\alpha_1, \alpha_2, \ldots, \alpha_M) | \alpha_j \in R^K, \ 1 \leq j \leq M, \text{ satisfy (3.16)-(3.18)}\}. \tag{3.44}$$

For $u \in \mathcal{U}$ and $(\alpha_1, \alpha_2, \ldots, \alpha_M) \in \mathcal{H}$, let

$$v_j(\alpha_1, \alpha_2, \ldots, \alpha_M) = \sum_{i=1}^{M} i\alpha_i^j, \ j = 1, 2, \ldots, K, \tag{3.45}$$

$$w(s, \alpha_1, \alpha_2, \ldots, \alpha_M) = \sum_{i=1}^{M} \sum_{j=1}^{K} i\alpha_i^j \chi_{[j-1,j)}(s), \tag{3.46}$$

$$\bar{\Phi}_K(y) = \Phi_K((y_1, y_2, \ldots, y_n)^T), \tag{3.47}$$

$$\bar{g}(s, y, \alpha_1, \ldots, \alpha_M) = g(y_{n+1}, (y_1, \ldots, y_n)^T, w(s, \alpha_1, \ldots, \alpha_M)), \tag{3.48}$$

and

$$g_0(s, u, y, \alpha_1, \alpha_2, \ldots, \alpha_M) = u\bar{g}(s, y, \alpha_1, \alpha_2, \ldots, \alpha_M), \tag{3.49}$$

where $\alpha^j = (\alpha_1^j, \alpha_2^j, \ldots, \alpha_M^j)^T$, $j = 1, 2, \ldots, K$. Now define

$$
\begin{aligned}
&\Psi(y, \alpha_1, \alpha_2, \ldots, \alpha_M) \\
&= \sum_{j=1}^{K-1} \Phi_j(v_j(\alpha_1, \alpha_2, \ldots, \alpha_M), v_{j+1}(\alpha_1, \alpha_2, \ldots, \alpha_M)) + \bar{\Phi}_K(y),
\end{aligned} \tag{3.50}
$$

and

$$
\begin{aligned}
&\bar{J}(u, \alpha_1, \alpha_2, \ldots, \alpha_M) \\
&= \Psi(y(K|u, \alpha_1, \alpha_2, \ldots, \alpha_M), \alpha_1, \alpha_2, \ldots, \alpha_M) \\
&+ \int_0^K g_0(s, u(s), y(s|u, \alpha_1, \alpha_2, \ldots, \alpha_M), \alpha_1, \alpha_2, \ldots, \alpha_M) ds.
\end{aligned} \tag{3.51}
$$

We introduce the following optimal control problem:
Problem (P_2): Minimize $\bar{J}(u, \alpha_1, \alpha_2, \ldots, \alpha_M)$, subject to system dynamics (3.39)- (3.40) and constraints $u \in \mathcal{U}$ and $(\alpha_1, \alpha_2, \ldots, \alpha_M) \in \mathcal{H}$.

We will use the term *knot set* (Teo et al (1991) and Jennings et al (1991)) to denote the set of switching points of a piecewise constant control. Note that the knot set for controls in $\mathcal{U}$ is $\{0, 1, \ldots, K\}$. By the definition of $\mathcal{U}$, $u \in \mathcal{U}$ if and only if (3.29) and (3.28) hold. ¿From (3.28)-(3.33) and (3.38), we obtain the following lemma, where T is the final time of the system in Problem P_2.

Lemma 3.2 *Under the assumption that u is a piecewise constant function with knot set $\{0, 1, \ldots, K\}$, the constraint $u \in \mathcal{U}$ is equivalent to*

$$u(s) \geq 0 \tag{3.52}$$

and

$$y_{n+1}(K) - T = 0, \tag{3.53}$$

where T is the final time of the system in Problem P_2.

Theorem 3.4 *Problem (P_1) and Problem (P_2) are equivalent in the sense that*

$$(\xi^*, \alpha_1^*, \alpha_2^*, \ldots, \alpha_M^*)$$

is a solution of Problem (P_1) if and only if

$$(u^*, \alpha_1^*, \alpha_2^*, \ldots, \alpha_M^*)$$

is a solution of Problem (P_2), where $u^* = U(\xi^*)$. Furthermore,

$$\hat{J}(\xi^*, \alpha_1^*, \alpha_2^*, \ldots, \alpha_M^*) = \bar{J}(u^*, \alpha_1^*, \alpha_2^*, \ldots, \alpha_M^*). \tag{3.54}$$

Proof: From previous discussions, it is clear that $(\xi^*, \alpha_1^*, \alpha_2^*, \ldots, \alpha_M^*)$ (not necessarily optimal here) satisfies (3.15)-(3.18) if and only if $U(\xi^*) \in \mathcal{U}$, and $(\alpha_1^*, \alpha_2^*, \ldots, \alpha_M^*) \in \mathcal{H}$. Furthermore, from (3.14), (3.51) and Theorem 3.3, we can easily verify that

$$\hat{J}(\xi^*, \alpha_1^*, \alpha_2^*, \ldots, \alpha_M^*) = \bar{J}(u^*, \alpha_1^*, \alpha_2^*, \ldots, \alpha_M^*). \tag{3.55}$$

The results of the theorem follow readily. □

Theorem 3.5 *Problem (P) and Problem (P_2) are equivalent in the sense that (ξ^*, v^*) is a solution of Problem (P) if and only if $(u^*, \alpha_1^*, \alpha_2^*, \ldots, \alpha_M^*)$ is a solution of Problem (P_2), where (ξ^*, v^*) and $(u^*, \alpha_1^*, \alpha_2^*, \ldots, \alpha_M^*)$ are related through (3.20), (3.27) and (3.29). Furthermore,*

$$J(\xi^*, v^*) = \bar{J}(u^*, \alpha_1^*, \alpha_2^*, \ldots, \alpha_M^*). \tag{3.56}$$

Proof: The theorem is a direct consequence of Theorem 3.2 and Theorem 3.4. □

From Theorem 3.5, in order to solve Problem (P), we need only to solve Problem (P_2). Through (3.20), (3.27) and (3.29), the solution for Problem (P) can then be obtained from the solution of Problem (P_2). ¿From Lemma 3.2, we see that Problem (P_2) is in a form solvable by the optimal control software package MISER3.2, Jennings et al (1991), with the knot set taken as $\{0, 1, \ldots, K\}$. However, since the feasible set for the parameter vectors is a finite (hence discrete) set, directly solving the problem may cause numerical difficulties. Now we define

$$G(\alpha_1, \alpha_2, \ldots, \alpha_M) = \sum_{j=1}^{K} q(\alpha^j). \tag{3.57}$$

Note that, assuming (3.16) and (3.17), the constraint (3.18) can be replaced by

$$G(\alpha_1, \alpha_2, \ldots, \alpha_M) = 0 \tag{3.58}$$

and the function G is nonnegative. To make the feasible region of the system parameters connected, we use penalty function method, taking away the constraints (3.18) from Problem (P_2) and pasting the function G to the cost functional. Based on the above discussion, we propose the following algorithm for solving Problem (P_2).

Algorithm.
Step 1. Choose an initial penalty constant $\lambda_1 > 0$, an increment constant $d > 0$ and an accuracy number $\varepsilon > 0$.
Step 2. Find an initial guess $(u^0, \alpha_1^0, \alpha_2^0, \ldots, \alpha_M^0)$ satisfying (3.16)-(3.17). Set i=1.
Step 3. Use a suitable optimal control algorithm (e.g MISER3.2) to find a solution $(u^i, \alpha_1^i, \alpha_2^i, \ldots, \alpha_M^i)$ to the problem

$$\begin{aligned}
&\min \ \bar{J}(u, \alpha_1, \alpha_2, \ldots, \alpha_M) + \lambda_i G(\alpha_1, \alpha_2, \ldots, \alpha_M) \\
&\textit{subject to system dynamics (3.39)-(3.40) and constraints} \\
&\alpha_i^j \geq 0, \ 1 \leq j \leq K, \ 1 \leq i \leq M; \ \sum_{i=1}^{M} \alpha_i^j - 1 = 0, \ j = 1, 2, \ldots, K; \\
&u(s) \geq 0, \ 0 \leq s \leq K \text{ and } y_{n+1}(K) - T = 0,
\end{aligned}$$

In solving the above problem, the knot set is always taken to be $\{0, 1, \ldots, K\}$ and the initial guess is $(u^{i-1}, \alpha_1^{i-1}, \alpha_2^{i-1}, \ldots \alpha_M^{i-1})$.
Step 4. If $G(\alpha_1^i, \alpha_2^i, \ldots, \alpha_M^i) \leq \varepsilon$, stop. Otherwise, set $i = i + 1$ and $\lambda_i = \lambda_{i-1} + d$, goto Step 3.

4 GRADIENT FORMULAE

To solve Problem (P_2), we need the gradient formulae for the functions Ψ, g_0, G, and ϕ as appeared in the algorithm given in the previous section. We summarize the results in the following as a theorem.

Theorem 4.1 *The gradients for the functions ϕ, G, $k = 1, 2, \ldots, K$, g_0, and Ψ are as follows.*

$$\frac{\partial \phi}{\partial y}(s, u, y, \alpha_1, \alpha_2, \ldots, \alpha_M) = \begin{bmatrix} u \cdot \sum_{i=1}^{M} \eta(s, \alpha_i) \frac{\partial p_i(y)}{\partial y} \\ \theta \end{bmatrix}, \tag{4.1}$$

$$\frac{\partial \phi}{\partial u}(s, u, y, \alpha_1, \alpha_2, \ldots, \alpha_M) = \begin{bmatrix} \sum_{i=1}^{M} \eta(s, \alpha_i) p_i(y) \\ 1 \end{bmatrix}, \tag{4.2}$$

$$\frac{\partial \phi}{\partial \alpha_i^j}(s, u, y, \alpha_1, \alpha_2, \ldots, \alpha_M) = \begin{bmatrix} u \chi_{[j-1,j)}(s) p_i(y(s)) \\ 0 \end{bmatrix},$$
$$i = 1, 2, \ldots, M; \; j = 1, 2, \ldots, K, \tag{4.3}$$

$$\frac{\partial G}{\partial \alpha_i^j}(\alpha_1, \alpha_2, \ldots, \alpha_M) = i^2 - 2i \left(\sum_{l=1}^{M} l \alpha_l^j \right),$$
$$1 \le i \le M, \; 1 \le j \le K, \tag{4.4}$$

$$\frac{\partial g_0}{\partial y}(s, y, u, \alpha_1, \alpha_2, \ldots, \alpha_M) = u \cdot \frac{\partial \bar{g}}{\partial y}(s, y, \alpha_1, \alpha_2, \ldots, \alpha_M), \tag{4.5}$$

$$\frac{\partial g_0}{\partial u}(s, y, u, \alpha_1, \alpha_2, \ldots, \alpha_M) = \bar{g}(s, y, \alpha_1, \alpha_2, \ldots, \alpha_M), \tag{4.6}$$

$$\frac{\partial g_0}{\partial \alpha_i^j}(s, y, u, \alpha_1, \alpha_2, \ldots, \alpha_M) = u \frac{\partial g}{\partial w} \cdot i \chi_{[j-1,j)}(s), \tag{4.7}$$

$$\frac{\partial \Psi}{\partial y} = \frac{\partial \Phi_K(y_1, y_2, \ldots, y_n)^T}{\partial y}, \tag{4.8}$$

$$\frac{\partial \Psi}{\partial \alpha_i^1} = \frac{\partial \Phi_1}{\partial v_1} i, \; i = 1, 2, \ldots, M, \tag{4.9}$$

$$\frac{\partial \Psi}{\partial \alpha_i^j} = \left(\frac{\partial \Phi_{j-1}}{\partial v_j} + \frac{\partial \Phi_j}{\partial v_j} \right) \cdot i, \; j = 2, 3, \ldots, K, \; i = 1, 2, \ldots, M, \tag{4.10}$$

where, in (4.1), θ is the $(n+1)$-dimensional zero row vector.

Proof: All the gradient formulae in the theorem can be obtained by using the chain rule and by following the definitions of the corresponding functions. We omit the details. □

We note that the integrand of the constraint (3.53) is a linear function of u, its gradient is obvious and hence is not included in Theorem 4.1.

5 EXAMPLE

In this section, we use the transformation method given in Section 3 to solve a numerical problem.

Example 5.1 *Our numerical example is specified as follows:*
The number of candidate subsystems is: $M = 3$,
The number of subsystems to be chosen is: $K = 2$,
The 3 candidate subsystems are given by specifying their right hand side:

$$f_1(t,x) = 1, \ 0 \le t \le 2,$$
$$f_2(t,x) = -1, \ 0 \le t \le 2,$$
$$f_3(t,x) = 2t, \ 0 \le t \le 2,$$

The initial condition for the overall system is: $x(0) = 0$,
The cost functional is

$$J(\xi, v) = x(2|\xi, v)^2 + \int_0^2 \left(sin(\frac{\pi}{2}t) - x(t|\xi, v)\right)^2 dt,$$

where $\xi = t_1 \in [0,2]$ *and* $v = (v_1, v_2) \in \{1,2,3\} \times \{1,2,3\}$.

The problem may now be stated as:

$$\min \quad J(\xi, v) \tag{5.1}$$

subject to

$$\dot{x}(t) = \begin{cases} f_{v_1}(t,x), & 0 < t \le t_1, \\ f_{v_2}(t,x), & t_1 \le t < 2, \end{cases}$$
$$x(0) = 0,$$

and

$$0 \le t_1 \le 2, \ (v_1, v_2) \in \{1,2,3\} \times \{1,2,3\}. \tag{5.2}$$

In view of Section 3, the problem can be transformed into an optimal control problem with system parameters. Let $u(s) = \tau_1$, for $0 \le s < 1$, and $= \tau_2$, for $1 \le s \le 2$, where $\tau_1 = t_1$, $\tau_2 = 2 - t_1$, and t_1 is the variable switching time. By the transformations and discussions given in Section 3, it is routine to show that the above problem is equivalent to the following problem.

$$\min \quad \bar{J}(u, z) \tag{5.3}$$

subject to system dynamics

$$\dot{y} = \phi(s, u, y, z), \ y(0) = (0,0)^T, \tag{5.4}$$

and constraints

$$0 \le u(s) \le 2, \ \ 0 \le s \le 2,$$
$$0 \le z_i \le 1, \ \ i = 1, 2, \ldots, 6,$$
$$y_2(2) - 2 = 0,$$
$$z_1 + z_2 + z_3 - 1 = 0,$$
$$z_4 + z_5 + z_6 - 1 = 0,$$
$$q_1(z) \equiv z_1 + 4z_2 + 9z_3 - (z_1 + 2z_2 + 3z_3)^2 = 0,$$
$$q_2(z) \equiv z_4 + 4z_5 + 9z_6 - (z_4 + 2z_5 + 3z_6)^2 = 0,$$

where s is the scaled time, u is the control, $z = (z_1, z_2, \dots, z_6)^T$ is the parameter vector consisting of all the α_i^j's, $y = (y_1, y_2)^T$ is the state, and the functions $\bar{J}$ and $\phi = (\phi_1, \phi_2)^T$ are given as follows:

$$\bar{J}(u, y, z) = [y_1(2)]^2 + \int_0^2 u(s) \left[sin(\frac{\pi}{2} y_2(s)) - y_1(s) \right]^2 ds,$$

$$\phi_1(s, u, y, z) = \begin{cases} u(z_1 - z_2 + 2z_3 y_2), & 0 \le s < 1, \\ u(z_4 - z_5 + 2z_6 y_2), & 1 \le s < 2, \end{cases}$$

$$\phi_2(s, u, y, z) = u.$$

This transformed problem is solved using the algorithm of this paper. In Step 3 of the algorithm, the optimal control software MISER3.2 is used to solve the problem obtained by the penalty function method. The computed results are listed in the following table.

TABLE 1. COMPUTED RESULTS FOR EXAMPLE 5.1

	u	z	penalized constraints	objective $\bar{J}$
initial values	$\tau_1 = 0.5$ $\tau_2 = 1.5$	0 1 0 0 0 1	$q_1 = 0$ $q_2 = 0$	4.11177
computed optimal values	$\tau_1 = 1.03461$ $\tau_2 = 0.96539$	1.0 0.0 0.0 0.0 1.0 0.0	$q_1 = 0.0$ $q_2 = 0.0$	0.03611

The optimal solution to the original problem is easily obtained from the computed solution of the transformed problem. The optimal switching time is $t_1^* = 1.03461$, the optimal system is $\dot{x} = F(t, x)$, $0 \le t \le 2$, where $F(t, x) = f_1$ for $0 \le t < t_1^*$, and $= f_2$ for $t_1^* \le t \le 2$.

The computed optimal state x of the original problem is plotted against the original time in Figure 5.1. A better understanding of the original problem can be made from this figure. From the original problem, it is easy to see that, once the switching time t_1 is given, there are only 9 possibilities for the choice of the approximating curve. Based on this observation, it is clear that the computed result is correct.

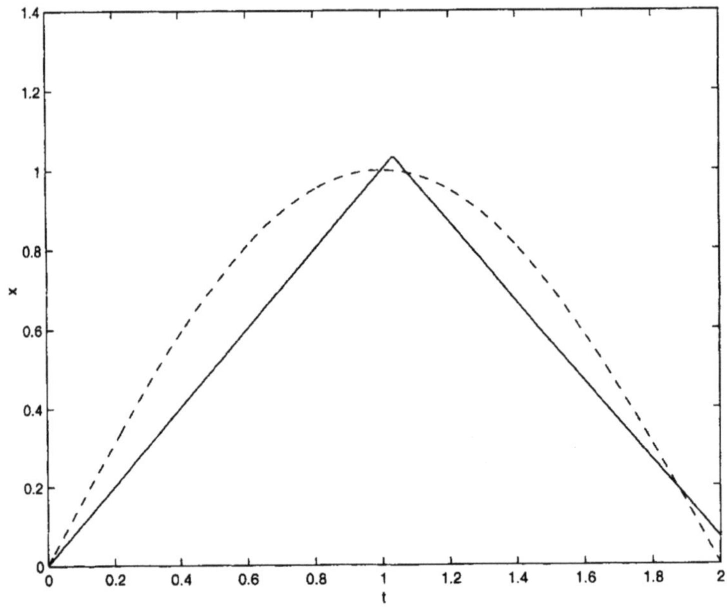

Figure 5.1 Computed Trajectory of the State y_1

6 CONCLUSIONS

In this paper, we considered a class of optimal switching control problems which is mixed continuous-discrete in nature. We note that the overall system to be constructed has sudden changes in the dynamics (at the switching times). Hence, the optimal control problem involves system discontinuities. To approach the problem, a set of scalar parameters (*i.e.* the α_i^j's) are introduced to transform the discrete 'variables' into continuous variables. By doing so, the sacrifice is the increase of the number of decision variables and the introduction of new constraints.

To deal with the switching times, a time scaling and a control function are introduced to eliminate the effect of the discontinuities in the system dynamics. The control function is a piecewise constant function with knot set corresponding to the switching times of the original problem, and the system discontinuities of the problem coincide with the knots of the control function. Thus, the problem is solvable by the optimal control software MISER3.2.

On this basis, a numerical algorithm is developed for the solution of the transformed problem. A numerical example is solve using the proposed algorithm. Computation results show that algorithm proposed in this paper is efficient.

Acknowledgments

This research was partially supported by a research grant from the Australian Research Council.

References

Evans, L. C. and Friedman, A. (1979), Optimal stochastic switching and the Dirichlet problem for the Bellman equation, *Trans. Amer. Math. Soc.*, Vol. 253, pp. 365-389.

Jennings, L. S., Fisher, M.E., Teo, K.L. and Goh, C.J. (1991), MISER 3 Solving optimal control problems–an update, *Advances in Engineering Software and Workstations*, Vol. 13(4), pp. 190-196.

Jennings, L. S. and Teo, K. L. (1990), A computational algorithm for functional inequality constrained optimization problems, *Automatica*, Vol. 26, pp. 371–375.

Li, X. and Yong, J. (1995), *Optimal Control Theory for Infinite Dimensional Systems*, Birkhäuser, Boston.

Teo, K.L., Goh, C.J. and Wong K.H. (1991), *A Unified Computational Approach to Optimal Control Problems*, Longman Scientific and Technical, London.

Teo, K. L. and Jennings, L.S. (1989), Nonlinear optimal control problems with continuous state inequality constraints, *Journal of Optimization Theory and Applications*, Vol. 63, No. 2, pp. 1-12.

Teo, K. L., Rehbock, V. and Jennings, L. S. (1993), A new computational algorithm for functional inequality constrained optimization problems, *Automatica*, Vol. 29(3), pp. 789–792.

Wang, S. Teo, K. L. and Lee, J. (1996), A new approach to nonlinear mixed programming problems, Preprint.

Yong, J. (1989), Systems governed by ordinary differential equations with continuous, switching and impulse controls, *Appl. Math. Optim.*, Vol. 20, pp. 223-235.

Yong, J. (1991), Existence of the value for a differential game with switching strategies in a Banach space, *System Sci. & Math. Sci.*, Vol. 4, pp. 321-340.

16 OPTIMIZATION BY WAY OF THE TRAJECTORY FOLLOWING METHOD

Thomas L. Vincent

Department of Aerospace and Mechanical Engineering,
University of Arizona,
Tucson, AZ 85721, USA

Abstract: The trajectory following method is a relatively old concept for solving the nonlinear programming problem. It is based on the idea of solving continuous differential equations whose equilibrium solutions satisfy the necessary conditions for a minimum. The method is "trajectory following" in the sense that an initial point is moved along a trajectory generated by the differential equations to the solution point. The fact that this method has not been popular is understandable. It may require many function evaluations for a solution and it may suffer from the same slow convergence difficulties that the "gradient method" is known to have. However with the advent of fast computers and efficient integration solvers, these difficulties may no longer exist for many problems of practical interest. Here the focus is on ease of use rather than number of function evaluations as a performance criteria. It is shown that the trajectory following method, used in conjunction with a good "stiff" integration routine, can provide a tool that is easy to use for solving a wide variety of optimization problems including linear programming, quadratic programming and global optimization problems.

Key words: optimization, trajectory following, gradient flow.

X. Yang et al. (eds.), Progress in Optimization, 239–254.

1 INTRODUCTION

The trajectory following method (Arrow and Hurwicz (1977), Goh (1978), Kelly (1962)) is based on the idea of solving continuous differential equations whose equilibrium solutions satisfy the necessary conditions for a minimum. We have recently shown that one of the major objections to this method (possible slow convergence rate) is overcome with the advent of fast computers and efficient integration solvers which can handle "stiff" systems (Vincent and Grantham (1981), Vincent and Teo (1999), Zang and Vincent (1999)). This method is still likely to require many more function evaluations than other methods, however, if one considers its ease of use and applicability to a wide range of optimization problems as commensurate performance criteria, the trajectory following method deserves a second look.

The original idea behind a **trajectory following** method is to form a set of differential equations from the gradient of the cost function. Quite naturally, it is also referred to as a **gradient flow** method Yan et al (1994). However, as will be seen below, use of the trajectory following method need not be restricted to using only the cost function gradient.

1.1 Minimizing $G(\mathbf{u})$ subject to no constraints

Consider the problem of minimizing the scalar function $G(\mathbf{u})$ of the n_u component control variables $\mathbf{u}$ subject to no constraints. The function G and its derivative with respect to $\mathbf{u}$ are assumed to be continuous. The unconstrained trajectory following algorithm is given by

$$\dot{\mathbf{u}} = \begin{cases} 0 & \text{if norm}\left(\frac{\partial G}{\partial \mathbf{u}}\right) < \varepsilon \\ -\left[\frac{\partial G}{\partial \mathbf{u}}\right]^{\mathsf{T}} & \text{otherwise} \end{cases}. \tag{1.1}$$

In general there may be many equilibrium points associated with this set of differential equations. If the initial value of $\mathbf{u}$ is chosen to be any point other than an equilibrium point, then motion using (1.1) will always be toward a local minimal point. The reason for this is that

$$\dot{G} = -\frac{\partial G}{\partial \mathbf{u}}\left[\frac{\partial G}{\partial \mathbf{u}}\right]^{\mathsf{T}} < 0$$

along any such trajectory which makes maximal points and saddle points unstable. It follows that if G has a unique proper minimum at $\mathbf{u}^*$, then every trajectory from all points in the control space (other than $\mathbf{u}^*$) must asymptotically approach $\mathbf{u}^*$.

Since a minimal solution is approached asymptotically, some accuracy criteria must be specified to stop the integration process. Here integration is stopped when the norm of the gradient is less than some specific value. Unless stated otherwise, the gradient is assumed to be evaluated numerically using the central difference method.

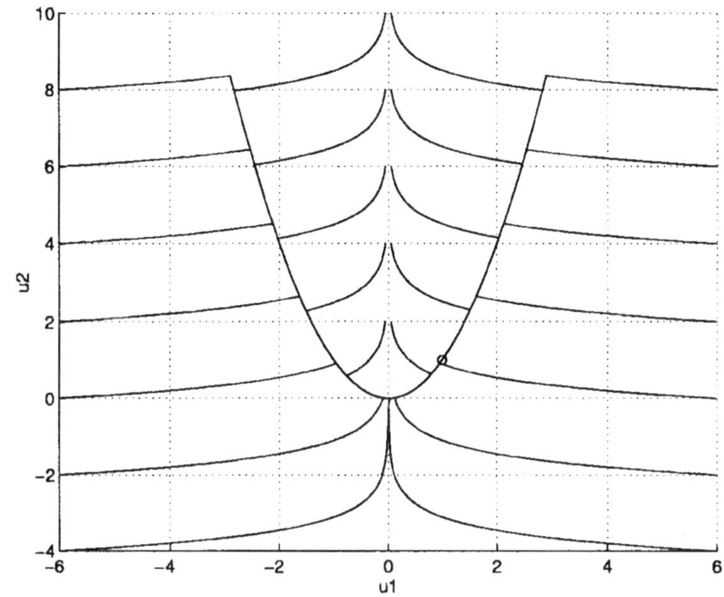

Figure 1.1 Using trajectory following to solve the banana problem.

Rosenbrock's banana function. For some problems such as Rosenbrock's banana function (Grace (1992)), where

$$G = 100\left(u_1^2 - u_2\right)^2 + \left(1 - u_1\right)^2,$$

the algorithm given by (1.1) results in a **stiff system** of differential equations. In such a situation, one must use a stiff integration solver (e.g. Matlab's Gear or ode15s) in order to obtain satisfactory results.

Figure 1.1 illustrates the trajectories obtained using (1.1) from various initial conditions. All trajectories flow into the parabolic shaped curve and then follow this curve asymptotically approaching the solution point located at $u_1 = u_2 = 1$. The parabola is the bottom of a very steep valley which would be evident if G were added to make a three dimensional plot. Lines of constant cost are somewhat banana shaped and are bunched along the parabola (Vincent and Grantham (1997), p. 146).

The ability to solve this problem is very much dependent on the integration routine used. For example if a Runge-Kutta method is used (e.g. Matlab's ode45) then all trajectories would appear to stop as soon as they reach the parabola. While the Runge-Kutta routine continues to make progress down the valley, it is so slow, one is not tempted to see how long it takes to satisfy the norm criteria. On the other hand, all 24 of the trajectories illustrated in Figure 1.1 takes less than 1.5 minutes to generate using a 120mh Pentium pc and Matlab/Simulink Gear integration routine.

2 TRAJECTORY FOLLOWING ALGORITHM FOR SIMPLE CONSTRAINTS

Various methods can be used to handle constraints. We will use the same approach as in Vincent and Grantham (1997). When minimizing $G(\mathbf{u})$ subject to simple inequality constraints, of the form

$$\mathbf{u}_{\min} \leq \mathbf{u} \leq \mathbf{u}_{\max}, \tag{2.1}$$

the trajectory following algorithm is modified according to

$$\dot{u}_i = \begin{cases} 0 & \text{if} u_i = u_{i_{\max}} \text{ and } \frac{\partial G}{\partial u_i} \leq 0 \\ 0 & \text{if } u_i = u_{i_{\min}} \text{ and } \frac{\partial G}{\partial u_i} \geq 0 \\ 0 & \text{if } \;\text{norm}\left(\frac{\partial G}{\partial \mathbf{u}}\right) < \varepsilon \\ -\frac{\partial G}{\partial u_i} & \text{otherwise,} \end{cases} \tag{2.2}$$

where $i = 1, \ldots, n_u$. Note that this algorithm can handle initial conditions which do not satisfy the constraints, as well as any problems associated with integrating past a boundary if the integration routine includes a test for the inequality conditions being satisfied and setting u_i to the appropriate boundary value for any inequality not satisfied. In this way the first two conditions of (2.2) will be automatically satisfied. Also we are assuming that the integration code has the ability to detect when all components of $\dot{u}_i = 0$ and will stop the integration accordingly. If every component of the control variable $\mathbf{u}$ is bounded by a simple constraint, then the constraint set $\mathcal{U}$ defined by these constraints is compact and it is guaranteed that a global minimum exists in $\mathcal{U}$ (Vincent and Grantham (1981)). If there are no other local minimums, then the trajectory generated by (2.2) will asymptotically approach the global minimal point starting from almost all $\mathbf{u} \in \mathcal{U}$. The exceptions would be any interior points of $\mathcal{U}$ for which $\partial G/\partial \mathbf{u} = \mathbf{0}$.

2.1 Forward/backward integration

If it is known that G has several local minimal points, the above algorithm can be modified to search for all of them (Vincent and Grantham (1997)). This process is facilitated by searching for all local minimal points and all maximal points at the same time. To start the process an initial point is chosen and (2.2) is integrated forward in time to find the first local minimal candidate. A new initial point is then randomly chosen in the neighborhood of the first local minimal candidate and the system (2.2) is then integrated backward in time to find the first local maximal candidate. A new initial point is randomly chosen in the neighborhood of this maximal point with forward integration again used. This process continues, using forward-backward integration, until some stopping criteria is satisfied. Clearly, there is no guarantee that this process will find all local minimal and maximal points given a specified number of searches, but it can find a lot of them, some points many times over.

Unless the number of minimum points were known and that number were small, the suggested forward/backward integration procedure is not a likely

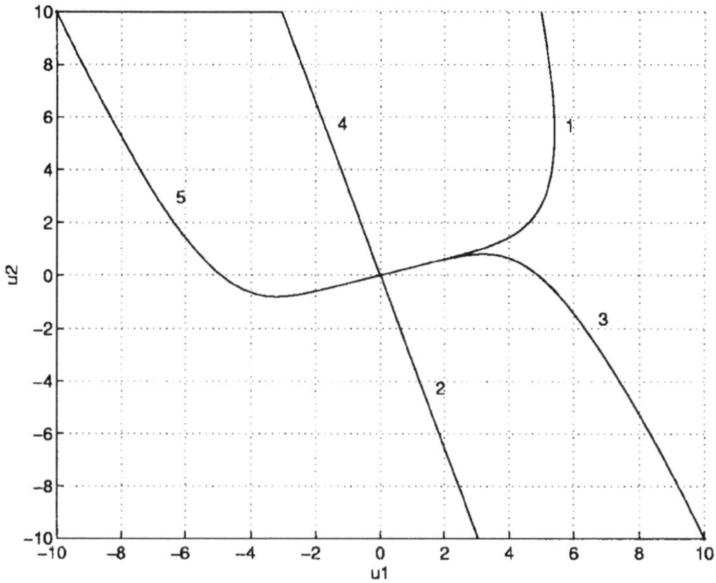

Figure 2.1 Finding both minimum and maximum points.

method for finding the global minimum. In section 4, a trajectory following method specifically directed toward finding global minimums in one, two or three dimensions is presented.

Quadratic Cost Example. A two-dimensional quadratic cost function subject to simple inequality constraints will either have one global minimum point with one or more local maximal points or one global maximal point with one or more local minimal points. Consider the function and simple constraints

$$G = u_1^2 - 2u_1u_2 + 4u_2^2$$

$$-10 \leq u_i \leq 10.$$

All the local minimal and maximal points can easily be found using the trajectory following method with forward/backward integration. Rather than automating this process using a random function generator as suggested above, the iterative process was done by hand with new initial conditions chosen to obtain the trajectories shown in Figure 2.1. The first forward trajectory, labeled 1, starts from a point on the boundary and finds a local minimum at the origin. The first backward trajectory, labeled 2, starts at a point very near the origin and finds a local maximum at $u_1 = 10$, $u_2 = -10$. The next forward trajectory 3 again finds the minimum at the origin, and so on. In this way the two global maximums are found along with the unique global minimum (3 times!). Note that when the trajectories run into simple inequality constraint, they abruptly turn and follow the constraint.

3 NON-SIMPLE CONSTRAINTS

Constraints which can not be expressed as (2.1) require some modification of the algorithm. Courant (1943) proposed dealing with the constraint problem by introducing a new cost function composed of the original cost function plus a penalty function. The penalty function is composed of the square of a constraint multiplied by a positive constant. A minimum is then sought using the new unconstrained cost with the value of the constant arbitrarily large. Many variants of this penalty function approach have been proposed (e.g. using a penalty function to guarantee that the inequality constraints are always satisfied) (Fiacco and McCormick (1968)). Here Courant's original penalty function method is used which means, in particular, that constraints may be (slightly) violated at the solution points obtained.

The notation $\mathbf{h}(\mathbf{u}) \geq \mathbf{0}$ is used to denote n_h non-simple inequality constraints and $\psi(\mathbf{u}) = \mathbf{0}$ to denote n_ψ equality constraints. Both of these constraint types are included in the trajectory following algorithm through the use of penalty functions. In particular, consider the new function

$$W = G + \beta_h \sum_{i=1}^{n_h} \gamma_i h_i^2 + \beta_\psi \sum_{i=1}^{n_\psi} \psi_i^2,$$

where β_h and β_ψ are large positive numbers and γ_i is chosen such that $\gamma_i = 0$ if $h_i \geq 0$ and $\gamma_i = 1$ if $h_i < 0$. It is assumed that the function W and its derivative with respect to $\mathbf{u}$ are continuous in $\mathbf{u}$.

Since each term added to G results in a positive penalty, a minimum point for W would be equal to a minimum point for G, subject to the constraints, only when $\beta_h \to \infty$ and $\beta_\psi \to \infty$. Since both G and W are continuous in $\mathbf{u}$, a minimum point for W should remain close to a minimum point for G subject to the constraints, provided that the penalty parameters are taken sufficiently large. There is no way of telling ahead of time how large these parameters should be. After a solution has been found, using this method, the constraints should always be evaluated to see how accurately they have been satisfied. If need be, the parameters can be increased in value and a new run made using as starting conditions the solution obtained from the previous run.

In this case the trajectory following algorithm is obtained by simply replacing G in (2.2) with W.

$$\dot{u}_i = \begin{cases} 0 & \text{if} \, u_i = u_{i_{\max}} \text{ and } \frac{\partial W}{\partial u_i} \leq 0 \\ 0 & \text{if } u_i = u_{i_{\min}} \text{ and } \frac{\partial W}{\partial u_i} \geq 0 \\ 0 & \text{if } \operatorname{norm}\left(\frac{\partial W}{\partial \mathbf{u}}\right) < \varepsilon \\ -\frac{\partial W}{\partial u_i} & \text{otherwise} \end{cases} \tag{3.1}$$

where $i = 1, \ldots, n_u$. It should be noted that even if G by itself would not make (3.1) a stiff system, large penalty values used to form W certainly will.

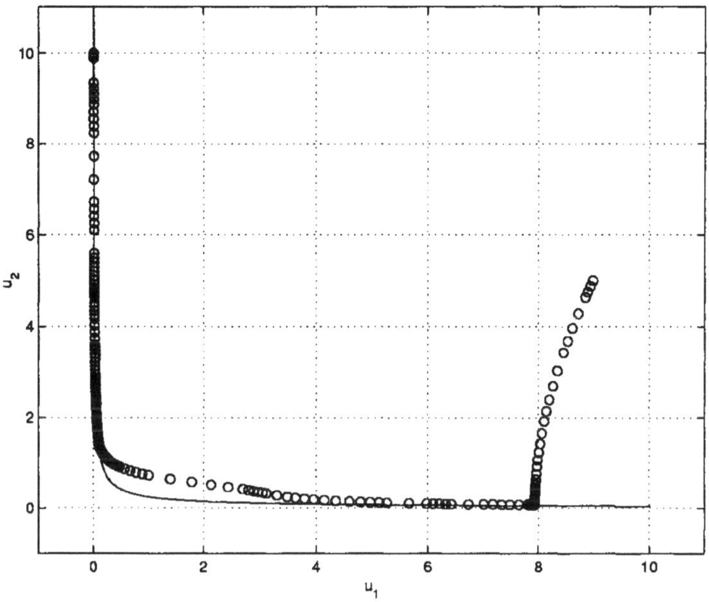

Figure 3.1 Circles are integration steps and the solid line is the equality constraint.

3.1 Non-simple example

The following problem is from Reklaitis et al (1983), p. 621. The cost function

$$G = u_1^2 u_2^4,$$

is subject to the constraints

$$\begin{aligned} 0.25 - (u_1)^{\frac{3}{4}} u_2 &= 0 \\ u_1 u_2^4 + u_2^2 \sqrt{u_1} - 1 &\geq 0 \end{aligned}$$

$$\begin{aligned} 0 \leq \ & u_1 \leq 10 \\ 0 \leq \ & u_2 \leq 10. \end{aligned}$$

Consider using the trajectory following algorithm (3.1) to find all minimal and maximal points for this problem using $\varepsilon = 10^{-5}$, with $\beta_h = \beta_\psi = 200$ (unless the inequality constraint is inactive, in which case $\beta_\psi = 500$). Figure 3.1 illustrates the results. Starting at the initial point $u_1 = 9$, $u_2 = 5$, and integrating forward in time, the trajectory quickly moves to satisfy the equality constraint which is illustrated in this figure by the solid line. The circles show the progression of the trajectory following algorithm at every integration step. Note how the trajectory gets pulled off of the equality constraint by the penalty associated with the inequality constraint. However, when the norm condition is satisfied, the solution $u_1 = 0.100959$, $u_2 = 1.394652$, with $G_{\min} = 0.386062$ is obtained with the inequality constraint active. A new initial condition is then chosen close to this solution and the trajectory following algorithm (3.1) integrated backward in time until the norm condition is again satisfied at the solution point given by $u_1 = 0.007416$, $u_2 = 10.00000$, with $G_{\max} = 0.542615$. Note that since u_2 is on the boundary for the maximal solution, its value is exact.

4 LINEAR/QUADRATIC PROGRAMMING

While efficient code for solving either the linear or quadratic programming problem is available, there are advantages to re-examining the solution to these problems using trajectory following. First, the trajectory following method can be programmed in a few lines of code to handle problems of large size. Perhaps more important, the same code, with minor alterations, can easily handle problems that are not strictly linear programming or quadratic programming. For example one might have a problem that is strict except for the fact that the cost function is the sum of a linear and quadratic function or perhaps some of the constraints are quadratic equality constraints.

Both the linear and quadratic programming problem will be examined simultaneously by using the cost function

$$G = \eta_1 \mathbf{c}^{\mathsf{T}} \mathbf{u} + \eta_2 \mathbf{u}^{\mathsf{T}} \mathbf{P} \mathbf{u}$$

with simple inequality constraints of the form

$$\mathbf{u}_{\min} \leq \mathbf{u} \leq \mathbf{u}_{\max}$$

and non-simple inequality constraints of the form

$$\mathbf{h}(\mathbf{u}) = \mathbf{A}\mathbf{u} + \mathbf{b} \geq \mathbf{0}. \tag{4.1}$$

where η_1 and η_2 are chosen according to the problem to be solved. The trajectory following algorithm (3.1) is still applicable, however, for the linear/quadratic problem there is no point in calculating the gradient used in (3.1) numerically. The following steps are used to instead.

1. First calculate the gradient of G

$$\text{Grad } G = \eta_1 \mathbf{c} + \eta_2 \mathbf{P} \mathbf{u}$$

2. Test to see if the non-simple constraints are satisfied

$$\mathbf{L} = \mathbf{h} > 0$$

   this is a (Matlab) logic statement which will create a column vector $\mathbf{L}$ which will have a 1 or 0 at every row element depending on whether the corresponding row in the vector $\mathbf{h}$ is greater than zero (1 = true) or less than or equal to zero (0 = false).

3. Form the gradient of the W function according to the following (Matlab) logic statements

(a) Find out if **L** has any zero elements by comparing the length of the vector with the sum of its elements

$$\text{if } \mathrm{sum}(L) < \mathrm{length}(L)$$

(b) then include them in the gradient of W according to the method described above. That is, for every component h_j which is less than zero, add β times the gradient of that component.

for j = 1:length(L)
if L(j) == 1
Grad W = Grad G + β * h(j,:) * A(j,:)

The trajectory following algorithm for this case is again given by (3.1) using the notation $\frac{\partial W}{\partial u_i}$ to denote the ith component of Grad W.

4.1 Linear/Quadratic Example

Consider the two dimensional problem,

$$\mathbf{u} = \begin{bmatrix} u_1 \\ u_2 \end{bmatrix}$$

in which the cost function to be minimized

$$G = \mathbf{c}^\top \mathbf{u} + 2\mathbf{u}^\top \mathbf{P}\mathbf{u}$$

is a sum of both linear terms and quadratic terms where

$$\mathbf{c} = \begin{bmatrix} 1 \\ -2 \end{bmatrix} \quad \text{and} \quad \mathbf{P} = \begin{bmatrix} 1 & 0 \\ 0 & 1 \end{bmatrix}$$

Let the inequality constraints be given by (4.1) where

$$\mathbf{A} = \begin{bmatrix} 1 & 1 \\ 1 & -1 \\ -1 & 1 \\ -1 & -1 \end{bmatrix} \quad \text{and} \quad \mathbf{b} = \begin{bmatrix} 1 \\ 1 \\ 1 \\ 1 \end{bmatrix}.$$

When there exists a potential for several inequality constraints to be active at the same time (in particular, when applying this method to problems of high dimension), computer time can be saved by solving the problem in a sequential fashion (Fiacco and McCormick (1968)). Using this method here, a trajectory is first generated using the initial conditions $u_1 = 1$ and $u_2 = -1$ along with the parameters $\beta_h = 100$ and $\varepsilon = .002$. After this solution is stopped by the norm condition, it is used as a starting condition for a new trajectory using $\beta_h = 1000$ and $\varepsilon = 10^{-5}$. These two trajectories are illustrated in Figure 4.1 with the integration steps indicated by the circles. The inequality constraints define a feasible region inside the diamond. Note that it is not necessary to

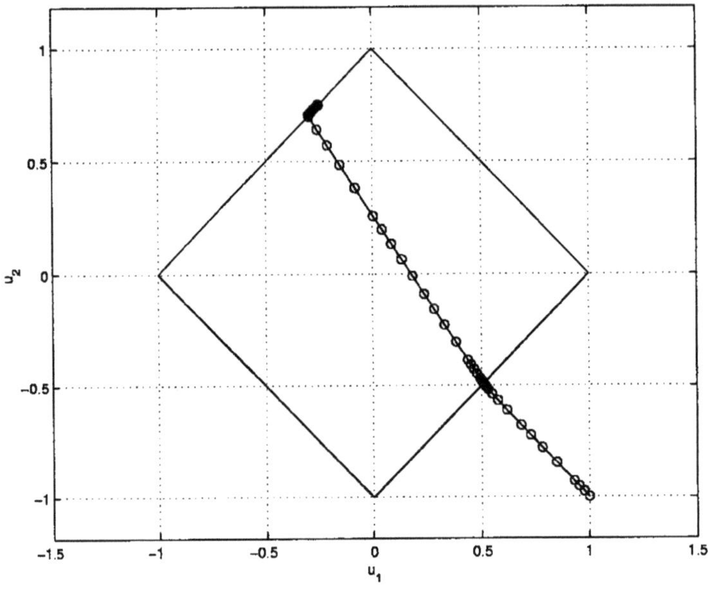

Figure 4.1 Linear/Quadratic Problem.

choose an initial condition which lies inside this constraint set. Since one of the inequality constraints is violated with the initial condition, the trajectory following algorithm produces a trajectory almost perpendicular to this constraint. As soon as the constraint is satisfied the trajectory moves perpendicular to lines of constant cost until it hits the upper left hand constraint. At this point the trajectory violates the constraint by a small amount, but changes direction to follows it until the first norm condition is satisfied. At this point $u_1 = -0.253124$, $u_2 = 0.751826$, yielding $h = 0.00495049$, and $G = -0.498147$. This process is seen more clearly in Figure 4.2 which zooms in on the crossing point. The first trajectory ends right where the second integration run bends the trajectory back toward the equality constraint. When the second norm condition is satisfied the solution is given by $u_1 = -0.250253$, $u_2 = 0.750246$, yielding $h = 0.0004995$, and $G = -0.499753$.

If an inequality constraint is active, the trajectory following method as described here will always slightly violate the constraint. This fact is very handy in being able to identify which constraints are active. If the inequality constraints must be satisfied exactly, this can be done after a solution has been obtained by simply using the constraint as an equality constraint to solve for the variables. In the above example, if u_2 is held fixed and $h = 0$ is used to solve for u_1 one obtains $u_1 = -0.249754$, $u_2 = 0.750246$, yielding $h = 0$, and

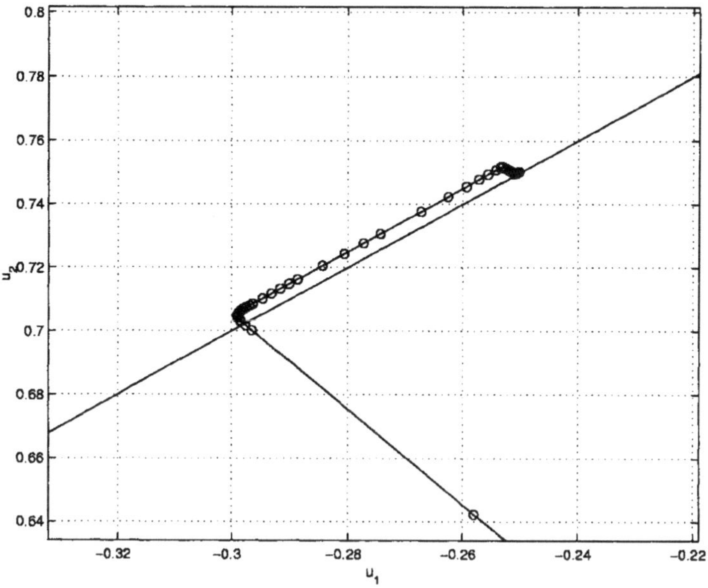

Figure 4.2 Zooming in on the solution to the Linear/Quadratic Problem.

$G = -0.499753$. To six significant figures, this correction does not change the minimum cost solution obtained.

5 GLOBAL MINIMIZATION

If the dimension of the problem is small and there are few local maximum or minimum then one may find a global minimum by finding all local minima and maxima using the forward/backward method mentioned above. However this can be tedious if there are many maxima and minima. Another method for problems of low dimensions (i.e. one) with many maxima and minima was suggested by Liu and Teo (1999). They use a bridging technique to move from one local minimum to another one of lower value by "bridging" between the two local minima. This method requires construction of a bridging function and to date, this method has not been extended beyond one dimensional problems. Construction of the bridging function becomes very difficult in dimensions greater than one.

By using trajectory following, it is possible to utilize the concept of bridging between one local minimum and another one of lower value but without the difficulties introduced by having to construct a bridging function (Vincent and Teo (1999)). This is done by using the trajectory following algorithm, modified by using a flooding method, so that while "bridging" between two local minimums, the system is driven along a prescribed trajectory rather than according to a gradient.

5.1 *Flooding Method*

The idea behind the flooding technique is most easily visualized if we think in terms of global maximum rather than global minimum. Consider a mountain landscape contained in a given square area with many peaks and valleys. Pick any starting point in this landscape and use the trajectory following algorithm to climb up the landscape until a local peak is reached. Now flood the landscape with water up to the level of this peak. Clearly the only other peaks of interest are those which rise above the water. However, because of fog we can't see any other peaks, and, if there are any, we must find them by some searching technique. Continuing with the analogy, we divide the area into a grid of North-South lines, then starting at one of the grid lines on the South boundary we drive a motor boat from South to North. If along the way we hit land, we get out and use the trajectory following algorithm to climb to a new peak. The landscape is then flooded up to this level. We get back into the boat, locate where we left the grid line and continue our search along the line until the North boundary is met. If this process is continued for all grid lines we will be guaranteed to find the highest peak provided that the distance between the grid lines is sufficiently small.

The flooding method requires a grid and two different ways of driving the system of differential equations. When searching for a local minimum the trajectory following algorithm (3.1) is used. However when searching along a grid line, the system of differential equations are driven along the line at a velocity proportional to the difference between the value of the function and the last minimum value found. In this way a graceful stop can be made when "land" is discovered.

The first thing that needs to be done is define a box within which the search will take place. This is done by means of simple constraints of the form

$$\mathbf{u}_{\min} \leq \mathbf{u} \leq \mathbf{u}_{\max}$$

Starting at the point where $u_i = u_{i_{\min}}$ for $i = 1, \ldots, n_u$ the trajectory following algorithm (3.1) is used to find a local minimum candidate, W^*. The results are then saved and we then go back to the starting point and use

$$\begin{aligned}
\dot{u}_1 &= \begin{cases} 0 & \text{if } \; |W - W^*| < \epsilon \\ 0 & \text{if } \; u_1 \geq u_{1_{\max}} \\ W - W^* & \text{otherwise} \end{cases} \qquad (5.1) \\
\dot{u}_j &= 0 \quad \text{for } \; j = 2, \ldots, n_u
\end{aligned}$$

to drive the system along a grid line until either the norm condition is satisfied or the end of the grid line is reached. If the norm condition is satisfied before reaching the end of a grid line, then at that point, the system dynamics is switched from (5.1) to (3.1). Once integration is stopped by the trajectory following algorithm (3.1) the results are saved, and the new value for W^* is used to resume integration using (5.1) starting where the system was last stopped on the grid line. This process continues until all the grid lines have been investigated. Note that as soon as the global minimum has been found using

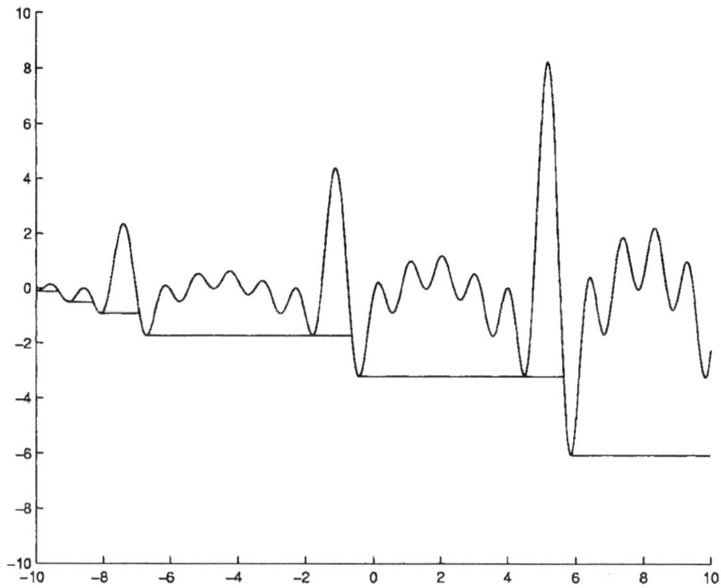

Figure 5.1 Flooding technique in one dimension.

this procedure, all remaining integration will be by means of (5.1) which is generally faster than (3.1).

Clearly the success of this method in finding a global minimum will depend on how the grid lines are chosen. In particular, they must be close enough together so that the system will not be driven past a global minimum without it being detected.

One Dimensional Example. We will use one of the examples in Liu and Teo (1999). We seek a global minimum for

$$G = -e^{0.1u} \sum_{i=1}^{5} \sin\left((i+1)\,u + i\right)$$

subject to

$$-10 \leq u \leq 10.$$

For one dimensional problems we need only one grid line. Consequently the flooding technique is guaranteed to find the global minimum along this line. Figure 5.1 illustrates use of the algorithm.

The actual function with its many maxima and minima is plotted along with the solution trajectory as generated by the algorithm. The flooding technique first uses (3.1) to find a local minimum at $u = -9.9925$ with $G = -0.1182$ near the left boundary. At this point the system is then driven according to (5.1) until reaching the point $u = -9.3197$ (the corresponding trajectory is the left most straight line segment). At this point (3.1) is again used to find the next local minimum at $u = -9.0232$. This process continues until the right hand boundary is reached. Note that there were two near misses along the way. However both of these local minimums had values sightly higher than the last local minimum found. The following table summarizes the six local minimums obtained in the order that they were found. It took approximately 30 seconds to generate these solutions using Matlab's integrator ode15s on a 120MH Pentium lap top

u	G
−9.9925	−0.11820
−9.0232	−0.4934
−8.0767	−0.9074
−6.7152	−1.7229
−0.4320	−3.2295
5.8511	−6.0536

Clearly the global minimum would have been the first local minimum found if we had of started from the right hand side.

Two Dimensional Example. The following two dimensional example is from Vincent and Teo (1999). Minimize

$$G = \frac{1}{500}\left(u_1^2 + u_2^2\right) - \frac{\cos u_1}{\sqrt{1}}\frac{\cos u_2}{\sqrt{2}}$$

subject to

$$\begin{aligned} -10 &\leq u_1 \leq 10 \\ -10 &\leq u_2 \leq 10. \end{aligned}$$

In this case we must choose a grid with appropriate spacing. For this problem we note that

$$\frac{\partial G}{\partial u_1} = \frac{1}{250}u_1 + \frac{\sqrt{2}}{2}(\sin u_1 \cos u_2)$$

$$\frac{\partial G}{\partial u_2} = \frac{1}{250}u_2 + \frac{\sqrt{2}}{2}(\cos u_1 \sin u_2)$$

Over the range given, $\frac{\partial G}{\partial u_2}$ can never be greater than 0.7471. Thus if we form grid lines parallel to the x_1 axis, we can guarantee not to miss a local minimum point with a value smaller by h where

$$h = \frac{d}{2}\tan(0.7471)$$

For example if we use $d = 1$, then $h = 0.4631$. Let us assume that this is acceptable and solve the problem using the flooding technique based on a grid size with $d = 1$. Figure 5.2 illustrates the results of the various grid runs.

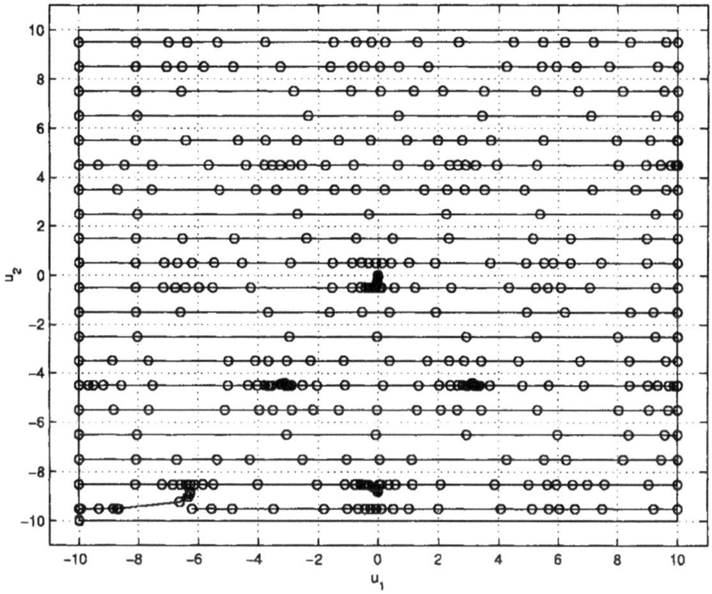

Figure 5.2 Global optimization in two dimensions.

The following table summarizes the six local minimums obtained in the order that they were found. It took approximately 60 seconds to generate these solutions using Matlab's integrator ode45 on a 120MH Pentium lap top. For this problem a "stiff integrator" was not needed.

u_1	u_2	G
-9.9974	-10.0000	0.9927
-6.2597	-8.8152	-0.7647
0.0000	-8.8151	-0.8433
3.1276	-4.4076	0.9412
-3.1300	-4.4076	-0.9412
0.0000	0.0000	-1.0000

The solution process was started on the lower left corner of the feasible set and the first local minimum was found at a boundary point. Each circle in Figure 5.2 represents a solution point for the integrator. It is fairly clear where the local minimum solutions are located as the solutions points bunch up near these points. Note that the solutions need not lie on the grid lines, but after a minimal solution is found the routine takes the system back to a grid line. After the global minimum was found at the origin, all other trajectories were generated using only (5.1)

6 CONCLUSIONS

In terms of ease of use and the ability to handle a wide range of problems the trajectory following algorithm (3.1) provides a convenient optimization tool for nonlinear problems of moderate size (e.g. $n_u = 10$). Analytical derivatives are not required, nor is it necessary to start the algorithm at a feasible point. For linear/quadratic problems, the algorithm is used with exact derivatives and may be applied to problems of larger dimension (e.g. $n_u = 100$). The algorithm may also be used, in conjunction with a flooding technique, to find a global minimum for problems of smaller dimension (e.g. $n_u = 3$). No claim is made in terms of efficiency as measured by the number of function evaluations. However it does provide an attractive programming alternative for many difficult practical problems of modest dimension.

References

Arrow, K.J. and Hurwicz, L. (1977), Gradient methods for constrained maxima, in *Studies in Resource Allocation Processes*, pp. 146-153, Cambridge University Press, Cambridge.

Courant, R., (1943), Variational methods for the solution of problems of equilibrium and vibrations, *Bull. Am. Math. Soc.*, Vol. 49, pp. 1-23.

Fiacco, A.V. and McCormick, G.P. (1968), *Nonlinear Programming: Sequential Unconstrained Minimization Techniques*, Wiley, New York.

Goh, B.S. (1978), Global convergence of some differential equations algorithms for solving equations involving positive variables, *BIT*, Vol. 18, pp. 84-90.

Grace, A. (1992), *Optimization Toolbox for Use with MATLAB*, The Mathworks, Inc., Matick, MA.

Kelly, H.J. (1962), Method of gradients, in *Optimization Techniques*, pp. 205-254, Academic Press, New York.

Liu, Y. and Teo, K.L. (1999), A bridging method for global optimization, *In Review*.

Reklaitis, G.V.A, Ravindran, V.A. and Ragsdell, K.M. (1983), *Engineering Optimization Methods and Applications*, Wiley, New York.

Yan, W.Y., Teo, K.L. and Moore, J.B. (1994), A gradient flow approach to computing (LQ) optimal output feedback gains, *Optimal Control Applications and Methods*, Vol. 15, pp. 67-75.

Vincent, T.L. and Grantham, W.J. (1981), *Optimality in Parametric Systems*, Wiley-Interscience, New York.

Vincent, T.L. and Grantham, W.J. (1997), *Nonlinear and Optimal Control Systems*, Wiley, New York.

Vincent, T.L. and Teo, K.L. (1999), Global optimization using trajectory following, *In Review*.

Zang, Z. and Vincent, T.L. (1999), Continuous-time envelope-constrained filter design, *In Review*.

17 SOLVING HAMILTON-JACOBI-BELLMAN EQUATIONS BY AN UPWIND FINITE DIFFERENCE METHOD

S. Wang,

Centre for Applied Dynamics and Optimization,
Department of Mathematics and Statistics,
The University of Western Australia,
Nedlands, WA 6907, Australia

F. Gao

School of Mathematics & Statistics,
Curtin University of Technology,
GPO Box U1987, Perth 6845, Australia

and K.L. Teo

Department of Applied Mathematics,
The Hong Kong Polytechnic University,
Kowloon, Hong Kong

Abstract: In this paper we present an upwind finite difference method for the numerical approximation of viscosity solutions of a two dimensional Hamilton-Jacobi-Bellman (HJB) equation arising from a class of optimal feedback control problems. The method is based on an explicit finite difference scheme and it is shown that the method is stable under certain constraints on the step lengths of the discretization. Numerical results, performed to verify the usefulness of the method, show that the method gives accurate approximate solutions to both of the control and the value function.

Key words: HJB equation, optimal feedback control, dynamic programming, upwind finite difference method, stability.

X. Yang et al. (eds.), Progress in Optimization, 255–268.

1 INTRODUCTION

Consider the following optimal control problem:

$$\begin{aligned} \min \quad & \int_s^1 L(t, \boldsymbol{x}(t), \boldsymbol{u}(t))dt + h(\boldsymbol{x}(1)), \\ \text{subject to} \quad & \dot{\boldsymbol{x}}(t) = \boldsymbol{f}(t, \boldsymbol{x}(t), \boldsymbol{u}(t)), \quad \text{a.e. } t \in [s, 1] \\ & \boldsymbol{x}(s) = \boldsymbol{y}, \end{aligned}$$

over a set of admissible controls $U \subset \mathbb{R}^m$, where $\boldsymbol{f} = (f_1, f_2, \ldots, f_n)$ and $(s, \boldsymbol{y}) \in [0,1] \times \mathbb{R}^n$. It is well known that, via the dynamic programming approach, this problem can be formulated as a Hamilton-Jacobi-Bellman (HJB) or dynamic programming equation of the form

$$-\frac{\partial v}{\partial t} + \sup_{\boldsymbol{u} \in U} \left[-\nabla v \cdot \boldsymbol{f}(t, \boldsymbol{x}(t), \boldsymbol{u}(t)) - L(t, \boldsymbol{x}, \boldsymbol{u}) \right] = 0 \quad (t, \boldsymbol{x}) \in [0,1) \times \mathbb{R}^n \tag{1.1}$$

with the initial condition

$$v(1, \boldsymbol{x}) = h(\boldsymbol{x}) \tag{1.2}$$

where v is called the value function and ∇v denotes the gradient of v with respect to $\boldsymbol{x}$. There are two unknown functions in this equation – the value function v and the optimal control $\boldsymbol{u}$. Because of the importance of this initial value problem in optimal feedback control, it has been discussed in the literature for many years. Solutions to (1.1) and (1.2) are normally continuous but non-smooth in $\mathbb{R}^n$. For the construction of optimal feedback controls via the dynamic programming approach together with the associated verification techniques (Zhou (1993)), non-smooth analysis (Clarke (1993); Ekeland (1974)) is required. The classic theory does not have a firm foundation, as it is based on the smoothness assumption of the value function v in (1.1), which is not satisfied even for some simple examples. To handle such non-smooth phenomena, the concept of viscosity solutions is used in Lions (1983). This concept comes originally from the solution of the equation

$$\varepsilon \nabla^2 v - \frac{\partial v}{\partial t} + \sup_{\boldsymbol{u} \in U} \left[-\nabla v \cdot \boldsymbol{f}(t, \boldsymbol{x}, \boldsymbol{u}) - L(t, \boldsymbol{x}, \boldsymbol{u}) \right] = 0, \ (t, \boldsymbol{x}) \in (0,1) \times \mathbb{R}^n \tag{1.3}$$

governing a viscous fluid flow, where $\varepsilon > 0$ is the viscosity of the fluid. The solution to (1.3) is smooth and converges to the viscosity solution of (1.1) as $\varepsilon \to 0$ (cf., for example, Yong (1992)). Let $G : \Omega \times \mathbb{R}^1 \times \mathbb{R}^n \mapsto \mathbb{R}^1$ be a continuous function. Mathematically, v is said to be a viscosity solution to the equation

$$G(\boldsymbol{x}, v(\boldsymbol{x}), \nabla v(\boldsymbol{x})) = 0$$

if

$$\begin{aligned} G(\boldsymbol{x}, v(\boldsymbol{x}), \boldsymbol{p}) \leq 0 \quad & \text{for all} \quad \boldsymbol{p} \in D^+_{\boldsymbol{x}} v(\boldsymbol{x}), \\ G(\boldsymbol{x}, v(\boldsymbol{x}), \boldsymbol{p}) \geq 0 \quad & \text{for all} \quad \boldsymbol{p} \in D^-_{\boldsymbol{x}} v(\boldsymbol{x}), \end{aligned}$$

where $D^+_{\boldsymbol{x}}v(\boldsymbol{x})$ and $D^-_{\boldsymbol{x}}v(\boldsymbol{x})$ denotes respectively the super- and sub-differentials of v at $\boldsymbol{x}$ defined by

$$D^+_{\boldsymbol{x}}v(\boldsymbol{x}) := \left\{ \boldsymbol{p} \in \mathbb{R}^n : \lim_{\boldsymbol{y}\to\boldsymbol{x}} \sup_{\boldsymbol{y}\in\Omega} \frac{v(\boldsymbol{y}) - v(\boldsymbol{x}) - \boldsymbol{p}\cdot(\boldsymbol{y}-\boldsymbol{x})}{|\boldsymbol{y}-\boldsymbol{x}|} \leq 0 \right\},$$

$$D^-_{\boldsymbol{x}}v(\boldsymbol{x}) := \left\{ \boldsymbol{p} \in \mathbb{R}^n : \lim_{\boldsymbol{y}\to\boldsymbol{x}} \inf{}_{\boldsymbol{y}\in\Omega} \frac{v(\boldsymbol{y}) - v(\boldsymbol{x}) - \boldsymbol{p}\cdot(\boldsymbol{y}-\boldsymbol{x})}{|\boldsymbol{y}-\boldsymbol{x}|} \geq 0 \right\}.$$

It can be shown that the viscosity solution to an HJB equation is unique and equal to the classic (smooth) solution if it exists (cf, for example, Yong (1992) and Li & Yong (1995)). Although viscosity solutions have been discussed for many years, most of them are from a theoretical point of view (cf., for examples, Lions (1983); Li & Yong (1995); Zhou (1990); Zhou (1993); Zhou & Sethi (1993)). Since, in general, this equation is not analytically solvable, numerical approximations to this equation are needed in practice. However, there are only one or two numerical algorithms proposed for the numerical solution of this equation in the literature (cf., for example, Maslov (1987)).

In this paper we present a finite difference method for (1.1) and (1.2) in the case that $n = 1$, i.e., two dimensional case in space and time. The method is based on an explicit upwind finite difference scheme proposed originally for solving equations of the form (1.3) with $\varepsilon << 1$ (cf., for example, Peyret & Taylor (1983)). The stability of this method is analyzed and it is shown that under some conditions on the step lengths the method is stable. To verify the usefulness of the method, several test problems are solved. The numerical results show that the method is accurate and stable.

Although the method is discussed in two dimensions, the idea can be used in any dimensions.

2 THE METHOD

We now consider the following initial value problem

$$-v_t + \sup_{\boldsymbol{u}\in U} \left[-f(t,x,\boldsymbol{u})v_x - L(t,x,\boldsymbol{u}) \right] = 0 \quad (t,x) \in [0,1) \times [a,b], \tag{2.1}$$

$$v(1,x) = h(x) \tag{2.2}$$

where a and b are two real constants satisfying $a < b$ and U is a given set of admissible controls. We assume that $f(t,x,\boldsymbol{u})$ and $L(t,x,\boldsymbol{u})$ are continuous and differentiable with respect to $\boldsymbol{u}$. We also assume that $f(t,x,\boldsymbol{u})$ is bounded in $\mathbb{R}^1$.

To discuss the discretization of (2.1) we first assume that $x_i = a + i\Delta x$ for $i = 0, 1, \ldots, M$ and $t_j = 1 - j\Delta t$ for $j = 0, 1, \ldots, N$ are two partitions along the x and t axes respectively, where $\Delta x = (b-a)/M$, $\Delta t = 1/N$ and M and N are two positive integers. Consider the 6-point stencil with the vertices (x_{i-1}, t_j), (x_i, t_j), (x_{i+1}, t_j), (x_i, t_{j+1}), (x_i, t_{j+1}) and (x_{i+1}, t_{j+1}), shown in Figure 2.1.

On this stencil we approximate (2.1) by the following difference equations:

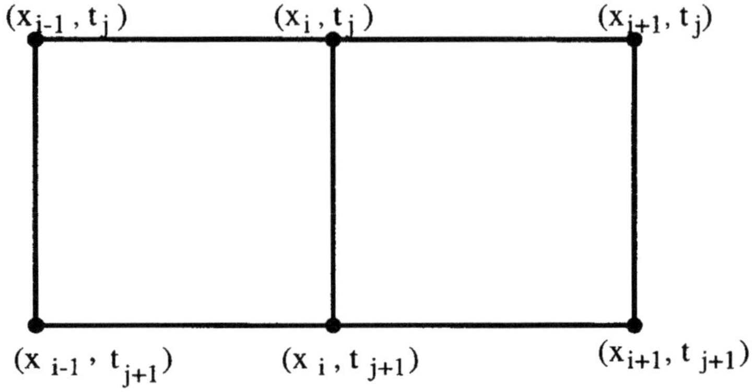

Figure 2.1 The 6-point finite difference stencil

$$\frac{v_i^{j+1} - v_i^j}{\Delta t} + \frac{1 + \mathrm{sign}(f_i^j)}{2} f_i^j \frac{v_{i+1}^j - v_i^j}{\Delta x} + \frac{1 - \mathrm{sign}(f_i^j)}{2} f_i^j \frac{v_i^j - v_{i-1}^j}{\Delta x} + L_i^j = 0, \tag{2.3}$$

$$\boldsymbol{u}_i^{j+1} = \arg \sup_{\boldsymbol{u} \in U} \left[- f(t_{j+1}, x_i, \boldsymbol{u}) \frac{v_i^{j+1} - v_{i-1}^{j+1}}{\Delta x} - L(t_{j+1}, x_i, \boldsymbol{u}) \right] \tag{2.4}$$

for $i = j+1, j+1, \ldots, M-j-1$ and $j = 0, 1, \ldots, N-1$, where $\mathrm{sign}(z)$ denotes the sign of z and

$$f_i^j = f(t_j, x_i, \boldsymbol{u}_i^j), \quad L_i^j = L(t_j, x_i, \boldsymbol{u}_i^j). \tag{2.5}$$

This is the upwind finite difference scheme which reduces to the forward difference scheme when $f_i^j > 0$ and to the backward difference schemes when $f_i^j < 0$. Similar schemes are often used for solving convection-diffusion equations (cf., for example, Peyret & Taylor (1983)). We will show in the next section that this upwind scheme is stable under some conditions. Obviously, (2.3) is an explicit scheme. Based on this scheme, we propose the following algorithm for the numerical solution of (2.1) and (2.2):

Algorithm A:

Initialize $\{v_i^0\}_{i=0}^M$ and $\{\boldsymbol{u}_i^0\}_{i=0}^M$ using the initial condition (2.2).
For $j = 0$ to $N-1$ do
  For $i = j+1$ to $M-j-1$ do

$$v_i^{j+1} = (1 + |f_i^j|\gamma) v_i^j - \frac{1 + \mathrm{sign}(f_i^j)}{2} \gamma f_i^j v_{i+1}^j + \frac{1 - \mathrm{sign}(f_i^j)}{2} \gamma f_i^j v_{i-1}^j - \gamma L_i^j$$

$$\tag{2.6}$$

$$u_i^{j+1} = \arg\sup_{u\in U}\Big[-f(t_{j+1},x_i,u)\frac{v_i^{j+1}-v_{i-1}^{j+1}}{\Delta x} - L(t_{j+1},x_i,u)\Big], \tag{2.7}$$

with $\gamma = \Delta t/\Delta x$.

end do i

end do j

Remark 2.1. Although the original problem is defined on a rectangular domain, the computational domain may be trapezoid depending on the sign pattern of $f(x,t,u)$. This is because of the explicit nature of the scheme. Thus, in order to solve the problem on a rectangular region we need to choose a trapezoid region containing the original one as a subset.

Remark 2.2. The above scheme can be easily extended to non-uniform meshes. In this case, we need to replace Δx and Δt by $\Delta x_i := x_{i+1} - x_i$ and $\Delta t_j := t_{j+1} - t_j$ respectively, and modify the first term on the right side of (2.6) accordingly.

3 STABILITY OF THE SCHEME

We now discuss the stability of the difference scheme (2.3) or (2.6). Because the scheme is explicit, it is stable under some conditions on the step lengths Δt and Δx. In the rest of this section we shall derive these conditions.

From the construction of the scheme we see that (2.6) is equivalent to

$$v_i^{j+1} = (1+|f_i^j|\gamma)v_i^j - \gamma f_i^j v_{i+1}^j - \gamma L_i^j \tag{3.1}$$

when $f_i^j > 0$ and to

$$v_i^{j+1} = (1+|f_i^j|\gamma)v_i^j + \gamma f_i^j v_{i-1}^j - \gamma L_i^j \tag{3.2}$$

when $f_i^j < 0$. Both of these are of the form

$$\boldsymbol{v}^{j+1} = A^j\boldsymbol{v}^j + \boldsymbol{b}^j \tag{3.3}$$

for $j = 0,1,...,N-1$. From this recursive expression we have

$$\begin{aligned}\boldsymbol{v}^{j+1} &= A^j(A^{j-1}\boldsymbol{v}^{j-1} + \boldsymbol{b}^{j-1}) + \boldsymbol{b}^j \\ &\vdots \\ &= \left(\prod_{k=0}^{j} A^k\right)\boldsymbol{v}^0 + \sum_{k=0}^{j}\left(\prod_{l=0}^{k} A^l\right)\boldsymbol{b}^k.\end{aligned}$$

Similarly, this equality is also satisfied by the errors at each time step j. In this case $\boldsymbol{v}^0$ and $\boldsymbol{v}^{j+1}$ represent the initial and $(j+1)$th approximation errors in the solution of (3.3), and $\boldsymbol{b}^k$ is the approximation error in L at the kth time step. Let ρ denote the maximum of the spectral radii , or equivalently the maximum of all the eigenvalues of $A^k, k = 0, 1, ..., j$. Then, from the above equality we have that there exists a matrix norm, $||\cdot||$, such that

$$\begin{aligned} ||v^{j+1}|| &\leq \left(\prod_{k=0}^{j} ||A^k||\right) ||v^0|| + \sum_{k=0}^{j} \left(\prod_{l=0}^{k} ||A^l||\right) ||b^k|| \\ &\leq \rho^j ||v^0|| + \sum_{k=0}^{j} \rho^k ||b^k|| \end{aligned}$$

(cf. Ortega (1990), §4.2). Obviously, if $\rho \leq 1$, then the initial error $||v^0||$ will not grow with respect to j, but the approximation errors $||b^k||, k = 1, 2, ..., j$ are accumulative. In this case, the finite difference equation (3.3) is stable for a finite number of time steps. However, in the case that $\rho < 1$, the last term of the above inequality is bounded for all j, and thus the scheme is asymptotically stable (cf. Ortega (1990), p.75). Based on this discussion, we now derive the sufficient asymptotic stability condition for (2.6).

For each $j = 0, 1, \ldots, N-1$, we assume, for discussion brevity, that f_i^j changes its sign pattern only once for $i = j, j+1, \ldots, M-j$ and a fixed j. The situation that $\{f_i^j\}_j^{M-j}$ changes sign pattern more than once can be treated in a similar way. This discussion includes the following two cases.

Case 1. We assume that $f_i^j > 0$ for $i = j+1, j+1, \ldots, I$ and $f_i^j < 0$ for $i = I+1, I+2, \ldots, M-j-1$, where $j+1 < I < M-j-1$ is a fixed integer. In this case, from (3.1) and (3.2) we see that ((2.6) is of the form (3.3) where

$$\begin{aligned} v^{j+1} &= (v_{j+1}^{j+1}, v_{j+2}^{j+1}, \ldots, v_{M-j-1}^{j+1})^T, \\ v^j &= (v_{j+1}^j, v_{j+2}^j, \ldots, v_{M-j-1}^j)^T, \\ b^j &= \gamma(L_{j+1}^j, L_{j+2}^j, \ldots, L_{M-j-1})^T \end{aligned}$$

and A^j is a tridiagonal matrix defined by

$$A^j = \begin{pmatrix} d_{j+1}^j & -a_{j+1}^j & & & & & & & \\ 0 & d_{j+2}^j & -a_{j+2}^j & & & & & & \\ & \ddots & \ddots & \ddots & & & & & \\ & & 0 & d_I^j & -a_I^j & & & & \\ & & & -a_{I+1}^j & d_{I+1}^j & 0 & & & \\ & & & & \ddots & \ddots & \ddots & & \\ & & & & & -a_{J-2}^j & d_{J-2}^j & 0 \\ & & & & & & -a_{J-1}^j & d_{J-1}^j \end{pmatrix} \tag{3.4}$$

with $a_i^j = \gamma|f_i^j|$, $d_k^j = 1 + a_k^j$ for all $k = j+1, ..., J-1$ and $J = M-j$. All undefined entries in (3.4) are zeros. Now the asymptotic stability of (3.3) (Von Neumann condition) requires that all the eigenvalues of A^j lie on the interval $(-1, 1)$, i.e. the solutions to

$$\det(A^j - \lambda I) = 0$$

are all in the interval $[-1, 1]$, where I denotes the identity matrix of order $M - 2(j+1)$. From (3.4) we have

$$\det(A^j - \lambda I) = (1 + a^j_{j+1} - \lambda)\dots(1 + a^j_{I-1} - \lambda)(1 + a^j_{I+2} - \lambda)\dots (1 + a^j_{M-j-1} - \lambda)\begin{vmatrix} 1 + a^j_I - \lambda & -a^j_I \\ -a^j_{I+1} & 1 + a^j_{I+1} - \lambda \end{vmatrix},$$

and thus the above equation has the solutions

$$\lambda_i = 1 + a^j_i \quad \text{for} \quad i = j+1, \dots, I-1, I+2, \dots, M-j-1 \tag{3.5}$$

and two other solutions λ_I and λ_{I+1} satisfying

$$(1 + a^j_I - \lambda)(1 + a^j_{I+1} - \lambda) - a^j_I a^j_{I+1} = 0.$$

Solving this quadratic equation we have

$$\lambda_I = 1 \quad \text{and} \quad \lambda_{I+1} = 1 + a^j_I + a^j_{I+1}. \tag{3.6}$$

Thus, using (3.5) and (3.6) we see that the asymptotic stability condition $|\lambda| < 1$ becomes

$$|1 + a^j_i| < 1 \quad \text{for} \quad i = j+1, \dots, I-1, I+2, \dots, M-j-1,$$
$$|1 + a^j_I + a^j_{I+1}| < 1.$$

Recall that $a^j_i = |f^j_i|\Delta t/\Delta x$, we have that $a^j_i < 0$ because $\Delta t < 0$ and $\Delta x > 0$. Thus, the above becomes

$$-2 < \frac{|f^j_i|\Delta t}{\Delta x} < 0 \quad \text{for} \quad i = j+1, \dots, I-1, I+2, \dots, M-j-1,$$
$$2 < \frac{(|f^j_I| + |f^j_{I+1}|)\Delta t}{\Delta x} < 0,$$

or

$$|\Delta t| < \frac{2\Delta x}{|f^j_i|} \quad \text{for} \quad i = j+1, \dots, I-1, I+2, \dots, M-j-1, \tag{3.7}$$

$$|\Delta t| < \frac{2\Delta x}{|f^j_I| + |f^j_{I+1}|}. \tag{3.8}$$

In the above we have assume that all the denominators are not zero. In the case that some of them are zeros, (3.5) or (3.6) is satisfied for any Δt and any Δx.

Because $f^j_I > 0$ and $f^j_{I+1} < 0$, $f = 0$ at a point in the segment from (x_I, t^j) to (x_{I+1}, t^j). Thus, both $|f^j_I|$ and $|f^j_{I+1}|$ are of order $O(\Delta x)$, and so the right hand side of (3.8) is of order $O(1)$. Therefore, if we choose Δx and Δt such that

$$|\Delta t| < \frac{2\Delta x}{f_{\text{sup}}}, \tag{3.9}$$

then both (3.7) and (3.8) are satisfied when Δx is sufficiently small, where

$$f_{\sup} = \sup_{(x,t,\boldsymbol{u})\in(a,b)\times(0,1)\times U} f(x,t,\boldsymbol{u}).$$

The inequality (3.9) is a sufficient condition for the stability of the upwind scheme presented in the previous section.

Case 2. Assume that $f_i^j < 0$ for $i = j+1, j+1, \ldots, I$ and $f_i^j > 0$ for $i = I+1, I+2, \ldots, M-j-1$. In this case the matrix form of (3.1) and (3.2) is the same as (3.3) where

$$\begin{aligned} \boldsymbol{v}^{j+1} &= (v_j^{j+1}, v_{j+1}^{j+1}, \ldots, v_{M-j}^{j+1})^T, \\ \boldsymbol{v}^j &= (v_j^j, v_{j+1}^j, \ldots, v_{M-j}^j)^T, \\ \boldsymbol{b}^j &= \gamma(0, L_{j+1}^j, L_{j+2}^j, \ldots, L_{M-j-1}, 0)^T \end{aligned}$$

and A^j is a tridiagonal matrix defined by

$$A^j = \begin{pmatrix} 0 & 0 & & & & & & \\ -a_{j+1}^j & 1+a_{j+1}^j & 0 & & & & & \\ & \ddots & \ddots & \ddots & & & & \\ & & -a_I^j & 1+a_I^j & 0 & & & \\ & & & 0 & 1+a_{I+1}^j & -a_{I+1}^j & & \\ & & & & \ddots & \ddots & \ddots & \\ & & & & & 0 & 1+a_J^j & -a_J^j \\ & & & & & & 0 & 0 \end{pmatrix} \tag{3.10}$$

with $a_i^j = \gamma|f_i^j|$ and $J = M-j-1$. All undefined entries in (3.10) are zeros. Obviously, from (3.3 and (3.10) we have $v_j^{j+1} = v_{M-j}^{j+1} = 0$. These values are auxiliary and will not be used in the rest of the scheme. Notice that (3.10) is a block-diagonal matrix of the form

$$A^j = \begin{pmatrix} A_{11}^j & 0 \\ 0 & A_{22}^j \end{pmatrix},$$

where A_{11} and A_{22} are lower and upper triangular matrices respectively. Thus, it is easy to show that the eigenvalues of the matrix (3.10) are

$$\begin{aligned} &\lambda_i = 1 + a_i^j \quad \text{for} \quad i = j+1, j+2, \ldots, M-j-1, \\ &\lambda_j = \lambda_{M-j} = 0. \end{aligned}$$

The non-zero solutions are in exactly the same form as those in (3.5), and so the result for the previous case applies to this one, i.e., the scheme is stable if (3.9) is satisfied.

We remark that, as mentioned in the previous section, the above stability analysis also cover the case that the mesh is not uniform. In this case we need only to replace all Δx and Δt by Δx_i and Δt_j respectively.

Comments on the accuracy of the method

The stability of the scheme under the condition (3.9) guarantees that the error due to the discretization at each time step will not grow in the time stepping procedure. Thus the error is dominated by the approximation of the derivatives and the functions in (2.1). Since the discretization scheme is based on the forward or backward difference scheme depending on the sign of the characteristic function $f(t,x,u)$, using a standard argument we can show that the approximation error in the sup norm is of order $O(|\Delta t| + \Delta x)$ in the case that $||v'||_\infty$ is bounded. In general, the approximation error in the L_2 norm is bounded by the same as above, because $v'(x)$ is always square-integrable. This will be illustrated numerically in the next section. Also, when (3.9) is satisfied, the approximate error at any time step will not propagate in the time stepping procedure.

4 NUMERICAL EXPERIMENTS

To verify the efficiency of the method discussed in the previous sections, some numerical experiments were carried out. All computations were performed in Fortran double precision on a Unix workstation.

Test 1. Consider the following control problem

$$\begin{array}{rll} \text{minimize} & -x(1), & \\ \text{subject to} & \dot{x}(t) = x(t)u(t), & \text{a.e. } t \in [s,1] \\ & x(s) = y, & \\ \text{control} & u(\cdot) : [0,1] \mapsto \{r \in \mathbb{R} : 0 \le r \le 1\}. & \end{array}$$

This control problem is from Zhou (1993) and the corresponding HJB initial value problem is

$$-v_t + \sup_{0 \le u \le 1} [-xuv_x] = 0, \tag{4.1}$$

$$v(1,x) = -x. \tag{4.2}$$

The exact value function v is

$$v(t,x) = \begin{cases} -xe^{1-t}, & \text{if } x > 0, \\ -x, & \text{if } x \le 0. \end{cases}$$

We now solve the problem by the previous method in the region $D_0 = [-1,1] \times [0,1]$, i.e. $-1 \le x \le 1$ and $0 \le t \le 1$. The number of nodes N on the t-axis and the computational region are determined using the following iterative procedure.

For any given number of nodes M ($\Delta x = 2/(M-1)$) in the x-axis, we find an N_1 using M, $f_{\text{sup},0}$ and the inequality (3.9), where $f_{\text{sup},0}$ denotes the maximum value of f on D_0. We then extend the solution region D_0 to a new region D_1 by adding the $2(N_1 - 1)$ nodes $x = \pm(1 + j\Delta x)$ for $i = 1, 2, \ldots, N_1 - 1$. We now determine an N_2 such that (3.9) is satisfied using the new value

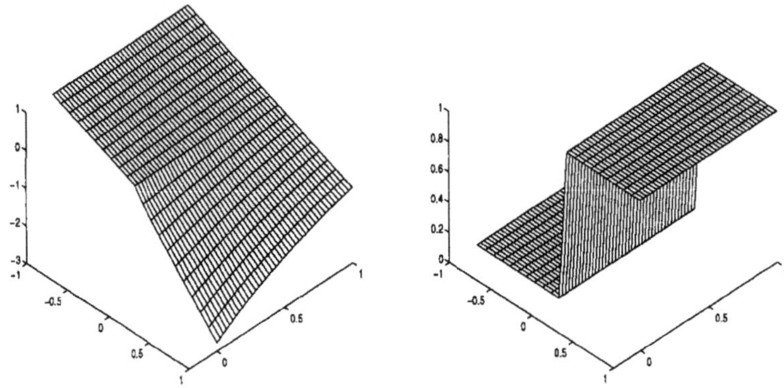

(a) Value function (b) Control function

Figure 4.1 Computed value and control functions for Test 1.

$f_{\text{sup},1}$ on the region D_1. If $N_2 = N_1$ (or equivalently $f_{\text{sup},1} = f_{\text{sup},0}$), then stop. Otherwise repeat the above process. This procedure may not converge if f becomes unbounded as $x \to \infty$ or/and $-\infty$. However, since we are only interested in the solution in the region D_0, we can simply truncate f when (t, x) is far away from D_0. For example, in the numerical solution of (4.1) and (4.2), we replace the function x in the coefficient of the second term in (4.1) by

$$f(t,x) = \begin{cases} -4, & \text{if } x < -4, \\ x, & \text{if } -4 \le x \le 4, \\ 4, & \text{if } x > 4. \end{cases} \tag{4.3}$$

The problem has been solved for various values of M and Table 4.1 is a list of the errors in the maximum norm of the numerical solutions in the region D_0 defined above. From the table we see that the numerical solution converges to the exact one at the rate of order one. This is in agreement with our comments in Remark 3.2. The value function v and the control u obtained using $\Delta x = 1/10$ are plotted in Figure 4.1. The step length Δt is chosen to be $1/41$ automatically by the computer code.

Table 4.1 Computed errors in the maximum norm for Test 1

M	11	21	41	81	161
Error	1.758e-1	8.685e-2	4.319e-2	2.154e-2	1.075e-2

The second test problem is

Test 2. Consider the following control problem

$$\begin{array}{ll} \text{minimise} & -x^2(1), \\ \text{subject to} & \dot{x}(t) = u(t), \quad \text{a.e. } t \in [s,1] \\ & x(s) = y, \\ \text{control} & u(\cdot) : [0,1] \mapsto \{r \in \mathbb{R} : -1 \le r \le 1\}. \end{array}$$

This problem is from Yong (1992) and the corresponding HJB initial value problem is

$$\begin{aligned} -v_t + \sup_{-1 \le u \le 1} [-uv_x] &= 0, \\ v(1,x) &= -x^2 \end{aligned}$$

with the exact value function

$$v(t,x) = \begin{cases} -[x + (1-t)]^2, & \text{if } x \ge 0, \\ -[x - (1-t)]^2, & \text{if } x < 0 \end{cases}$$

$$-\left[|x| + (1-t)\right]^2.$$

For this example we have that $\max|f| = \max|u| = 1$ in (3.9). Table 4.2 is a list of maximum errors at the mesh points and the discrete L_2-norms for different values of M. From the table we see that, for this case, the approximate solution converges to the exact one at the order of h in the discrete maximum norm and at the order of $h^{1.5}$ in the discrete L_2 norm. The results using $\Delta x = 1/10$ are plotted in Figure 4.2. From the figure we see that v is not smooth along the line $x = 0$.

Table 4.2 Computed errors in the maximum norm for Test 2

error \ M	10	20	40	80	160
in $\|\|\cdot\|\|_\infty$	4.222e0	2.053e-2	1.013e-1	5.032e-2	2.508e-2
in $\|\|\cdot\|\|_2$	1.337e-1	4.158e-2	1.378e-2	4.716e-3	1.640e-3

The last test problem is

Test 3. Consider the following control problem

$$\begin{array}{ll} \text{minimise} & \dfrac{1}{4}\displaystyle\int_s^1 u^2(t)dt - x(1), \\ \text{subject to} & \dot{x}(t) = x(t)u(t), \quad \text{a.e. } t \in [s,1] \\ & x(s) = y, \\ \text{control} & u(\cdot) : [0,1] \mapsto \{r \in \mathbb{R} : -1 \le r \le 1\}. \end{array}$$

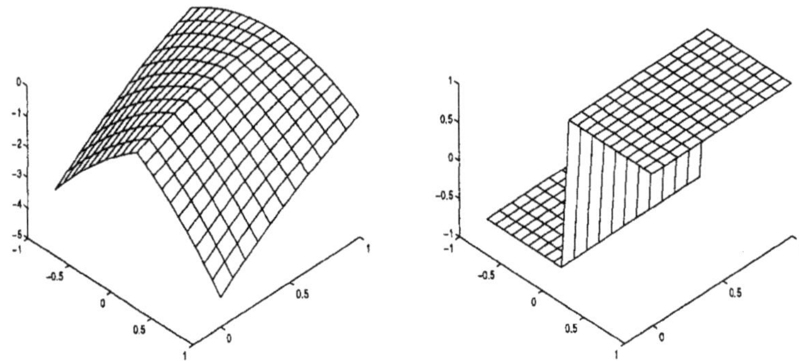

(a) Value function (b) Control function

Figure 4.2 Computed value and control functions for Test 2.

This control problem is similar to Test 1, but it has different cost function and range for u. The corresponding HJB equation is

$$
\begin{aligned}
-v_t + \sup_{-1 \leq u \leq 1} \left[- xuv_x - \frac{u^2}{4} \right] &= 0, \\
v(1, x) &= -x.
\end{aligned}
$$

The exact solution to this problem is unknown. We now use the method discussed in the previous sections to solve the problem in the range $-1 \leq x \leq 1$ and $0 \leq t \leq 1$. We choose a uniform mesh along the x-axis with $M = 21$ nodes, i.e. $\Delta x = 0.1$. Using the truncated function f in (4.3) and the meshing technique for the t range given above we obtained that $N = 42$. The computed results for u and v are plotted in Figure 4.3. From this figure we see that both of the control and the value function are smooth in the solution domain.

5 CONCLUSIONS

In this we paper presented an upwind finite difference method for the numerical solution of a two dimensional Hamilton-Jacobi-Bellman (HJB) equation governing a class of optimal feedback control problems. The method is based on an explicit upwind finite difference scheme and it has been shown that the method is stable under certain constraints on the step lengths of the discretization. Numerical experiments were performed to show the effectiveness of the method. The numerical results show that the method gives accurate approximations to both of the control and the value function for problems having both smooth and non-smooth controls.

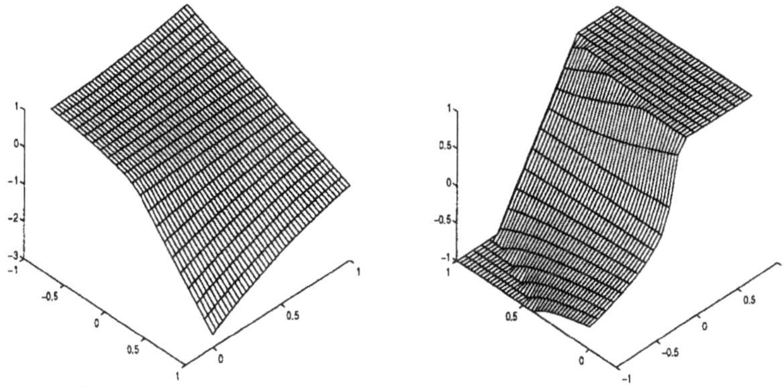

(a) Value function (b) Control function

Figure 4.3 Computed value and control functions for Test 3.

Acknowledgments

The authors gratefully acknowledge the support of the Australian Research Council.

References

Clarke, F.H. (1993), *Optimization and Nonsmooth Analysis*, Wiley, New York.

Ekeland, I. (1974), On the variational principle, *J. Mathematical Analysis and Applications*, Vol.47, pp. 324–353.

Li, X. and Yong, J. (1995), *Optimal Control Theory for Infinite Dimensional Systems*, Birkhäuser, Boston-Basel-Berlin.

Lions, P.L. (1983), Optimal control of diffusion process and Hamilton-Jacobi-Bellman equations, part 2: viscosity solutions and uniqueness, *Comm. Partial Differential Equations*, Vol. 11, pp. 1229–127.

Maslov, V.P., On a new principle of superposition for optimization problems, *Russian Maths. Surveys*, Vol. 42:3, pp. 43–84.

Ortega, J.M. (1990), *Numerical Analysis: A Second Course*, SIAM, Philadelphia.

Peyret, R. and Taylor, T.D. (1983), *Computational Methods for Fluid Flow*, Springer-Verlag, New York.

Yong, J. (1992), *The Method of Dynamic Programming and Hamilton-Jacobi-Bellman Equations*, Shanghai Science & Technology Publisher, Shanghai (in Chinese).

Zhou, X.Y. (1990), Maximum principle, dynamic programming and their connection in deterministic control, *J. Optimization Theory and Applications*, Vol. 65, pp. 363–373.

Zhou X.Y. (1993), Verification theorems within the framework of viscosity solutions, *J. Mathematical Analysis and Applications*, Vol. 177, pp. 208–225.

Zhou, X.Y. and Sethi, S. (1993), A sufficient condition for near optimal stochastic controls and its applications to manufacturing systems, *J. Applied Mathematics and Optimization*, Vol. 177, pp. 208–225.

18 AN EFFICIENT APPROXIMATION METHOD FOR A CLASS OF CONTINUOUS LINEAR PROGRAMS

K. H. Wong, M. I. Kasiama and C. Myburgh

Department of Mathematics,
The University of the Witwatersrand,
South Africa

Abstract: A general class of continuous linear programs, which involves an integral objective function and a Volterra type constraint inequality with convolution kernel is considered. By using a method similar to that of Lukas and Teo (1991), the above problem is converted into a sequence of optimal control problems not involving any integral constraint, each of which can be solved by the control parameterisation method. Convergence analysis of the above approximation scheme is considered.

Key words: polynomial evaluation, iteration scheme

X. Yang et al. (eds.), Progress in Optimization, 269–284.

1 INTRODUCTION

In this paper, we use optimal control theory to develop an efficient algorithm for solving a general class of continuous linear programs. The problem under consideration was first considered by Bellmann (1957) in an attempt to model some economic processes which he called Bottleneck processes. Since then, the problem has been studied by a number of authors whose work can be loosely divided into two areas, those concerned with establishing strong duality theorems and those with computational methods.

All these authors gave strong duality results with varying algebraic restrictions on the problem. On the computational side, two approaches can be found. The simpler of the two is to solve the problem via a series of discrete approximations, called discretisations, to the original continuous linear programs (CLP). Such is the approach taken by Burie and Abrham (1973). However, these methods fail to address the infinite-dimensional nature of the problem and these have two disadvantages. First, the computational times involved can be very large as a result of solving a very large finite-dimensional linear program, and second, the problem may have an optimal solution of a simple form that is obscured in the discretised version. These observations have led a number of authors to attempt to solve the problem directly.

The approach used in this paper involves the conversion of the continuous linear program into a sequence of optimal control problems which involve integrodifferential equations. Our problem involves an integral objective function and a Volterra type constraint inequality with convolution kernel. We use a method similar to that used by Lukas and Teo [5] to approximate the kernel so that the Volterra type constraint will give rise to a simple continuous constraint not involving any kernel, together with a system of ordinary differential equations.

On this basis, we obtain a sequence of optimal control problems not involving any integral constraint, each of which can be solved by the control parameterisation method. Convergence analysis is carried out to justify the proposed approximation scheme.

2 STATEMENT OF THE PROBLEM

We consider the problem (P) :

$$\text{Maximise } \; f(x) = \int_0^T c^T(t)x(t)dt \tag{2.1}$$

$$\text{subject to } \; g(x(t),t) = B^T(t)x(t) + \int_0^t K^T(t-s)x(s)ds - b(t) \leq 0 \tag{2.2}$$

$$-d \leq \dot{x}(t) \leq d, t \in [0,T] \tag{2.3}$$

$$-e \leq x(0) \leq e \tag{2.4}$$

where $x(t) \in \mathbb{R}^n$ is a continuous function, $c(t) \in \mathbb{R}^n$, $B(t) \in \mathbb{R}^n$, $K(t) \in \mathbb{R}^n$, $d \in \mathbb{R}^n$, $e \in \mathbb{R}^n$.

For the sake of simplicity, we shall here restrict our discussion to a single constraint. To proceed further, we assume throughout that the following condition is satisfied:

(A1) $K(t) \in C^3[0,T]$, $B(t)$, $c(t)$, $b(t)$ are all continuous functions.

Let $F = \{x : [x_1, \ldots, x_n]^T$ is a continuous function defined on $[0,T]$ such that the constraints (2.2)-(2.4) are satisfied$\}$. Thus the problem (P) may be re-stated as follows: Find an $\hat{x} \in F$ such that $f(x)$ is maximised over all $x \in F$.

3 APPROXIMATE PROBLEM

To begin, we firstly let

$$y_i(t) = \int_0^t K_i(t-s)x_i(s)ds, i = 1, \ldots, n \tag{3.1}$$

so that

$$\int_0^t K^T(t-s)x(s)ds = \sum_{i=1}^{n} y_i(t) \tag{3.2}$$

To solve problem (P), we shall construct a sequence of approximated problems (P^N) each of which can be solved by the method of control parameterisation. We start by approximating the kernel $K_i(t)$ by an expansion of the form

$$K_i(t) \approx K_{i,N}(t) \tag{3.3}$$

where

$$K_{i,N}(t) = \sum_{m=1}^{N} \alpha_i^m k^m(t) \tag{3.4}$$

Here the basis functions k^m solve a system of differential equations with constant coefficients of the form

$$\dot{k}^m(t) = \sum_{l=1}^{N} a^{m,l} k^l(t) \tag{3.5}$$

By substituting $K_{i,N}$ for K_i, equation (3.1) is approximated by

$$y_i(t) = \int_0^t K_{i,N}(t-s)x_i(s)ds \tag{3.6}$$

Because of the special form of $K_{i,N}$, this equation is equivalent to a particular system of ordinary differential equations.

For each positive integer m, $y^m(t/x)$ satisfies the following differential equation corresponding to each $x \in F$:

$$\dot{y}_i^m(t) = k^m(0)x_i(t) + \sum_{l=1}^{N} a^{m,l} y_i^l(t), \quad y_i^m(0) = 0 \tag{3.7}$$

Now, let us define

$$y_i^N(t/x) = \sum_{m=1}^{N} \alpha_i^m y_i^m(t/x) \tag{3.8}$$

The following result can be verified directly.

Lemma 3.1 *$y_i^N(t/x)$ satisfies the following system*

$$y_i^N(t/x) = \int_0^t K_{i,N}(t-s)x_i(s)ds \tag{3.9}$$

The original system (3.1) has been approximated by a system of ordinary differential equations. This can be solved by any of the available techniques. An important advantage of this representation is clearly that it requires only one kernel evaluation at $t = 0$, which can be done exactly. If the given kernel $K(t)$ has a known exact representation of the form (3.4), the system (3.7) defines the exact solution of (3.1).

There are obviously many ways of approximating $K_i(t)$ in the form of (3.4). We shall use a polynomial approximation of degree $N-1$ represented by its Chebyshev series

$$K_{i,N}(t) = \sum_{m=1}^{N} \alpha_i^m \bar{T}^{m-1}(t) \tag{3.10}$$

Here $\bar{T}^m$ is the shifted Chebyshev polynomial of degree m defined by

$$\bar{T}^m(t) = T^m(2t/T-1) = \cos\left(m\cos^{-1}(2t/T-1)\right), t \in [0,T] \tag{3.11}$$

where $T^m(t)$ is the usual Chebyshev polynomial on $[-1,1]$.

Lemma 3.2 *The functions $k^m(t) = \bar{T}^{m-1}(t)$ solve the system*

$$\dot{k}^m(t) = \frac{4(m-1)}{T}\begin{cases} 0 & m=1 \\ k^{m-1}+k^{m-3}+\cdots+k^3+\frac{k^1}{2} & m \geq 2 \text{ even} \\ k^{m-1}+k^{m-3}+\cdots+k^2 & m \geq 2 \text{ odd.} \end{cases} \tag{3.12}$$

Proof: The proof is by induction using the well–known recurrence relationship for Chebyshev polynomials,

$$T^{j+1}(t) = 2tT^j(t) - T^{j-1}(t) \tag{3.13}$$

□

Thus the basis functions $k^m(t) = \bar{T}^{m-1}(t)$ solve a system of the form (3.5), where the values of $a^{m,l}$ in (3.5) are

$$a^{m,l} = \begin{cases} 0 & l \geq m \\ [2(m-1)/T]\left[1-(-1)^{m-l}\right] & l \neq 1 \\ [(m-1)/T]\left[1+(-1)^m\right] & l = 1 \end{cases} \tag{3.14}$$

Also,

$$k^m(0) = \bar{T}^{m-1}(0) = T^{m-1}(-1) = (-1)^{m-1} \tag{3.15}$$

There are several methods of finding a good polynomial approximation of $K_i(t)$. We used the polynomial P_i^{N-1} of degree $N-1$ interpolating $K_i(t)$ at the Chebyshev points on $[0,T]$, i.e. the zeroes of $\bar{T}^N$:

$$t_m = T/2 + (T/2)\cos[(2m-1)\pi/2N], m = 1,\dots,N \tag{3.16}$$

It is well known that P_i^{N-1} can be written as

$$P_i^{N-1}(t) = \sum_{m=1}^{N} \alpha_i^m \bar{T}^{m-1}(t) \tag{3.17}$$

where α_i^m is easily computed from

$$\alpha_i^m = \begin{cases} \frac{1}{N}\sum_{j=1}^{N} K_i(t_j), & m = 1 \\ \frac{2}{N}\sum_{j=1}^{N} \bar{T}^{m-1}(t_j), & m = 2,\dots,N \end{cases} \tag{3.18}$$

This directly gives approximation (3.3).

The important reason for using $P_i^{N-1}(t)$ is that it is known to approximate a smooth function $K_i(t)$ very well in the maximum norm.

Having introduced an approximation of $K_i(t)$, we are now in a position to replace the problem (P) by a sequence of approximated problems (P^N). The problem (P^N) is obtained from that of (P) by replacing the constraint (2.2) by the following constraint:

$$g^N(x(t),t) = B^T(t)x(t) + \int_0^t \left[K^N(t-s)\right]^T x(s)ds - b(t) + \frac{1}{N^2} \le 0 \tag{3.19}$$

Problem (P^N)

The approximate problem (P^N) is equivalent to the following problem:

$$\text{Maximise } f(x) = \int_0^T c^T(t)x(t)dt \tag{3.20}$$

$$\text{subject to } \dot{y}_i^m(t) = k^m(0)x_i(t) + \sum_{l=1}^{N} a^{m,l} y_i^l(t) \tag{3.21}$$

$$y_i^m(0) = 0, i = 1,\dots,n \text{ and } m = 1,\dots,N \tag{3.22}$$

$$g^N(x(t),t,y(t)) = B^T(t)x(t) + \sum_{i=1}^{n}\sum_{m=1}^{N} \alpha_i^m y_i^m(t) - b(t) + \frac{1}{N^2} \le 0 \tag{3.23}$$

$$-d \leq \dot{x}(t) \leq d, t \in [0, T] \tag{3.24}$$

$$-e \leq x(0) \leq e \tag{3.25}$$

where $x(t)$ is a continuous function.

4 ANALYSIS OF THE METHOD

In this section, our aim is to investigate some convergence properties relating to the approximation of problem (P) by a sequence of problems (P^N).

Let

$$R_i^N(t) = K_i(t) - \left(P_i^N(t)\right), t \in [0, T] \tag{4.1}$$

denote the error of the approximation. We assume that $K_i(t)$ is approximated by the polynomial $P_i^N(t)$ defined in the above section.

It is clear that if $K_i(t) \in C^3[0, T]$, then the error R_i^N satisfies

$$\|(R_i^N)\|_\infty \leq O\left(\frac{c(N)}{N^3}\right) \quad \text{as } N \to \infty \tag{4.2}$$

where

$$c(N) = \left(\frac{2}{\pi}\right) \log N + 2 \tag{4.3}$$

In other words, if $K_i(t)$ is smooth, then it is known that $\|R_i^N\|_\infty$ is small even for moderate degree $N - 1$. In fact, $\|R_i^N\|_\infty$ is not much larger than

$$E_{N-1} = \min\left\{\|K - Q\|_\infty : Q \in \mathcal{P}^{N-1}\right\} \tag{4.4}$$

the smallest possible error achievable over the the space $\mathcal{P}^{N-1}$ of polynomials of degree $N - 1$ (which is achieved by the minimax polynomial). We can also note that $c(N)$ grows slowly and for $N = 20$, $c(N) \leq 4$. The following known result gives a computable bound on $\|R^N\|_\infty$.

Let $\dot{F}$ be the interior of F defined by

$$\dot{F} = \{x \in F : g(x(t), t) < 0, t \in [0, T]\} \tag{4.5}$$

Let F^N be the set of all continuous functions such that constraints (3.23)-(3.25) are satisifed.

To proceed further, we need the following assumptions:

(B1) $\dot{F}$ is non-empty.

(B2) For each $x \in F$, there exists $\hat{x} \in \dot{F}$ such that $(1 - \alpha)x + \alpha\hat{x} \in \dot{F}$ for $0 < \alpha \leq 1$.

Lemma 4.1 *There exists $N_1 \geq 0$ such that $F^N \subset F$ for all $N \geq N_1$*

Proof: Firstly, we note that for all $x \in F^N$ we have

$$|x(t)| \leq \sqrt{n}(Td + e) \tag{4.6}$$

where $|\cdot|$ denotes the usual Euclidean norm in $\mathbb{R}^n$ defined by

$$|x(t)| = \left(\sum_{i=1}^{n} x_i^2(t)\right)^{\frac{1}{2}} \tag{4.7}$$

Thus, from (3.19), (2.2), (4.2) and (4.6), we obtain the following equation

$$\begin{aligned} g(x(t),t) - g^N(x(t),t) &= \int_0^t \left[K(t-s) - K^N(t-s)\right]^T x(s)ds - \frac{1}{N^2} \\ &\leq O\left(\frac{c(N)}{N^3}\right) - \frac{1}{N^2} \end{aligned} \tag{4.8}$$

Thus from (4.8), (4.3), it is clear that there exists $N_1 > 0$ such that for all $N \geq N_1$, we have

$$g(x(t),t) \leq g^N(x(t),t) \tag{4.9}$$

for all $x \in F^N$ and for all $t \in [0,T]$.

The conclusion of the lemma then follows easily from (4.9) and the definition of F^N and F. □

Theorem 4.1 *Let x^* be an optimal function for the problem (P) and let x^N be an optimal function for the problem (P^N). Then*

$$\lim_{N\to\infty} f(x^N) = f(x^*) \tag{4.10}$$

In particular, if $x^* \in \dot{F}$, then there exists $\hat{N} \geq 0$ such that for all $N \geq \hat{N}$, x^N is also an optimal function for the problem (P).

Proof: From Lemma 4.1 and the definition of x^N, it is clear that there exists $N_1 \geq 0$ such that

$$x^N \in F \tag{4.11}$$

for all $N \geq N_1$. Thus from the definition of x^*, we have

$$f(x^*) \geq f(x^N) \tag{4.12}$$

for all $N \geq N_1$. Thus,

$$f(x^*) \geq \lim_{N\to\infty} \sup f(x^N) \tag{4.13}$$

On the other hand, from assumption (B2), there exists $\hat{x} \in \dot{F}$ such that

$$x_\alpha = (1-\alpha)x^* + \alpha\hat{x} \in \dot{F}, 0 < \alpha \leq 1 \tag{4.14}$$

From (3.19), (2.2), (4.2), (4.3) and the fact that $|x_\alpha|$ is bounded, we get

$$|g^N(x_\alpha(t),t) - g(x_\alpha(t),t)| \leq O\left(\frac{1}{N^2}\right) \tag{4.15}$$

Given any $\epsilon > 0$, there exists $\hat{\alpha} \in (0, 1]$ such that

$$x_{\hat{\alpha}} \in \dot{F} \tag{4.16}$$

and

$$|f(x_{\hat{\alpha}}) - f(x^*)| \leq \epsilon \tag{4.17}$$

Thus from (4.15) and (4.16), there exists $N_{\hat{\alpha}} \geq 0$ such that whenever $N \geq N_{\hat{\alpha}}$, we have

$$x_{\hat{\alpha}} \in F^N \tag{4.18}$$

The above equation together with the definition of x^N implies

$$f(x_{\hat{\alpha}}) \leq f(x^N) \tag{4.19}$$

for all $N \geq N_{\hat{\alpha}}$. Hence from (4.17) and (4.19), we obtain

$$f(x^*) - \epsilon \leq f(x^N) \tag{4.20}$$

for all $N \geq N_{\hat{\alpha}}$. Thus,

$$f(x^*) - \epsilon \leq \lim_{N \to \infty} \inf f(x^N) \tag{4.21}$$

Since ϵ is arbitrary, we have

$$f(x^*) \leq \lim_{N \to \infty} \inf f(x^N) \tag{4.22}$$

From (4.13) and (4.22), we have

$$f(x^*) \leq \lim_{N \to \infty} f(x^N) \tag{4.23}$$

Finally if $x^* \in \dot{F}$, then we can take $x_{\hat{\alpha}} = x^*$ and $\hat{N} = N_{\hat{\alpha}}$ to obtain

$$f(x^*) = f(x^N) \tag{4.24}$$

for all $N \geq \hat{N}$. The conclusion of this theorem follows easily from (4.11) and (4.24). □

5 PIECEWISE LINEAR CONTINUOUS APPROXIMATION

In this section, we shall use linear function approximation approach similar to Jennings et all (1991) so that the problem (P^N) can be converted into a sequence of finite dimensional optimisation problems which can be easily solved by some existing non-linear programming software.

We can generate piecewise linear function $x(t)$ as follows:

Firstly we let

$$\dot{x}(t) = u(t) \tag{5.1}$$

$$x(0) = \xi \tag{5.2}$$

Let $\{I_p\}_{p=1}^{\infty}$ be a sequence of partitions of the interval $[0, T]$ such that I_p has n_p elements, I_{p+1} is a refinement of I_p and $\|I_p\| \to 0$ as $p \to \infty$, where $\|I_p\|$ denotes the length of the largest element of I_p. Let

$$u(t) = \sum_{k=1}^{n_p} \sigma_k^p \chi_{I_{p,k}}(t), t \in [0, T] \tag{5.3}$$

where

$$\chi_{I_{p,k}}(t) = \begin{cases} 1 & \text{if } t \in I_{p,k} \\ 0 & \text{otherwise} \end{cases} \tag{5.4}$$

Let U^p be the set of all step-functions consistent with the partition I_p. Let

$$\theta^p = \begin{bmatrix} \sigma_p \\ \xi \end{bmatrix} \tag{5.5}$$

where

$$\sigma^p = [\sigma_1^{pT}, \sigma_2^{pT}, \ldots, \sigma_n^{pT}] \in U^p \tag{5.6}$$

Then each piecewise linear function $x(t)$ consistent with the partition I_p can be uniquely identified with the vector $\theta^p \in U^p \times \mathbb{R}^n$, and vice versa.

Let F^p be the subset of F consisting of all piecewise linear functions x(t) consistent with the partition I_p. The first approximated problem (P^p) is stated as follows:

Find $\hat{x} \in F^P$ such that $f(x)$ is maximised over all $x \in F^p$.

In order to prove some convergence results given later, we need to make two assumptions similar to assumption (B1) and (B2).

Let

$$\dot{F}^p = \{x \in F^p : g(x(t), t) < 0, t \in [0, T]\} \tag{5.7}$$

To proceed further, we need the following assumptions:

(C1) $\dot{F}^p$ is non-empty.

(C2) For each $x \in F^p$, there exists $\bar{x} \in F^p$ such that $(1 - \alpha)x + \alpha\bar{x} \in \dot{F}^p, 0 < \alpha \leq 1$.

However in the new approximate problem $(P^{N,p})$ described below, we need to consider θ as the control vector and $x(t)$ as some components of the state variable.

Let

$$z(t) = \left[y_1^1(t), \ldots, y_1^N(t), y_2^1(t), \ldots, y_2^N(t), \ldots, y_n^1(t), \ldots, y_n^N(t), x^T(t)\right]^T \tag{5.8}$$

be the new state variable.

Then the new differential equation takes the form

$$\frac{dz}{dt} = h(\theta, z, t) \tag{5.9}$$

$$z(0) = \hat{h}(\theta) \tag{5.10}$$

where

$$h_{m+iN}(\theta, z(t), t) = k(0)x_i(t) + \sum_{l=1}^{N} a^{m,l} y_i^l(t), i = 0, \ldots, n-1, m = 1, \ldots, N \tag{5.11}$$

and

$$\begin{aligned} h_{i+Nn}(\theta, z(t), t) &= \sum_{k=1}^{n_p} \sigma_{i,k}^p \chi_k^p(t), i = 1, \ldots, n & (5.12) \\ \hat{h}_i(\theta) &= 0, i = 1, \ldots, mN & (5.13) \\ \hat{h}_{Nn+i}(\theta) &= \xi_i, i = 1, \ldots, n & (5.14) \\ & & (5.15) \end{aligned}$$

Define

$$Z^p = \{\theta \in U^p \times \mathbb{R}^n : -e \leq \xi \leq e, -d \leq \sigma \leq d\} \tag{5.16}$$

Let $z(\cdot/\theta)$ be the solution of the system corresponding to each $\theta \in Z^p$. Then constraint (3.23) becomes:

$$\hat{g}^N(z(t/\theta), t) = B^T(t)x(t/\theta) + \sum_{i=1}^{n} \sum_{m=1}^{N} \alpha_i^m y_i^m(t/\theta) - b(t) + \frac{1}{N^2} \leq 0, t \in [0, T] \tag{5.17}$$

Let $F^{N,p}$ be the subset of Z^p such that the constraint (5.17) is satisfied. The approximated problem $(P^{N,p})$ can be stated as follows:
Problem $(P^{N,p})$

Given the system (5.9)-(5.14), find a vector $\theta^{N,p} \in F^{N,p}$ such that the cost functional

$$\hat{f}(\theta) = \int_0^T c^T(t)x(t/\theta)dt \tag{5.18}$$

is maximised over $F^{N,p}$.

Remark 5.1 The reason that we have to treat $x(t)$ as the state variable rather than the control variable is because the control parameterisation algorithm devised in Teo and Jennings (1989) cannot work efficiently when a control variable appears in the continuous inequality (5.17).

6 CONSTRAINT APPROXIMATION

Let the state inequality (5.17) when restricted to Z^p be written as

$$G^N(\theta) = \int_0^T \max\left\{0, \hat{g}^N\left(z(t/\theta), t\right)\right\} dt = 0 \tag{6.1}$$

which is however not smooth.

Now replace the non-smooth functions

$$\max\{0, \hat{g}(z(t/\theta), t)\} \tag{6.2}$$

by the smooth ones given by

$$\hat{g}_\epsilon^N(z(t/\theta), t) \tag{6.3}$$

where

$$\hat{g}_\epsilon^N(z(t/\theta), t) = \begin{cases} \hat{g}^N(z(t/\theta), t), & \hat{g}^N(z(t/\theta), t) > \epsilon \\ [\hat{g}^N(z(t/\theta), t) - \epsilon]^2/4\epsilon, & |\hat{g}^N(z(t/\theta), t)| \le \epsilon \\ 0 & \hat{g}^N(z(t/\theta), t) < -\epsilon \end{cases} \tag{6.4}$$

Let

$$G_\epsilon^N(\theta) = \int_0^T \hat{g}_\epsilon^N(z(t/\theta), t)dt \tag{6.5}$$

As mentioned in Teo and Jennings (1989), in order to overcome numerical difficulty, we need to solve the following approximation problem:

Problem $(P_{\epsilon,\gamma}^{N,p})$

This is problem $(P^{N,p})$ with the constraint (5.17) replaced by

$$G_\epsilon^N(\theta) \le \gamma \tag{6.6}$$

Lemma 6.1 *There exists a $\gamma(\epsilon) > 0$ such that for all γ, $0 < \gamma < \gamma(\epsilon)$, any feasible vector $\theta_{\epsilon,\gamma}^{N,p}$ of problem $(P_{\epsilon,\gamma}^{N,p})$, i.e.*

$$G_\epsilon^N(\theta_{\epsilon,\gamma}^{N,p}) \le \gamma \tag{6.7}$$

is also a feasible vector of the problem $(P^{N,p})$

Proof: The proof is similar to that of Lemma 8 of Teo and Jennings (1989). □

We can compute a sequence of sub-optimal parameters $\{\theta_{\epsilon,\gamma}^{N,p,*}\}$ to problem $(P^{N,p})$ with each of them in $F^{N,p}$. At this stage, the algorithm below, which is the same as that given in Teo and Jennings (1989), can be used to generate the solution of the problem.

Step 0 Choose initial values of $\epsilon(= 10^{-2})$ and $\gamma(= \epsilon^{T/4})$, say

Step 1 Solve $(P_{\epsilon,\gamma}^{N,p})$ to give $\theta_{\epsilon,\gamma}^{N,p,*}$

Step 2 If $\theta_{\epsilon,\gamma}^{N,p,*}$ is not feasible, set $\gamma = \gamma/2$ and go to Step 1. Otherwise, if $\epsilon > \bar{\epsilon}$, set $\epsilon = \epsilon/10$ and $\gamma = \gamma/10$ and got to Step 1. Otherwise stop.

Remark 6.1 Problem $(P_{\epsilon,\gamma}^{N,p})$ is essentially a nonlinear mathematical programming problem which can be solved by any standard software package such as NLPQL. However, in order to solve $(P_{\epsilon,\gamma}^{N,p})$, we need to know the gradient of the cost functional and that of the constraints with respect to θ. The method of finding this gradient is given in Teo and Jennings (1989).

7 CONVERGENCE RESULT

Our main concern in this section is to prove some results which justify the method described in the above section.

Lemma 7.1 *Let*

$$I_{p,k} = [t_{k-1}^p, t_k^p], k = 1, \dots, n_p \tag{7.1}$$

such that

$$0 = t_0^p < t_1^p \dots < t_{n_p}^p = T \tag{7.2}$$

For each function $x(t)$ and for each p, let

$$x^p(t) = \sum_{k=1}^{n_p} \left\{ x(t_{k-1}^p) + \frac{x(t_k^p) - x(t_{k-1}^p)}{t_k^p - t_{k-1}^p}(t - t_k^p) \right\} \chi_{I_{p,k}}(t) \tag{7.3}$$

where

$$\chi_{I_{p,k}}(t) = \begin{cases} 1 & \text{if } t \in I_{p,k} \\ 0 & \text{otherwise} \end{cases} \tag{7.4}$$

Then

$$\lim_{p \to \infty} \int_0^T |x^p(t) - x(t)| dt = 0 \tag{7.5}$$

Proof: Let $t_1 \in I_{p,k}$, then it is clear from the definition of the partition I_p that

$$t_k^p \to t_{k-1}^p \text{ as } p \to \infty \tag{7.6}$$

Similarly

$$t_1 \to t_{k-1}^p \text{ as } p \to \infty \tag{7.7}$$

Thus from (7.3), we have

$$\lim_{p \to \infty} x^p(t_1) = \lim_{p \to \infty} \left[x(t_{k-1}^p) + \frac{x(t_k^p) - x(t_{k-1}^p)}{t_k^p - t_{k-1}^p}(t_1 - t_k^p) \right] = x(t_1) \tag{7.8}$$

Thus the conclusion follows easily from the fact that $|x(t)|$ and $|x^p(t)|$ are bounded and the dominant convergence theorem. □

Lemma 7.2 *Let θ_p^N be an optimal vector for the problem P_p^N, let x_p^N be the piecewise linear function uniquely identified by θ_p^N, then there exists $N_1 \geq 0$ independent of p such that*

$$x_p^N \in F \tag{7.9}$$

for all $N \geq N_1$. That is for each $p > 0$, x_p^N is a feasible function for the problem (P) whenever $N \geq N_1$.

Proof: The proof follows directly from the fact that $x_p^N \in F^N$ and Lemma 4.1. □

Lemma 7.3 *Let x_p^*, x_p^N, x^N, x^* be, respectively, optimal functions for the problem (P^p), $(P^{N,p})$, (P^N) and the original problem (P), then*

$$\lim_{p\to\infty} f(x_p^*) = f(x^*) \tag{7.10}$$

$$\lim_{p\to\infty} f(x_p^N) = f(x^N) \tag{7.11}$$

Proof: In view of Lemma 7.1 and Assumptions (B1), (B2), (C1), (C2) and (C3), the proof is similar to that given in Theorem 4.2 of Teo and Jennings (1989). □

Theorem 7.1 *Let x_p^*, x_p^N, x^N, x^* be, respectively, optimal functions for the problem (P^p), $(P^{N,p})$, (P^N) and the original problem (P). Let θ_p^N be the optimal vector which represents x_p^N. Then for each $p > 0$, the sequence of vectors θ_p^N (considered as a sequence of N) has a subsequence which converges to $\hat{\theta}_p$, where $\hat{x}_p$, the piecewise linear function uniquely identified by $\hat{\theta}_p$, is also a feasible funtion for the problem (P). Moreover,*

$$f(x^*) = \lim_{N\to\infty}\lim_{p\to\infty} f(x_p^N) = \lim_{p\to\infty}\lim_{N\to\infty} f(x_p^N) = \lim_{p\to\infty} f(\hat{x}_p) \tag{7.12}$$

Proof: The sequence of vectors $\{\theta_p^N\}_{N=1}^{\infty}$ is obviously bounded in the Euclidean norm of $R^{(n_p+1)\times n}$. Thus, it has a subsequence which converges to θ_p^*. From Lemma 7.2, it is clear that $\hat{x}_p$ is a feasible function for the problem (P). Also,

$$\lim_{N\to\infty} \sup f(x_p^N) \geq f(\hat{x}_p) \geq \lim_{N\to\infty} \inf f(x_p^N) \tag{7.13}$$

From Lemma 7.3 and Theorem 4.1, we have

$$\lim_{N\to\infty}\lim_{p\to\infty} f(x_p^N) = \lim_{N\to\infty} f(x^N) = f(x^*) \tag{7.14}$$

By following exactly the same argument as that given in the proof of Theorem 4.1, we get

$$\lim_{N\to\infty} f(x_p^N) = f(x_p^*) \tag{7.15}$$

Hence, in view of the above equation and Lemma 7.3, we get

$$\lim_{p\to\infty}\lim_{N\to\infty} f(x_p^N) = \lim_{p\to\infty} f(x_p^*) = f(x^*) \tag{7.16}$$

From (7.13) and (7.15), we get

$$f(x_p^*) = f(\hat{x}_p) \tag{7.17}$$

Thus,

$$\lim_{p\to\infty} f(x_p^*) = \lim_{p\to\infty} f(\hat{x}_p) \tag{7.18}$$

Thus, from Lemma 7.3 and (7.18), we get

$$f(x^*) = \lim_{p\to\infty} f(\hat{x}_p) \tag{7.19}$$

This completes the proof of this theorem. □

8 EXAMPLE

In this section, we apply the approximation method to the following advertising problem.

Let $x(t)$ be the rate of advertising at time t. Then, the total amount of influence to the public at time t can be described by the integral equation

$$G(t) = \int_0^t Kx(s)e^{-\phi(t-s)^n} ds \tag{8.1}$$

This problem (Q) can be described as follows

$$\min \int_0^T x(t)c(t)dt \tag{8.2}$$

subject to

$$G(t) = \int_0^t Kx(s)e^{-\phi(t-s)^n} ds \geq g(t), t \in [0,T] \tag{8.3}$$

where the objective function is due to the total cost for advertising from $t = 0$ to $t = T$. $c(t)$ and $g(t)$ are given functions of t.

Let $(Q1)$ denote the problem obtained from (Q) by using $n = 2$, $\phi = 1/2$, $K = 1$, $c(t) = 1$ and $g(t) = t^2$. Let $(Q2)$ denote the problem obtained from (Q) by using $n = 2$, $\phi = 1/2$, $K = 1$, $c(t) = 1$ and $g(t) = \sqrt{(t)}/10$.

Both $(Q1)$ and $(Q2)$ have been solved using $N = 6$ and $N = 8$ and $n_p = 4$,8 and 16.

The results are given in Table 8.1 and Table 8.2 below:

Table 8.1

$f(x^*)$ for $(Q1)$	$n_p = 4$	$n_p = 8$	$n_p = 16$
$N = 6$	0.5263	0.5262	0.5261
$N = 8$	0.5263	0.5261	0.5260

The graphs of $x^*(t)$, $G(t)$ and $g(t)$ for problems $(Q1)$ and $(Q2)$ are plotted in Figures 8.1 and 8.2, respectively.

From these figures, it is clear that at the computed optimal solution, $G(t)$ and $g(t)$ are almost the same.

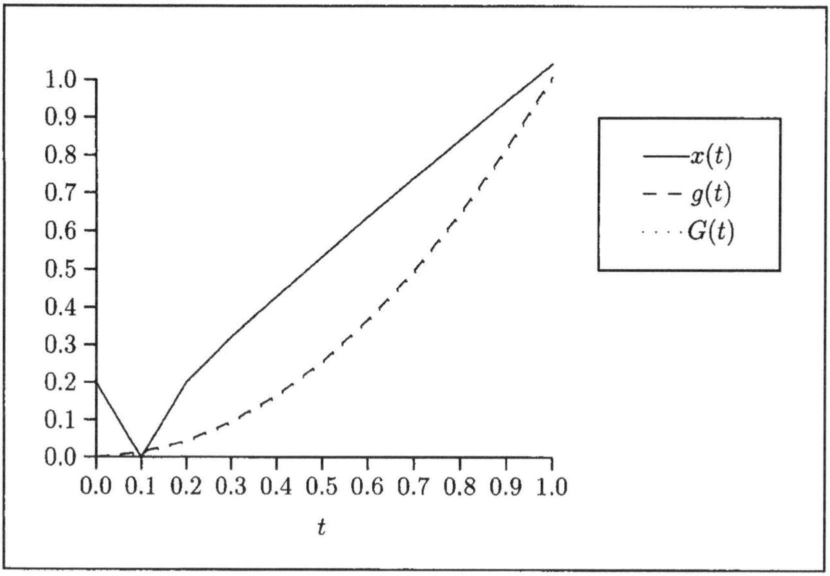

Figure 8.1 $x^*(t)$, $G(t)$ and $g(t)$ for problem $(Q1)$

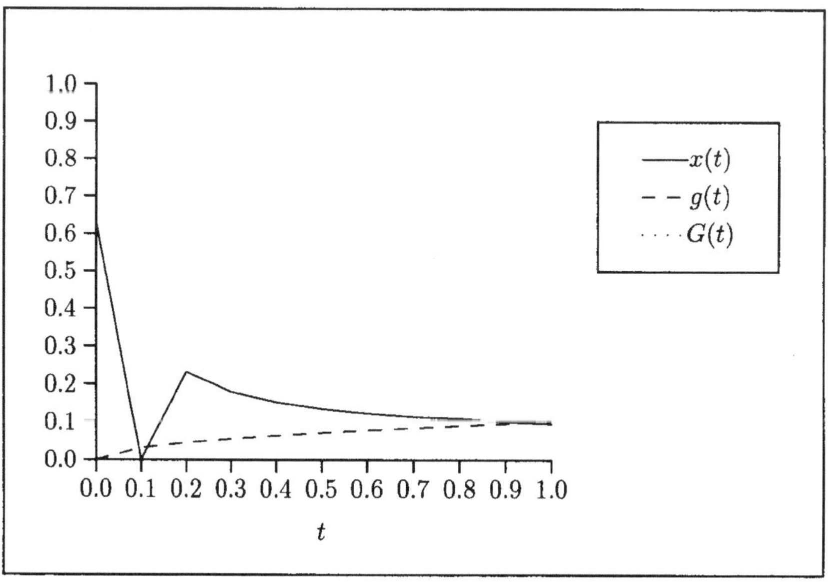

Figure 8.2 $x^*(t)$, $G(t)$ and $g(t)$ for problem $(Q2)$

Table 8.2

$f(x^*)$ for $(Q2)$	$n_p = 4$	$n_p = 8$	$n_p = 16$
$N = 6$	0.1506	0.1505	0.1504
$N = 8$	0.1506	0.1504	0.1503

References

Bellmann, R.E. (1957), *Dynamic Programming*, Princeton University Press, Princeton, N.J..

Burie, R.N. and Abrham, J. (1973), Numerical solutions to continuous linear programming problems, *J. Oper. Res.*, Vol. 17, pp. 107–117

Grinold, R.G. (1968), Continuous programming, Part one: linear objectives, *J. Math. Anal. Appl.*, Vol. 28, pp. 32–51.

Jennings, L.S., Teo, K.L., Fisher, M.E. and Goh, C.J. (1991), MISER3: solving optimal control problems - an update,, *Advances in Engineering Software*, Vol. 13, No. 4, pp. 190–196.

Levinson, N. (1966), A class of continuous linear programming problems, *J. Math. Anal. Appl.*, Vol. 16, pp. 73–83.

Lukas, M.A. and Teo, K.L. (1991), A computational method for a class of optimal control problems involving integrodifferential equations, *Opt. Control Appl. & Methods*, Vol. 12, pp. 141–162.

Teo, K.L., Goh, C.J. and Wong, K.H. (1991), *A Unified Computational Approach to Optimal Control Problems*, Longman Scientific and Technical.

Teo, K.L. and Jennings, L.S. (1989), Nonlinear optimal control problems with continuous state inequality constraints, *J. Optim. Theory Appl.*, Vol. 63, pp. 1–22.

Tyndall, W.F. (1965), A duality theorem for a class of continuous linear programming problems, *SIAM J. Appl. Math.*, Vol. 13, pp. 644–666.

IV Applications

19 CALIBRATION OF PARAMETERS FOR A COMBINED GRAVITY AND TRAFFIC ASSIGNMENT MODEL

Renlong Han

Department of of Information Management and Marketing,
The University of Western Australia,
Nedlands, Western Australia 6907,
Australia

Abstract: The transport planning process usually consists of four stages, trip generation, trip distribution, modal split and traffic assignment. This paper deals only with road traffic, so that there is no modal split. It describes a method for calibrating, from observed traffic counts, the parameters of a joint model, combining trip generation and distribution with traffic assignment. The objective is to minimise the total deviation of the values F estimated by the model $d(f, F)$, from the observed values f. A maximum likelihood model is formulated to calibrate the parameters and thus estimate the values F. The sampling distribution of traffic counts on each individual link is assumed to be Poisson. The global solution to calibrate the relationship cannot be found for this non-convex problem using conventional techniques such as hill-climbing. A genetic algorithm (GA) is introduced to reach a point close to the global optimum, and then Newton's method is used to search for an optimal or nearly optimal solution. Finally, three aggregate measures are applied to test the goodness of fit of the calibrated model. Tests for the individual parameters are performed using the Wald statistic.

Key words: traffic modelling, parameter calibration, maximum likelihood, Poisson distribution.

X. Yang et al. (eds.), Progress in Optimization, 287–304.

1 INTRODUCTION

The transport planning process usually consists of a number of stages. A widely used procedure is a four-step model which includes trip generation, trip distribution, modal split and traffic assignment. This paper deals only with road traffic, so that there is no modal split. It considers commonly used models for three stages, trip generation, trip distribution and traffic assignment, and derives a model combining them into a single entity.

The existing models for the estimation of origin and destination (OD) trip matrices are based on information regarding a target OD matrix, the number of trips observed on the links, the proportionate usage of the links for the trips in each OD pair being assumed given (Van Zuylen and Willumsen, 1980; Spiess, 1987; Cascetta and Nguyen, 1988; Brenninger-Gothe et al., 1989). However, in practice, a target OD matrix is rarely available or there is no OD data, or conventional methods of obtaining the information, e.g., home interviews and roadside interviews, are too expensive. The problem is to estimate a most likely OD trip matrix if there is no target matrix. In this situation, the objective function is a measure of the deviation of the values F estimated by the model, $d(f, F)$, from the traffic count observations f. Without constraints, this problem will have no unique solution and makes no sense. Although no target OD matrix is available, information on population and economic activity of zones can be used. This provides the basis for a joint model to estimate a most likely OD trip matrix.

The trip generation stage is to forecast the number of trips originating and terminating in different zones of the study area. These generation and attraction numbers are usually defined as functions of population, land use characteristics and social-economic activities (Patriksson, 1994). This paper only considers population and tourism as generation and attraction factors.

At the trip distribution stage, models are derived to estimate the number of trips which will be made between each OD pair in the study area. The models considered are gravity models so that the estimated pattern of trips depends on the three major factors: one is associated with the zone in which a trip begins, one with the zone in which it ends and the third with the separation between the zones (Evans, 1976). The measure of separation between a pair of zones is usually the total travel cost of traversing the links on the route between them. It is known that the cost of travelling along a link increases with the amount of traffic using the link. However, in this paper, we only consider cases of uncongested traffic, and so the increasing separation (costs) between any pair of zones with the increasing traffic using the links is assumed to be minimal and is ignored.

The purpose of traffic assignment is to provide an estimate of the amount of traffic which will use each link in a transport network under certain conditions. In general, this is done by taking the estimated trip demands obtained by solving a trip distribution model and allocating them to routes through the network according to some hypothesis about the way in which such a pattern of trip demands will be distributed between the various possible routes through the network (Evans, 1976).

The usual models for trip generation, trip distribution and traffic assignment can be equivalently formulated as optimisation problems subject to certain constraints. Conventional mathematical techniques such as goal programming and hill climbing have frequently been used to find solutions to them. However, multiple optimisation problems cannot be solved by these methods.

2 PROPOSED MODEL

For traffic demand forecasting, traditional four-step models are used in sequence, with the dependent variable of one stage being the explanatory variable of the next stage. For example, the trips between OD pairs are dependent in a gravity model at the trip generation and distribution stage, but they are explanatory variables in a traffic assignment model at the route choice stage. This paper combines these two models into a single model in order to calibrate the parameters and thus obtain OD matrix estimates from the observed link flows. In terms of the information available, a trip generation and distribution model is based on the gravity model.

$$q_{rs} = \alpha \frac{(P_r P_s)^{\beta}}{C_{rs}^{\gamma}} \tag{2.1}$$

where

q_{rs} is number of trips from origin r to destination s;
P_r, P_s is population of origin r and destination s;
C_{rs} is minimum separation (costs) between origin r and destination s;
α, β, γ are parameters.

An obvious weakness of such a gravity model in the tourism area is that a particular centre may have greater power to attract visits than its population alone would indicate (Taplin and Qiu, 1997). In addition, tourist traffic is usually round-trip. When a tourism zone is identified in the gravity model, whether as an origin or a destination, its population size is multiplied by a tourism attraction multiplier ξ expected to be greater than 1. If both origin and destination are tourist centres, both populations will be multiplied by ξ, i.e., ξ^2. Therefore, the above model is modified as follows.

$$q_{rs} = \alpha \frac{(\xi^{\varphi} P_r P_s)^{\beta}}{C_{rs}^{\gamma}} \tag{2.2}$$

where

ξ is the population multiplier associated with whether the zone is a recognised tourism destination;
$\varphi = 0, 1, 2$ means that none, one and both of the original and destined zones are recognised tourist centres.

Equilibrium assignment is based on the principle of stochastic user-equilibrium (SUE) that each traveller, in choosing his route, attempts to minimise his perceived journey cost, and at SUE, no traveller can improve his perceived journey cost by changing his route (Sheffi, 1985). The measured travel costs on links between each OD pair are assumed to be fixed values because we only consider situations where congestion effects are of minor importance, and assume the number of trips which use a specific link is independent of changes in the traffic volumes on that link.

A logit takes into account errors of judgement and the choice of non-minimum path routes for the sake of variety or to take in different scenery along the route. In other words, the distribution implied by the logit reflects a wider utility concept, so that some journeys are not at minimum cost. Therefore, the multi-nomial probabilistic logit model (Dial, 1971) is used for assignment of traffic to non-urban routes.

$$v^k_{rs} = q_{rs} \frac{e^{-\theta u^k_{rs}}}{\sum_{\forall k \in K_{rs}} e^{-\theta u^k_{rs}}} \tag{2.3}$$

where

v^{rs}_k is number of trips on the route k between origin r and destination s;
q_{rs} is number of trips from origin r to destination s;
u^k_{rs} is travel costs (distance or time) of route k between origin r and destination s;
θ is a parameter.

From the route flow, the link flow is calculated by:

$$F_i = \sum_{\forall r \in O} \sum_{\forall s \in S} \sum_{\forall k \in K_{rs}} v^k_{rs} \delta^k_i \tag{2.4}$$

where

v^{rs}_k is number of trips on the route k between origin r and destination s;
F_i is estimated traffic flow on link i;
δ^k_i is equal to 1 if link i is on route k, 0 otherwise.

Then models 2.2, 2.3, and 2.4 are combined; giving:

$$F_i = \sum_{\forall r \in O} \sum_{\forall s \in S} \sum_{\forall k \in K_{rs}} \frac{\delta^k_i \alpha (\xi^{\varphi} P_r P_s)^{\beta} e^{-\theta u^k_{rs}}}{C^{\gamma}_{rs} \sum_{\forall k \in K_{rs}} e^{-\theta u^k_{rs}}} \tag{2.5}$$

Once the joint model 2.5 is formulated, the problem of finding an algorithm for its solution must be solved by the application of suitable mathematical programming techniques.

2.1 Mathematical Formulation of the Proposed Model

The only traffic data available are traffic counts. So the objective function is to minimise the total difference between the traffic count observations f and the values F estimated by Model 2.5, min $d(f, F)$. Three closely related optimisation models are usually used to deal with this kind of problem. One model is based on least squares (Cascetta, 1984); another is based on entropy maximisation (Van Zuylen and Willumsen, 1980); and the third is the maximum likelihood model (Spiess, 1987; Qiu, 1995). The main difference between these models lies in how the distance between estimated and observed traffic counts is defined. It seems that no valid theoretical argument can be used to give preference to one of the three models (Brenninger-Gothe, 1989).

However, entropy maximisation assumes that observed link counts follow the multinomial distribution, and least-square methods were built on the basis that nothing is known about the sampling distribution. These two assumptions are not so reasonable as the Poisson distribution assumption for link counts. From the results of many trials, it has been found that entropy and least squares overestimate the importance of relatively big values and underestimate the importance of the small values.

The maximum likelihood method is based on an assumed sampling distribution. The normal and Poisson distributions are used in practice for traffic counts. For data sets with traffic counts ranging from 25 to 2,885 a day, the Poisson distribution seems to be preferable to the normal distribution for the following reasons. First, the normal distribution incorrectly permits negative values. Secondly, an additional parameter, variance must be calibrated, with one more row and column of second derivatives to be calculated. Finally, its first and second derivatives are more complicated than in the Poisson case.

Further, the Poisson is known to describe the variation in the traffic flow at a point on a road at a given time fairly well (Kirby, 1974). The Poisson distribution is such that its standard deviation is equal to the square root of the mean, and it seems reasonable that this distribution would realistically represent the effects of sampling method and day-to-day variability in the number of vehicles counted on roads. Therefore, in this paper, the Poisson distribution is used to formulate the maximum likelihood model.

The maximum likelihood procedure is formulated as follows:

The probability of obtaining the value of an actual link flow f_i is expressed in terms of the estimated value F_i by

$$\phi(f_i, F_i)$$

Let N denote the number of the links on which there are observed traffic counts. The compound probability of obtaining the base-year link flows f_i is then

$$\Phi = \prod_{i=1}^{N} \phi(f_i, F_i)$$

Defining the likelihood

$$L_1 = log\Phi = \sum_{i=1}^{N} log\phi(f_i, F_i)$$

The sampling distribution of the traffic count for each link is assumed to be Poisson, so that

$$\phi(f_i, F_i) = \frac{(F_i)^{f_i} e^{-F_i}}{f_i!}$$

Then the maximum likelihood estimator is

$$L_1 = \sum_{i=1}^{N}(f_i logF_i - F_i) - \sum_{i=1}^{N} log f_i!$$

By cancelling the last constant part $\sum_{i=1}^{N} log f_i!$, the above maximising problem becomes:

$$\max \qquad L = \sum_{i=1}^{N}(f_i logF_i - F_i) \tag{2.6}$$

$$\text{subject to} \qquad F_i \geq 0, \quad i = 1, \cdots, N \tag{2.7}$$

where f_i is the observed trip value and F_i is the estimated trip value calculated by Model 2.5. All traffic counts are strictly non-negative.

3 ALGORITHMS

After F_i in Model 2.6 is replaced by Model 2.5, five parameters $(\alpha, \beta, \gamma, \theta, \xi)$ are to be calibrated simultaneously to obtain the optimal solution F_i's. For this kind of non-convex, non-smooth optimisation problem, using traditional gradient methods such as Newton's method it is almost impossible to reach the global optimum. Therefore, a genetic algorithm (GA) is combined with Newton's method to search for a global or nearly global optimum.

3.1 Genetic Algorithms

GA as a computational technique was first developed and analysed by Holland early in the 1970s, but it was not widely used until about 10 years ago. GAs are robust search algorithms based on the mechanics of natural selection and natural genetics (Goldberg, 1989). The mechanics of evolution are simple yet powerful. The basic idea behind GAs is to generate randomly an initial pool of solutions, represented as string structures, and then, through continuous copying, swapping, and modifying of partial strings in a manner similar to natural evolution, to allow the solution pool to evolve toward better solutions.

An important and powerful aspect of GAs is their ability to handle non-convex, multiple optima problems. This makes GAs a potentially useful optimisation tool in cases where traditional methods such as hill-climbing might find only local optima. The main disadvantage of GAs is that they are unlikely to reach the exact global optimum. However, it is thought that GAs consistently yield solutions having a high probability of being near the global optimum.

3.2 Newton's Method

One of the most commonly used gradient methods to find the minimum of a function is Newton's method in which the iteration process is expressed as follows:

$$\mathbf{X}_{t+1} = \mathbf{X}_t - \rho_t \mathbf{H}_t^{-1} \mathbf{g}_t$$

where

$\mathbf{X}_{t+1}$ is a vector of parameters at the following iteration $t+1$;
$\mathbf{X}_t$ is a vector of parameters at the current iteration t;
$\mathbf{H}_t^{-1}$ is the inverse of the Hessian of the function, evaluated at $\mathbf{X}_t$;
$\mathbf{g}_t$ is the gradient vector of the function, evaluated at $\mathbf{X}_t$;
ρ_t is step size.

The key to using Newton's method is that the second derivatives matrix is computable. Usually it is complex and hard to get because it needs route enumeration. However, this study derived and computed the first and second derivatives of Model 2.5.

3.3 Quadratic Hill-climbing Method

Newton's method performed very poorly when initially programmed for Model 2.6. It failed to converge at all because the second derivatives matrix was not negative definite. This meant that the best solution (the final point) obtained from the genetic algorithm was still distant from the optimum and the function at that point not approximately quadratic, that is, not asymptotically convex.

The quadratic hill-climbing method (Greene, 1997) is used to solve the above problem. In any iteration, if Hessian H is not negative definite, it is replaced with

$$\mathbf{H}_\mu = \mathbf{H} - \mu\mathbf{I}$$

where μ is a positive number chosen large enough to make $\mathbf{H}_\mu$ negative definite.

In order to garantee the negative definiteness of $\mathbf{H}_\mu$, the value of μ is chosen to be larger than the largest positive eigenvalue of H.

4 MODEL TESTS

4.1 Aggregate Tests

Three measures of fit are applied as aggregate tests in the study case, the coefficient of determination, the normalised phi statistic, and the normalised mean absolute percentage error. Each of these is described in turn and all measures are used to test the goodness of fit.

Coefficient of Determination. One of the more widely used evaluation measures, especially for linear models, is the coefficient of determination or R^2 defined as:

$$R^2 = 1 - \frac{\sum_{i=1}^{N}(f_i - F_i)^2}{\sum_{i=1}^{N}(f_i - \bar{f})^2}$$

where f_i is the observed trip value; F_i is the estimated trip value and $\bar{f}$ is the mean trip value. Although this measure has been found to be insensitive to parameter calibration values by Wilson (1976), and Smith and Hutchinson (1981), it provides a useful descriptive measure (Greene, 1997). Therefore, it still is used to test the goodness of fit in this study.

Phi Statistic. The phi statistic is very similar to the information gain, or relative information, measure used by Smith and Hutchinson (1981).

The measure termed phi is given as:

$$\phi = \sum_{i=1}^{N} f_i \left| log \frac{f_i}{F_i} \right|$$

The larger the value of phi the poorer the model fit.

However, another version of the phi statistic was suggested by Smith and Hutchinson in their paper (1981), being defined as:

$$\phi = \sum_{i=1}^{N} \frac{f_i}{f} \left| log \frac{f_i}{F_i} \right|$$

where f is the total number of the observed traffic counts on the road links, that is, $f = \sum f_i$. A value of zero indicates that the observed and estimated values are identical. The results of some simulations using a calibrated gravity model reveals values in the range 0.35-0.62 when the randomly introduced error has a range of 60 to 90% (Smith and Hutchinson, 1981). This normalised phi statistic appears to be better than the original one for this case because there is a reference range and we can find how far the phi statistic is different from zero.

An important advantage of the phi statistic is that the error is weighted by the observed number or observed proportion of trips. The phi statistic automatically discounts the errors (Smith and Hutchinson, 1981).

Mean Absolute Percentage Error. The mean absolute error is the simplest measure tested and is defined as:

$$MAPE1 = \frac{\sum_{i=1}^{N} |f_i - F_i|}{N}$$

Another form of the mean absolute percentage error statistic was suggested by Smith and Hutchinson (1981), being the ratio of the mean absolute error to the total observed trips and defined as

$$MAPE2 = \sum_{i=1}^{N} \left| \frac{f_i - F_i}{f} \right|$$

The latter version of the mean absolute percentage error statistic, MAPE2, is more convenient to use for comparison because it turns an absolute difference into a relative difference. The closer to zero the value of MAPE2, the better the model fit.

4.2 Individual Parameter Tests

The most commonly used tests for maximum likelihood estimation of a parameter are likelihood ratio, Wald, and Lagrange multiplier tests. These three tests are asymptotically equivalent, and they behave well in a relatively large sample. As a consequence, the choice among them is typically made on the basis of ease of computation. The likelihood ratio test requires calculation of both restricted and unrestricted estimators, but the Wald test requires only the unrestricted estimator, while the Lagrange multiplier test requires only the restricted estimator (Greene, 1997). In the traffic problem, the model is so complex that it is a difficult and time-consuming task to make unrestricted estimates. Needless to say, it would be a great burden to make various restricted estimates. Therefore, the Wald statistic is preferred for this study.

The Wald Statistic. Let $\mathbf{X}$ be the vector of parameter estimates obtained without restrictions, that is

$$\mathbf{X} = (\alpha \ \beta \ \gamma \ \theta \ \xi)$$

A set of restrictions may be hypothesized as:

$$H_0 : \mathbf{c}(\mathbf{X}) = \mathbf{q}$$

where $\mathbf{c}$ is a set of restriction relationships between the parameters, and $\mathbf{q}$ is a set of constants.

Then the Wald statistic is

$$\mathbf{W} = [\mathbf{c}(\hat{\mathbf{X}}) - \mathbf{q}]'(Var[\mathbf{c}(\hat{\mathbf{X}}) - \mathbf{q}])^{-1}[\mathbf{c}(\hat{\mathbf{X}}) - \mathbf{q}]$$

where

$$Var[\mathbf{c}(\hat{\mathbf{X}}) - \mathbf{q}] = \mathbf{C} \ Var[\hat{\mathbf{X}}]\mathbf{C}', \ \mathbf{C} = \left[\frac{\partial \mathbf{c}(\hat{\mathbf{X}})}{\partial \hat{\mathbf{X}}'}\right]$$

Under H_0, in large samples, $\mathbf{W}$ has a chi-squared distribution with degrees of freedom equal to the number of restrictions (Greene, 1997).

The asymptotic covariance matrix of the parameter estimates is computed by evaluating the second derivatives matrix of the log-likelihood function at the maximum likelihood estimates (Greene, 1997), that is, the negative inverse of Hessian, given as

$$Var(\hat{\mathbf{X}}) = \left[-\frac{\partial^2 L(\hat{\mathbf{X}})}{\partial \hat{\mathbf{X}} \partial \hat{\mathbf{X}}'} \right]^{-1}$$

The diagonal elements of the covariance matrix provide the estimates of the asymptotic variances of all calibrated parameters.

Hypothesis Tests. The vector of parameter estimates obtained from the calibration is:

$$\hat{\mathbf{X}} = (1.6699802, 0.5908954, 1.7086951, 0.0472936, 2.6795139)$$

A set of restrictions for the proposed model can be hypothesized as follows:

- $\alpha = 1$: i.e., the hypothesis that the scalar parameter for the gravity model is not necessary in Model 2.2.

- $\beta = 0$ or $\beta = 1$: i.e., the hypotheses that the travel demand elasticity with respect to the product of populations between any OD pairs is 0 or 1. This means that the population product has no effect on travel demand or is proportional to it.

- $\gamma = 0$, $\gamma = 1$ or $\gamma = 2$: i.e., the hypotheses that the elasticity of travel demand with respect to the travel time between any OD pair is equal to 0, -1 or -2. The first hypothesis means the travel time does not affect travel demand; the second one means travel demand is just proportional to the reciprocal of the travel time; the last one tests whether Model 2.2 could be formulated as a naive gravity travel demand model analogous to Newton's Gravitational Law, in which the power of the distance between the two objects' centres of gravity is exactly two.

- $\theta = 0$: The hypothesis implies that the traveller is indifferent to travel time by the alternative routes between an OD pair, trips for the OD pair being spread equally over the rational routes connecting the zones.

- $\xi = 1$: i.e., the hypothesis that the population multiplier in Model 2.2 is 1, the zonal populations being homogeneous for the trip generation and attraction no matter whether the destination is a tourist centre.

- $\alpha = 1.67$, $\beta = 0.591$, $\gamma = 1.709$, $\theta = 0.047$ and $\xi = 2.68$: i.e., the joint hypothesis test. In order to conveniently use the calibrated model to forecast the traffic demand in practice, we test whether it is possible to cut the long floating point numbers to the simple ones.

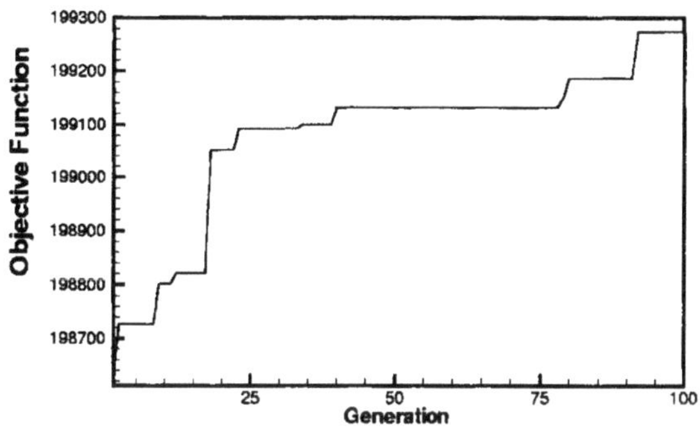

Figure 6.1 The Convergence of Objective Function Using GA

5 RESULTS

5.1 The Study Zones

The study area is the road network of Southern Wheatbelt Region of Western Australia. The roads to be dealt with are the Significant Roads in the draft Roads 2020 Regional Road Development Strategy 1997 (MRWA, 1997).

The region, Wheatbelt South, whose road network is the main focus of the research, is called the directly influenced area (DIA). The rest of Australia has economic and transport connections with the DIA and must be represented in the network for traffic assignment, but is considered as the indirectly influenced area (IIA). There are 39 zones of which 22 zones are in the DIA and 19 in the IIA, and 124 nodes and 191 links in the whole network. In addition, five of 39 zones are recognised as tourism destinations.

Other data include:

- * Observed vehicle counts on the 136 links of the DIA;
- * Link lengths and travel times for the 191 links in the whole network (DIA+IIA); and
- * Census populations for the 39 zones.

5.2 Results from the GA

The GA population size was 1,000, that is, 1,000 chromosomes in each generation, and it ran for 100 generations. Each chromosome takes roughly 3 seconds to run on a Fox Pro 200 PC computer. Thus, it took 1,000*100*3 seconds, i.e., about 83 hours, to get the results. Figure 6.1 shows the changes in the objective function during 100 generations.

Figure 6.1 shows that the objective function improved quickly during the initial 19 generations, slowly during the later generations. From Generation 40 to 78, there was no change. GA is a random search method and another GA run for the same problem, would give a different convergence trend. Nevertheless, the final solutions of the runs are likely to be near the optimal solution, after 100 generations with a population of 1,000 chromosomes.

Table 6.1 The Final Solution Using GA

Parameters		α	β	γ	θ	ξ
Best	Parameter value	1.200	0.623	1.739	0.046	1.704
	First derivative	-910.9	-23745.7	6086.4	5759.8	67.9
Solution	Objective function	199,273.58				

The best solution, the final point obtained using GA is listed in Table 6.1. It can be seen that the first derivatives with respect to different parameters are significantly different from zero, and thus this point can be thought to be still far from the maximum point. It is important to use Newton's method to drive the first derivatives towards zero.

Table 6.2 The Final Solution Using Newton's Method

Parameters	α	β	γ	θ	ξ
Value	1.6699802	0.5908954	1.7086951	0.0472936	2.6795139
First deriv.	-1.3E-11	-4.21E-10	1.16E-10	-8.9E-11	-3E-12
Variance	0.0155070	0.0000125	0.0001129	0.0000007	0.0075826
Std. Error	0.1245272	0.0035287	0.0106232	0.0008295	0.0870780
Obj. funcn.	199,403.79				

5.3 Results with Newton's Method

On the basis of the GA results in Table 1, Newton's method was used to keep searching for the optimal or near optimal solution. Because this method is still an incremental algorithm, it could not reach the best solution for a complex model like Model 2.5, but close to the best. In fact, it drove the first derivatives of the likelihood function nearly to zero as shown in Table 6.2. However, it could not bring them closer to zero because of the cumulative computing errors.

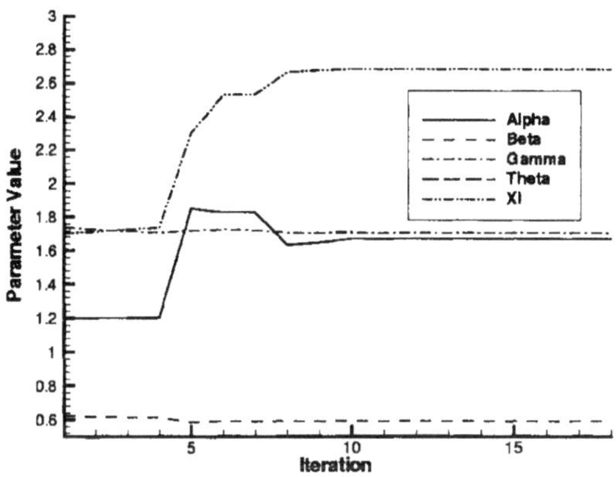

Figure 6.2 The Convergence of Five Parameters with Newton's Method

Figure 6.2 shows the convergence during 18 iterations of Newton's method. It can be concluded that all parameters converged well to reasonable values, but the parameters, β, θ and γ, are more stable than α and ξ. This means that the former three parameters in Model 2.5 are more reliable than the others.

It is concluded that the Newton's method worked extremely well in the study, giving the results in Table 6.2. The parameter values are plausible; the first derivatives converge to nearly zero; and the variances and standard errors are small.

The variance of the parameter θ is the smallest, relative to the size of θ, from which it can be inferred that θ is the most reliable among the five parameters. Next are β, γ, ξ, and, α, respectively. The variances of α and ξ are relatively much larger than the others, which explains why these two parameters are not so stable as the other three during the initial calibration process, with the values changing much more using Newton's method.

Figure 6.3 demonstrates the convergence of the first derivatives using Newton's method. All the first derivatives of the likelihood function with respect to each parameter converged to nearly zero very quickly, taking only 18 iterations; they could not diminish further because of cumulative computing errors. The Hessian matrix of the final iteration was found to be negative definite, and thus the final solution is indeed a maximum.

Figure 6.4 demonstrates the improving trend of the objective function with Newton's method. From Iteration 4 to 6, the objective function fluctuated sharply because the step size for Newton's Method changed to 1 from the original 0.5. After 18 iterations, the objective function converged to a stable

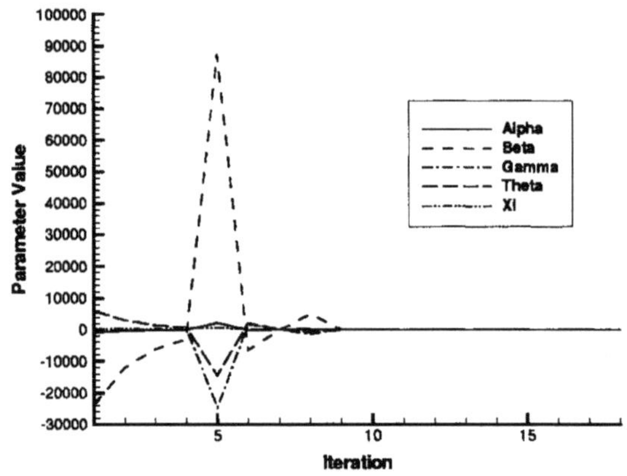

Figure 6.3 The Convergence of First Derivatives with Newton's Method

value, 199,403.79, starting from the final objective function value of 199,273.58 obtained with GA, an increase of about 0.065%.

5.4 Results of the Aggregate Tests

Table 6.3 The Results of the Aggregate Tests

Aggregate Tests	R^2	Normalised Phi	MAPE Statistic
Statistics Value	0.823	0.399	0.383

Table 6.3 gives the results of the aggregate tests. The R^2 of 0.823 is a general measure of goodness of fit. The normalised phi statistic of 0.399, and the normalised mean absolute percentage error statistic of 0.383, show that the calibrated joint model is suitable for use in travel demand forecasting.

5.5 Results of Hypothesis Tests

Table 6.4 shows the results of individual parameter hypothesis tests and the confidence intervals. The critical value for the Wald test at 5% level is 3.842. The Wald statistics for all null hypotheses but the last are much greater than 3.842. It can be concluded that all the null hypothesis tests for individual parameters at given values are rejected. However, the joint hypothesis test as defined to conveniently use the combined model is accepted. Therefore we can use Model 2.5 with rounded parameters for travel demand forecasting.

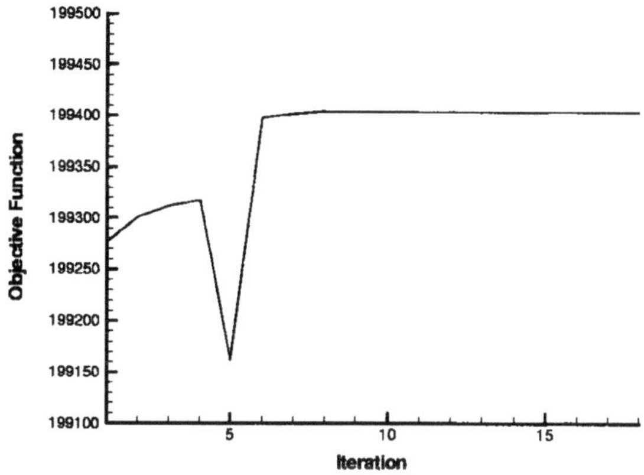

Figure 6.4 The Convergence of Objective Function with Newton's Method

Table 6.4 The Results of the Hypothesis Tests

Null Hypothesis		Wald Statistics	Evaluation Rej./Act.	Conf. Intervals (5%) L_Bound	Conf. Intervals (5%) U_Bound
α	1	28.946	Reject H_0	1.4259070	1.914054
β	0	28041.230	Reject H_0	0.5839792	0.5978116
	1	13441.400	Reject H_0		
γ	0	25871.294	Reject H_0	1.6878736	1.7295166
	1	4450.490	Reject H_0		
	2	751.942	Reject H_0		
θ	0	3250.930	Reject H_0	0.0456679	0.0489194
ξ	1	372.006	Reject H_0	2.5088410	2.8501867
J α	1.670	0.229	Accept H_0	n.a.	
O β	0.591				
I γ	1.709				
N θ	0.047				
T ξ	2.680				

6 CONCLUSION

6.1 Value of the Study

This paper showed that the optimisation problem for a road network can be solved using a genetic algorithm combined with Newton's method. For such a complicated model as Model 2.5, the second derivatives matrix is rarely derived. Furthermore, in order to make the second derivatives computable, we must enumerate all routes between any OD pair. This is a big task for a complicated network, which has inhibited previous studies.

This study also shows that a genetic algorithm can be applied in the field of transport planning to solve the multiple optima, non-convex optimisation problem, and reach a solution good enough to be applied in practice. Though one could not say that the best solution found by GA is at the global optimum, at least one can conclude that there is a high probablity that the point is on the global peak. In addition, so far it seems that no better methods than the combined GA and Newton's method can be used to solve this kind of non-smooth, non-convex multiple optimisation problem.

The final important contribution of this study is that it can estimate a most likely OD matrix needed in transportation planning even without a target OD matrix. This will be valuable in practice. Many countries still use traditional OD surveys to collect such information, and this uses a lot of human and financial resources.

6.2 Further Study

The calibrated parameters converged to reasonable values; and the first derivatives approached zero. However, the variances and standard errors are so small that they deserve further study. A convincing study of them would probably require simulation to determine whether the simulated variances were similar to those calculated.

Acknowledgments

The author would like to thank Professor John H E Taplin for his valuable advice and supervision. The author also would like to thank Main Roads Western Australia for supplying valuable data. Any error in the paper is the responsibility of the author. The author holds an Overseas Postgraduate Research Scholarship for his PhD study.

References

Brenninger-Gothe M., Jornsten, K. O. and Lundgren, J. T. (1989), Estimation of origin-destination matrices from traffic counts using multiobjective programming formulations, *Transportation Research*, Vol. 23B, No. 4, pp. 257-269.

Cascetta, E. (1984), Estimation of trip matrices from traffic counts and survey data: a generalized least squares estimator, *Transportation Research*, Vol. 18B, No. 4/5, pp. 289-299.

Cascetta, E. and Nguyen, S. (1988), A unified framework for estimating or updating origin/destination matrices from traffic counts, *Transportation Research*, Vol. 22B, No. 6, pp. 437-455.

Dial, R. B. (1971), A probabilistic multipath traffic assignment model which obviates path enumeration, *Transportation Research*, Vol. 5, pp. 83-111.

Evans, S. P. (1976), Derivation and analysis of some models for combining trip distribution and assignment, *Transportation Research*, Vol. 10, pp. 37-57.

Goldberg, D. E. (1989), *Genetic Algorithms in Searching, Operation and Machine Learning*, Addison Welsley Publishing Company, Don Mills.

Greene, W. H. (1997), *Econometric Analysis*, Prentice-Hall, New Jersey.

Kirby, H. R. (1974), Theoretical requirements for calibrating gravity models, *Transportation Research*, Vol. 8, pp. 97-104.

Main Roads, Western Australia (MRWA, 1997), *Road 2020 Regional Road Development Strategy (draft)*, Western Australia Municipal Association, Australia.

Patriksson, M. (1994), *The Traffic Assignment Problem: Models and Methods*, VSP,Utrecht.

Qiu, M. (1995), Simultaneous calibration of gravity and logit models of road traffic by a genetic algorithm, in *Proceeding of the IEEE International Conference on Evolutionary Computation*, IEEE Service Center, Piscataway NJ.

Sheffi, Y. (1985), *Urban Transportation Network: Equilibrium Analysis with Mathematical Programming Methods*, Prentice-Hall, New Jersey.

Smith, D. P. and Hutchinson, B. G. (1981), Goodness of fit statistics for trip distribution model, *Transportation Research*, Vol. 15A, pp. 295-303.

Spiess, H. (1987), A mximum likelihood model for estimating origin-destination matrices, *Transportation Research*, Vol. 21B, No. 5, pp. 395-412.

Taplin, J. H. E and Qiu, M. (1997), Car trip attraction and route choice in Australia, *Annalls of Tourism Research*, Vol. 24, No. 3, pp. 624-637.

Zuylen, Van. and Willumsen, (1980), The most likely trip matrix estimated from traffic counts, *Transportation Research*, Vol. 14B, pp. 281-293.

Wilson S. R. (1976), Statistical notes on the evaluation of calibrated gravity models, *Transportation Research*, Vol. 10, pp. 343-345.

20 A RESTRICTED VARIATION ARGUMENT TO DERIVE NECESSARY CONDITIONS FOR THE OPTIMAL CONTROL OF A TRAIN

Phil Howlett

Centre for Indutrial and Applicable Mathematics,
The University of South Australia,
The Levels, South Australia,
Australia

Abstract: A train travels from one station to the next along a smooth track with known gradient. The journey must be completed within a given time and it is desirable to minimise fuel consumption. We assume that only certain discrete control settings are possible. This conforms to the most common situation in diesel-electric locomotives. In the case where the track is flat and the train is controlled using a finite sequence of traction controls and a final brake control Cheng and Howlett (1992) have derived necessary and sufficient conditions that define the optimal switching times. In this paper we will show that these conditions can be derived (as necessary conditions) using an elegant argument of restricted variation.

Key words: train control, discrete control, optimal control.

X. Yang et al. (eds.), Progress in Optimization, 305–314.

1 INTRODUCTION

A train travels from one station to the next along a smooth track with known gradient. The journey must be completed within a given time and it is desirable to minimise the fuel consumption.

We assume that only certain discrete control settings are possible. This conforms to the most common situation in railway locomotives. In particular we note that diesel-electric locomotives such as the GM $JT26C-2SS$ and many other long-haul locomotives have 8 traction control settings. This model and other similar models are currently used by the *National Rail Corporation* in Australia to haul freight trains some 4,000 kilometres across the Australian continent between Brisbane and Perth. We normally assume that

- each traction control setting determines a constant rate of fuel supply;
- the power developed by the locomotive is directly proportional to the rate of fuel supply and the corresponding acceleration is non-negative; and
- each brake control gives constant negative acceleration.

The train is controlled using a finite sequence of traction phases and a final brake phase. During each phase the control setting is constant and for each control strategy there is a uniquely defined speed profile determined by the equations of motion. We will say that a strategy is feasible if the train arrives at the desired destination within the time allowed for the journey.

In general we wish to find a feasible strategy that minimises fuel consumption. However Howlett and Pudney (1995) and Howlett and Cheng (1997) show that such a strategy cannot normally be achieved using discrete control with a finite number of switches. Thus the intuitive problem is incorrectly posed. We consider instead a closely related problem that can be solved.

2 FORMULATION OF A GENERAL VEHICLE CONTROL PROBLEM WITH DISCRETE CONTROL

We consider a control problem with $t \in \mathbb{R}$ as the independent variable and with state variables $x = x(t) \in \mathbb{R}$ and $v = v(t) \in \mathbb{R}$ satisfying a simple autonomous system of differential equations in the form

$$\frac{dx}{dt} = v \tag{2.1}$$

and

$$\frac{dv}{dt} = F(x, v, u) \tag{2.2}$$

where $u \in U = \{-q, -q+1, \ldots, 0, \ldots, p-1, p\} \subseteq \mathbf{Z}$ is the control variable and p and q are fixed positive integers. In this problem t is the time, x is the position of the train, and v is the speed of the train. For each fixed sequence $\{u_{k+1}\}_{k=0,1,\ldots,n}$ of control settings and each partition

$$0 = t_0 < t_1 < \ldots < t_{n+1} \tag{2.3}$$

of the positive t-axis there is a corresponding control function $u : (0, t_{n+1}) \to U$ defined essentially by

$$u(t) = u_{k+1} \tag{2.4}$$

for $t \in (t_k, t_{k+1})$ and each $k = 0, 1, \ldots, n$. We denote the corresponding strategy by

$$S = S(\{[u_{k+1}; (t_k, t_{k+1})]\}_{k=0,1,\ldots,n}) \tag{2.5}$$

and use

$$\mathcal{S} = \mathcal{S}(\{u_{k+1}\}_{k=0,1,\ldots,n}) \tag{2.6}$$

to denote the collection of all strategies using the given sequence $\{u_{k+1}\}_{k=0,1,\ldots,n}$ of control settings. The value u_{k+1} is the control setting on the interval (t_k, t_{k+1}). The times $\{t_k\}_{k=1,2,\ldots,n}$ are called the switching times with t_0 the starting time and t_{n+1} the stopping time. We write $x_k = x(t_k)$ to generate a corresponding partition

$$0 = x_0 < x_1 < \ldots < x_{n+1} \tag{2.7}$$

of the positive x-axis, where the points $\{x_k\}_{k=1,2,\ldots,n}$ are called the switching points with x_0 the starting point and x_{n+1} the stopping point. We write $\tau_{k+1} = \Delta t_k$ and $\xi_{k+1} = \Delta x_k$ and also write $V_k = v(t_k)$ for the speed at the switching time t_k. If the cost rate of each control is defined by a function $f : U \to \mathbb{R}$ then the cost of the strategy S is given by

$$\begin{aligned} J(S) &= \int_0^{t_{n+1}} f[u(t)]dt \\ &= \sum_{k=0}^{n} f(u_{k+1})\tau_{k+1}. \end{aligned} \tag{2.8}$$

We can now state the problem concisely.
Problem Statement *For each fixed sequence $\{u_{k+1}\}_{k=0,1,\ldots,n}$ of control settings find the switching times $\{t_k\}_{k=1,2,\ldots,n}$ that minimise $J(S)$ subject to the initial condition $V_0 = 0$, the final condition $V_{n+1} = 0$ and the constraints $t_{n+1} \le T$ and $x_{n+1} \ge X$.*

3 EXISTENCE OF AN OPTIMAL STRATEGY

We suppose that the control sequence $\{u_{k+1}\}_{k=0,1,\ldots,n}$ is given. If there is a feasible strategy in the form

$$S = S(\{[u_{k+1}; (t_k, t_{k+1})]\}_{k=0,1,\ldots,n}) \tag{3.1}$$

for some partition

$$0 = t_0 \le t_1 \le \ldots \le t_{n+1} \le T \tag{3.2}$$

of the positive t-axis then we can argue that either there is a finite number of feasible strategies in which case we can select an optimal strategy or else there is an infinite number of feasible strategies in which case we can find a sequence of feasible strategies $\{S^{(p)}\}_{p=0,1,\ldots}$ given by

$$S^{(p)} = S(\{[u_{k+1}; (t_k^{(p)}, t_{k+1}^{(p)})]\}_{k=0,1,\ldots,n}) \tag{3.3}$$

where

$$0 = t_0^{(p)} \le t_1^{(p)} \le \ldots \le t_{n+1}^{(p)} \le T \tag{3.4}$$

and such that

$$J(S^{(p)}) \downarrow \inf J(S) \tag{3.5}$$

as $p \to \infty$ where the infimimum is taken over all feasible strategies using the given sequence of controls. We can now select a subsequence $\{S^{(p[q])}\}_{q=1,2,\ldots}$ of feasible strategies such that $t_k^{(p[q])} \to t_k^*$ as $q \to \infty$ for each $k = 1, 2, \ldots, n+1$. It is easy to see that the strategy

$$S^* = S(\{[u_{k+1}; (t_k^*, t_{k+1}^*)]\}_{k=0,1,\ldots,n}) \tag{3.6}$$

is an optimal strategy for the given sequence of controls.

4 THE EQUATIONS OF MOTION FOR A TYPICAL TRAIN CONTROL PROBLEM

For a typical train control problem we can write

$$F(x, v, u) = p(v, u) + g(x) - r(x, v) \tag{4.1}$$

where $p(v, u)$ is the driving force per unit mass, $g(x)$ is the component of gravitational acceleration in the direction of motion and $r(x, v)$ is the frictional resistance per unit mass opposing the motion. We assume that $v \ge 0$. We have

- $p(v, 0) = 0$;
- $p(v, u) = Af(u)/v > 0$ for $u > 0$;
- $p(v, u) = B(u) < 0$ for $u < 0$;
- $g(x) = g \sin\theta(x)$ where g is the acceleration due to gravity and θ is the angle of slope in the direction of motion;
- $r(x, v)$ is continuous in (x, v);
- $r(x, v) > 0$ when $v > 0$;
- $r(x, v_1) < r(x, v_2)$ when $v_1 < v_2$; and
- $\varphi(x, v) = vr(x, v)/A$ is convex in v for each x.

Solutions are given for problems of this general form in papers by Cheng and Howlett (1992); Cheng and Howlett (1993); Howlett (1990); Howlett (1996); Howlett and Cheng (1997); Pudney and Howlett (1994) and Pudney *et al* (1996). Asnis *et al* (1985) and Kraft and Schnieder (1981) have considered related train control problems. Howlett *et al* (1997) have considered the problem of finding an optimal driving strategy for a solar-powered car. This problem is closely related to the typical train control problem. In this paper we consider an argument of restricted variation to find necessary conditions for the optimal switching times. The elegant feature of the argument is that the number of degrees of freedom in the variational argument can be reduced to three; one more than the number of constraints. We restrict our attention to the special case of a train on a flat track with resistance dependent only on speed. In this case $F(x, v, u) = F(v, u)$ is given by

$$F(v, u) = \frac{Af(u)}{v} + B(u) - r(v) \tag{4.2}$$

where A is a positive real constant and $u \in [-q, p]$. We assume that $f(u) = 0$ for $-q \leq u \leq 0$ and $0 < f(u_1) < f(u_2) < f(p)$ when $0 < u_1 < u_2 < p$ and that $B(-q) < B(u_1) < B(u_2) < 0$ for $-q < u_1 < u_2 < 0$ and $B(u) = 0$ for $0 \leq u \leq p$.

The argument of restricted variation that we propose to use in this paper can also be used for a general train control problem but the algebra is significantly more complicated.

5 THE TRAIN CONTROL PROBLEM ON FLAT TRACK

We consider the train control problem with discrete control and a fixed sequence of control settings $\{u_{k+1}\}_{k=1,2,\ldots,n}$ as described above, on a flat track with the dynamics given by equation (4.2). In this case Cheng and Howlett (1992) have shown that the switching times $\{t_k\}_{k=1,2,\ldots,n}$ are optimal if and only if there exist real positive constants $\lambda \in \mathbb{R}$ and $\mu \in \mathbb{R}$ such that the switching speeds $\{V_k\}_{k=1,2,\ldots,n}$ satisfy the equations

$$\lambda V_k - \mu = \frac{f(u_k)F(V_k, u_{k+1}) - f(u_{k+1})F(V_k, u_k)}{F(V_k, u_{k+1}) - F(V_k, u_k)} \tag{5.1}$$

for each $k = 1, 2, \ldots, n$. A strategy satisfying the above conditions is said to be a strategy of optimal type. The distance and time constraints determine the appropriate values for λ and μ. In the case of a strategy with a control sequence $\{p, 0, p, 0, \ldots, p, 0, -q\}$ the equations (5.1) take a particularly simple form. If we define $\varphi(v) = vr(v)/A$ then we have

$$\lambda V_k - \mu = \varphi(V_k) \tag{5.2}$$

for each $k = 1, 2, \ldots, n-1$ and

$$\lambda V_n - \mu = 0. \tag{5.3}$$

Since the function φ is convex in practice we can see that the equation (5.2) has at most two solutions V and W with $0 \leq V \leq W$. For a strategy of optimal type we require two distinct solutions with $0 < V < W$. Under these circumstances equation (5.3) has a unique solution U with $0 < U < V < W$ given by

$$U = \frac{WV[r(W) - r(V)]}{Wr(W) - Vr(V)}.$$

The *power-coast* strategy is optimal if each power phase finishes at the upper critical speed W and if each coast phase (except the final coast phase) finishes at the lower critical speed V. The critical speed U is the speed at which the final coast phase finishes and at which braking begins. Howlett and Pudney (1995) show that any sensible strategy of continuous control can be approximated as closely as we please by a strategy of discrete control in the form $\{p, 0, p, 0, \ldots, p, 0, -q\}$ and hence it follows that a strategy of optimal type can be used to approximate the *idealised* minimum cost strategy. We note that it makes no sense in most circumstances to use negative values of the control variable in an energy-efficient strategy except to stop the train at the end of the journey. In many cases the minimum cost strategy can only be achieved using continuous control. On a flat track it is well known that the strategy of optimal type approaches an *idealised* speedholding strategy as the number of switches tends to infinity. These comments also apply when the track is not steep. More information about this topic can be found in Cheng and Howlett (1992), Cheng and Howlett (1993) and Howlett and Pudney (1995).

6 THE RESTRICTED VARIATION ARGUMENT TO DERIVE NECESSARY CONDITIONS FOR OPTIMAL SWITCHING TIMES

In this section we will show that the equations (5.1) can be derived by considering a differential variation of three successive switching times $\{t_{k-1}, t_k, t_{k+1}\}$ for an arbitrarily chosen integer $k \in (2, n-1)$ in a feasible strategy of optimal type. Such variations are the simplest variations that allow us to preserve feasibility. The time constraint is preserved if t_{n+1} remains fixed and the distance constraint is preserved if the area under the graph $v = v(t)$ is not changed. A restricted variation that retains feasibility is illustrated in Figure 6.1.

We have the following expressions for the differential increments of distance, time and cost. The distance differential is

$$d\xi^{(k)} = d\xi_{k-1} + d\xi_k + d\xi_{k+1} + d\xi_{k+2}, \tag{6.1}$$

the time differential is

$$d\tau^{(k)} = d\tau_{k-1} + d\tau_k + d\tau_{k+1} + d\tau_{k+2} \tag{6.2}$$

and the cost differential is

$$dJ^{(k)} = f(u_{k-1})d\tau_{k-1} + f(u_k)d\tau_k + f(u_{k+1})d\tau_{k+1} + f(u_{k+2})d\tau_{k+2}. \tag{6.3}$$

By defining a Lagrangean differential

$$d\mathcal{J}^{(k)} = dJ^{(k)} + \pi_k d\xi^{(k)} - \rho_k d\tau^{(k)} \tag{6.4}$$

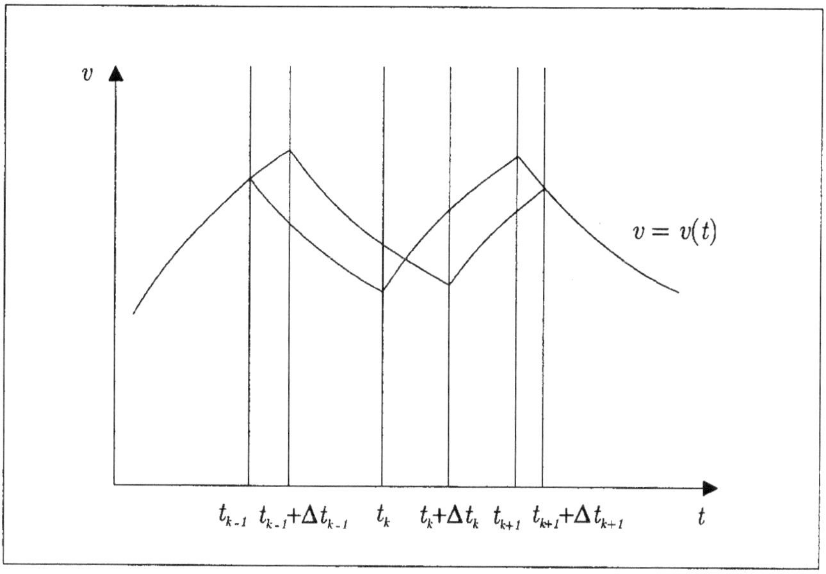

Figure 6.1 Graphical representation of a typical feasible variation.

where $\pi_k \in \mathbb{R}$ and $\rho_k \in \mathbb{R}$ are non-negative multipliers we can write the necessary Kuhn-Tucker conditions for a local constrained minimum as

$$d\mathcal{J}^{(k)} = 0 \tag{6.5}$$

or equivalently in the form

$$dJ^{(k)} = \rho_k d\tau^{(k)} - \pi_k d\xi^{(k)}. \tag{6.6}$$

The condition (6.6) can be interpreted geometrically as saying that the cost differential lies in the cone formed by the time differential and the negative distance differential. From the equation of motion we can see that

$$\tau_{k+1} = \int_{V_k}^{V_{k+1}} \frac{dv}{F(v, u_{k+1})} \tag{6.7}$$

and

$$\xi_{k+1} = \int_{V_k}^{V_{k+1}} \frac{vdv}{F(v, u_{k+1})}. \tag{6.8}$$

It follows that the distance differential can be re-written as

$$d\xi^{(k)} = \sum_{j=k-1}^{k+1} V_j \left[\frac{1}{F(V_j, u_j)} - \frac{1}{F(V_j, u_{j+1})} \right] dV_j, \tag{6.9}$$

the time differential becomes

$$d\tau^{(k)} = \sum_{j=k-1}^{k+1} \left[\frac{1}{F(V_j, u_j)} - \frac{1}{F(V_j, u_{j+1})} \right] dV_j, \tag{6.10}$$

the cost differential is given by

$$dJ^{(k)} = \sum_{j=k-1}^{k+1} \left[\frac{f(u_j)}{F(V_j, u_j)} - \frac{f(u_{j+1})}{F(V_j, u_{j+1})} \right] dV_j \tag{6.11}$$

and the Kuhn-Tucker conditions take the more explicit form

$$\begin{aligned} \sum_{j=k-1}^{k+1} \Bigg\{ & \left[\frac{f(u_j)}{F(V_j, u_j)} - \frac{f(u_{j+1})}{F(V_j, u_{j+1})} \right] \\ & -\pi_k V_j \left[\frac{1}{F(V_j, u_j)} - \frac{1}{F(V_j, u_{j+1})} \right] \\ & +\rho_k \left[\frac{1}{F(V_j, u_j)} - \frac{1}{F(V_j, u_{j+1})} \right] \Bigg\} dV_j = 0 \end{aligned} \tag{6.12}$$

for each feasible differential $dV^{(k)} = (dV_{k-1}, dV_k, dV_{k+1})$ of the variable $V^{(k)} = (V_{k-1}, V_k, V_{k+1})$. In Figure 6.2 we have illustrated a typical feasible differential in the case where $\xi(V) = X$ and $\tau(V) = T$ and where $d\xi^{(k)} = 0$ and $d\tau^{(k)} = 0$. Because we can find three independent feasible differentials to satisfy this equation it follows that the differential coefficients must each be zero. Therefore

$$\pi_k V_j - \rho_k = \frac{f(u_j)F(V_j, u_{j+1}) - f(u_{j+1})F(V_j, u_j)}{F(V_j, u_{j+1}) - F(V_j, u_j)} \tag{6.13}$$

for each $j = k-1, k, k+1$. If we write

$$\varphi_j(v) = \frac{f(u_j)F(v, u_{j+1}) - f(u_{j+1})F(v, u_j)}{F(v, u_{j+1}) - F(v, u_j)} \tag{6.14}$$

then by manipulating the three equations (6.13) we can see that

$$\pi_k = \frac{\varphi_{j+1}(V_{j+1}) - \varphi_j(V_j)}{V_{j+1} - V_j} \tag{6.15}$$

and

$$\rho_k = \frac{V_j \varphi_{j+1}(V_{j+1}) - V_{j+1}\varphi_j(V_j)}{V_{j+1} - V_j} \tag{6.16}$$

for each $j = k-1, k$. By knowing that these identities must hold for all $k = 2, 3, \ldots, n-1$ in a strategy of optimal type it follows that there are positive real numbers λ and μ with $\pi_k = \lambda$ and $\rho_k = \mu$ for all $k = 1, 2, \ldots, n$. This completes the argument. An argument of restricted variation can also be used to find the speed at which braking should begin.

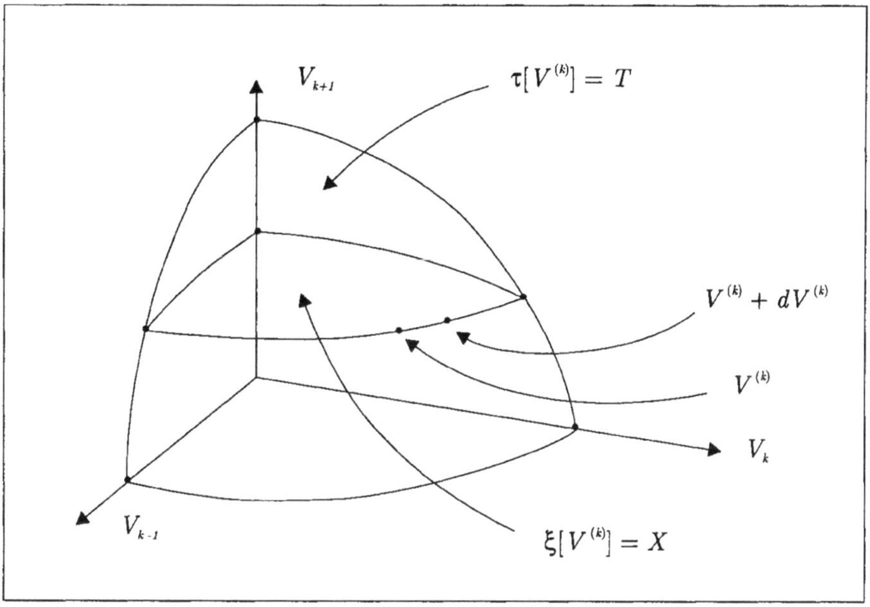

Figure 6.2 Geometric representation of a typical feasible differential $dV^{(k)}$.

References

Asnis, I.A., Dmitruk, A.V. and Osmolovskii, N.P. (1985), Solution of the problem of the energetically optimal control of the motion of a train by the maximum principle, *U.S.S.R. Comput.Maths.Math.Phys.*, Vol.25, No.6, pp. 37-44.

Cheng, J. and Howlett, P.G. (1992), Application of critical velocities to the minimisation of fuel consumption in the control of trains, *Automatica*, Vol.28, No.1, pp. 165-169.

Cheng, J. and Howlett, P.G. (1993), A note on the calculation of optimal strategies for the minimisation of fuel consumption in the control of trains, *IEEE Transactions on Automatic Control*, Vol.38, No.11, pp. 1730-1734.

Howlett, P.G. (1984), *The optimal control of a train*, Study Leave Report, School of Mathematics, University of South Australia.

Howlett, P.G. (1990), An optimal strategy for the control of a train, *J. Aust. Math. Soc.*, Ser. B, Vol. 31, pp. 454-471.

Howlett, P.G., Pudney, P.J. and Benjamin, B.R. (1992), Determination of optimal driving strategies for the control of a train, in *Computational Techniques and Applications: CTAC91*, edited by Noye, B.J., Benjamin, B.R. and Colgan, L.H., pp. 241-248.

Howlett, P.G., Cheng, J., and Pudney, P.J. (1995), Optimal strategies for energy-efficient train control, in *Control Problems in Industry* (Progress in Systems and Control Theory), edited by Lasiecka, I. and Morton, B., Birkhauser, pp. 151-178.

Howlett, P.G., Milroy, I.P. and Pudney, P.J. (1994), Energy-efficient train control, *Control Engineering Practice*, Vol.2, No.2, pp. 193-200.

Howlett, P.G. (1996), Optimal strategies for the control of a train, *Automatica*, Vol. 32, No. 4, pp. 519-532.

Howlett, P.G. and Pudney, P.J. (1995), *Energy-efficient Train Control*, Advances in Industrial Control, Springer-Verlag, London.

Howlett, P.G. and Cheng, J., (1997), Optimal driving strategies for a train on a track with continuously varying gradient, *J. Aust. Math. Soc.*, Ser. B, Vol. 38, pp. 388-410.

Howlett, P.G., Pudney, P.J., Gates, D. and Tarnopolskaya, T. (1997), Optimal driving strategy for a solar car on a level road, *IMA Journal of Mathematics Applied in Business and Industry*, Vol. 8, pp. 59-81.

Kraft, K.H. and Schnieder, E. (1981), *Optimale Trajektrien im Spurgebundenen Schnellverkehr* (Optimal Trajectories for Rapid Transit Systems), Regelungstechnik 29. Jahrgang.

Pudney, P.J. and Howlett, P.G. (1994), Optimal driving strategies for a train journey with speed limits, *J. Aust. Math. Soc.*, Ser. B, Vol. 36, pp. 38-49.

Pudney, P.J., Howlett, P.G., Benjamin, B.R. and Milroy, I.P. (1996), Modelling the operational performance of large systems: a railway example, in *Computational Techniques and Applications: CTAC-95*, (May, R.L. and Easton, A.K. Eds.), World Scientific, pp. 655-662.

21 DETERMINATION OF OPTIMAL BATCH SIZE FOR A MANUFACTURING SYSTEM

Ruhul Sarker and Charles Newton

School of Computer Science,
University College, The University of New South Wales,
ADFA, Northcott Drive, Canberra,
ACT 2600, Australia

Abstract: This paper considers two unconstrained non-convex optimization programs which represent joint raw material-production batch sizing problems in a real world situation. The batch sizing problems are introduced and two mathematical models are developed for two different scenerios. Then simple heuristics are developed to solve these models. Computational experiences of the problems are also discussed.

Key words: batch size, nonlinear function, non-convex, optimality, heuristic.

X. Yang et al. (eds.), Progress in Optimization, 315–327.

1 INTRODUCTION

This paper considers two unconstrained non-convex nonlinear programs which represent joint raw material-production batch sizing problems in real world manufacturing, production planning/scheduling, environment. The planning environment is introduced and the mathematical models are developed. The proposed model determines an economic manufacturing quantity and recommends a raw material ordering policy to deliver a fixed quantity of finished products at a regular interval. In this paper, first we estimate production batch sizes for a JIT (Just-in-Time) delivery system and then we incorporate a JIT raw material supply system. These models may be solved using sophisticated mathematical tools. However, we develope a simple heuristic considering the acceptability of these methods to the practitioners. The quality of solutions produced by the developed methods are discussed and a sensitivity analysis is provided.

The paper is organized as follows. Following the introduction, the problems of the current research are introduced and relevant literature is discussed. Then the mathematical formulation of two coordinated policies are presented. Then, the solution methodology is provided, followed by results and discussions. Finally, conclusions are provided.

2 PROBLEM STATEMENT

A desirable condition in long-term purchase agreements in a just-in-time (JIT) manufacturing environment is the frequent delivery of small quantities of items by suppliers/vendors so as to minimize inventory holding costs for the buyer. Considering a manufacturing system that procures raw materials from outside suppliers and processes them to convert into finished products for retailers/ customers. The manufacturer must deliver the products in small quantities to minimize the retailer's holding cost, and accept the supply of small quantities of raw materials to minimize his own holding costs. In the traditional JIT environment, the supplier of raw materials is dedicated to the manufacturing firm, and normally is located close by. The manufacturing lot size is dependent on the retailer's sales volume (market demand), unit product cost, set-up cost, inventory holding cost and transportation cost. The raw material purchasing lot size is dependent on raw material requirement in the manufacturing system, unit raw material cost, ordering cost and inventory holding cost. Therefore, the optimal raw material purchasing quantity may not be equal to the raw material requirement for an optimal manufacturing batch size. To operate the JIT manufacturing system optimally, it is necessary to optimize the activities of both raw material purchasing and production batch size simultaneously, taking all operating parameters into consideration. Unfortunately, until recently, most JIT studies in the literature have been descriptive [Aderohunmu et al(1995) and Chapman and Carter(1990), and most of the analytical studies do not take all the costs for both sub-systems into consideration [Ansari and Modarress (1987), Ansari and Hechel(1987), and O'Neal(1989)]. Therefore, the overall result may not be optimal.

A larger manufacturing batch size reduces the set-up cost component to the overall unit product cost. The products produced in one batch (one manufacturing cycle) are delivered to the retailer in m small lots (m retailer cycles per one manufacturing cycle) at fixed time intervals. So the inventory forms a saw tooth pattern during the production up-time and a stair case pattern during production down time in each manufacturing cycle. Likewise, the manufacturer receives n small lots of raw material, at regular intervals, during the production up-time of each manufacturing cycle. The raw materials are consumed at a given rate during the production up-time only. It is assumed that the production rate is greater than the demand rate. So the accumulated inventory during production up-time is used for making delivery during production down-time until the inventory is exhausted (see Figure 2.1). Production is then resumed and the cycle repeated.

Sarker et al [1995a, 1995b, & 1993] developed a model operating under continuous supply at a constant rate. In their paper, they considered two cases. In case I, the ordering quantity of raw material is assumed to be equal to the raw material required for one batch of the production system. The raw material, which is replenished at the beginning of a production cycle, will be fully consumed at the end of this production run. In case II, it is assumed that the ordering quantity of a raw material to be n times the quantity required for one lot of a product, where n is an integer. Though case I can be fitted in the JIT supply system, however, Case II is not favourable for the JIT environment.

An ordering policy, for raw materials to meet the requirements of a production facility under a fixed quantity, periodic delivery policy, has been developed by Sarker and Parija(1994), Jamal and Sarker(1993), Golhar and Sarker(1992) and Sarker and Golhar(1993). They considered that the manufacturer is allowed to place only one order for raw materials per cycle that is similar to above case-I. In this case a fixed quantity of finished goods (say x units) is to be delivered to the customer at the end of every L units of time (fixed interval). This delivery pattern forces inventory build-up in a saw-tooth fashion during the production up-time. The on hand inventory depletes sharply at a regular interval during the production down time until the end of the cycle time. The later one forms a stair case pattern. Recently, Sarker and Parija(1996) developed another model for purchasing a big lot of raw material that will be used in n consecutive production batches.

In this paper, we consider two problems. The first problem considers a JIT delivery system. This problem is somewhat similar to Sarker and Parija (1994), which considers a fixed quantity-periodic delivery policy. The raw material purchasing quantity is equal to the exact requirement of one manufacturing cycle. Chakravarty and Martin (1991) and Golhar and Sarker (1992) perceived that the production time may not be composed of exactly m full shipment periods. To satisfy the demand for the whole cycle time, the production run length may extend for a fraction of a shipment period in addition to m shipments. However, when the transportation cost is independent of the shipment quantity, the fractional shipment may not be cost effective. The retailer may not accept a fractional shipment either. So we consider the frequency of the finished prod-

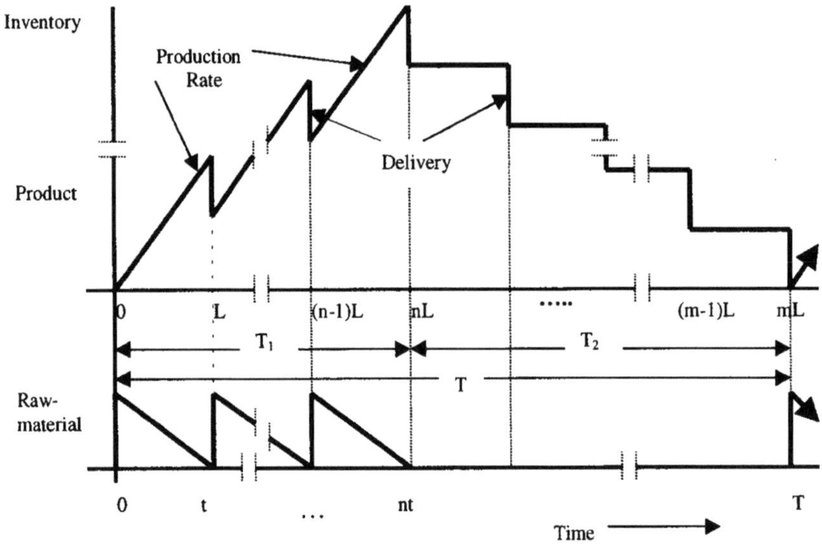

Figure 2.1 Inventory level with time for product (top), and raw material (bottom)

uct supply, m, as strictly integer. The transportation cost is also considered to cover the situation where the retailers are located at some significant distance from the manufacturer. The second problem considers a JIT system for both finished product delivery and raw material receiving. This problem is similar to the first problem except that the raw material requirements in one manufacturing cycle will be purchased in n small lots. Each lot of raw material will be consumed in L units of time during the production up-time.

The cost factors considered are ordering/setting up, holding and transportation costs. We have devised the total cost equation of the system with respect to the production quantity, finished product delivery frequency and/or raw material supply frequency and then solved the problem as an unconstrained optimization problem. The cost function is non-convex which will be discussed in a later section.

3 MODEL FORMULATION

The total quantity manufactured during the production time, T_1, must exactly match the demand for cycle time, T. It is assumed that T_1 must be less than or equal to T. In order to ensure no shortage of products, we assume that the production rate, P, is greater than the demand rate of the finished products, D_P. There is a demand of x units of finished goods at the end of every L time units (due to fixed interval batch supply) and the quantity produced during each period (of L time units) is PL, where $(PL-x)$ is positive. So the finished goods inventory buildup forms a saw-tooth pattern during the production period (see Fig. 1). The on-hand inventory depletes sharply at a regular interval after the production run till the end of the cycle time. The later one forms a stair case pattern.

In this paper, we consider two cases of JIT manufacturing environment. Case I: JIT Delivery System, where the delivery of finished goods is in small lots of x units at a regular time intervals and the raw material purchasing quantity is exactly equal to the requirement of raw material in a manufacturing cycle. Case II: JIT Supply and Delivery System, modifies the raw material supply pattern of Case I. In this case, the raw material required in one manufacturing cycle will be received in n small lots at regular time intervals during the production up-time. For simplicity, we assume that the length of time interval L is equal for both finished goods delivery and raw material receiving. However, the carrying of raw material is not permitted during production down-time in both cases, since it is not cost effective.

To simplify the analysis, the following assumptions are made:

1. There is only one manufacturer and only one raw material supplier for the one item.
2. The production rate is uniform and finite.
3. There are no shortages.
4. The delivery of the product is in a fixed quantity at a regular interval.
5. The raw material supply is available in a fixed quantity whenever required.
6. The producer is responsible to transport the products to the retailers' location.

The notations used in developing the cost functions are shown below:

D_P = demand rate of a product p, units per year
P_P = production rate, units per year (here, $P_P \leq D_P$)
Q_P = production lot size
H_P = annual inventory holding cost, \$/unit/year
A_P = setup cost for a product p (\$/setup)
f_r = amount/quantity of raw material r required in producing one unit of a product
D_r = demand of raw material r for the product p in a year, $D_r = f_r D_P$
Q_P = ordering quantity of raw material r
A_r = ordering cost of raw material r
x = shipment quantity to customer at a regular interval (units/shipment)
L = time between successive shipments = x/D_P
T = cycle time measured in year = Q_P/D_P
m = number of shipments during the cycle time = T/L
n = number of shipments during production up-time

C = transportation cost per shipment
T_1 = production up-time in a cycle
T_2 = production down-time in a cycle = $T - T_1$
IP_{avg} = average finished goods inventory

3.1 *Case I: JIT Delivery System*

The total cost function, for case I, can be expressed as follows:

$$TC_1 = \frac{D_r A_r}{Q_r} + \frac{D_p Q_r H_r}{2P} + \frac{D_p A_p}{Q_p} + IP_{avg} H_p + mC, \tag{3.1}$$

The first two terms in equation (3.1) represent the inventory cost for raw material, the next two terms represent the inventory costs for finished goods, and the last term represent the transportation cost of finished goods from the manufacturing place to the retailer. Since the raw materials are carried for only T_1 time units, the holding cost in second term, H_r, is rescaled by the factor $T_1/T = D_p/P$ to compute the raw material inventory carrying cost.

Unlike the assumption of a one-to-one conversion of raw materials to finished goods a raw material may be transformed to a new product through the manufacturing process at a different conversion rate, $f_r = D_r/D_p = Q_r/Q_p$. Therefore, substituting D_p/Q_p for D_r/Q_r, and $f_r.Q_p$ for Q_r, the above equation leads to

$$TC_1 = \frac{D_p A_r}{Q_p} + \frac{D_p f_r Q_p H_r}{2P} + \frac{D_p A_p}{Q_p} + IP_{avg} H_p + mC, \tag{3.2}$$

in which the total inventory costs for both raw materials and finished goods are expressed in terms of manufacturing batch size, Q_p. The average finished goods inventory per cycle can be expressed as follows:

$$IP_{avg} = Q_p(1 - \frac{D_p}{2P}) - (\frac{m-1}{2})x. \tag{3.3}$$

To simplify the problem, m is assumed to be integer. The detail of this expression can be found for integer m in Sarker and Uddin (1995) and for continuous m in Sarker and Parija (1994).

Using this relationship in (3.3), the total cost equation (3.1) may be written as

$$\begin{aligned} TC_1 \quad = \quad & \frac{D_p(A_r + A_p)}{Q_p} + \frac{D_p f_r Q_p H_r}{2P} + Q_p(1 - \frac{D_p}{2P})H_p \\ & -(\frac{m-1}{2})xH_p + mC, \end{aligned} \tag{3.4}$$

After simplification, equation (3.4) becomes

$$TC_1 = \frac{D_p(A_r + A_p)}{Q_p} + Q_p(\frac{D_p f_r H_r}{2P} - \frac{D_p H_p}{2P} + H_p) + m(C - \frac{xH_p}{2}) + \frac{xH_p}{2} \qquad (3.5)$$

The optimization problem for case I:

Problem I:

Minimize TC_1
Subject to $Q_p = mx$
$Q_p \geq 0$, and
m is integer and greater than one

3.2 Case II: JIT Supply and Delivery System

The total cost function, for case II, can be expressed as follows:

$$TC_2 = \frac{D_p A_p}{Q_p} + \frac{D_r A_r}{Q_r} + \frac{D_p f_r Q_r H_r}{2P} + Q_p(1 - \frac{D_p}{2P})H_p - (\frac{m-1}{2})xH_p + mC, \qquad (3.6)$$

The optimization problem for case II:

Problem II:

Minimize TC_2
Subject to $Q_p = mx$
$Q_p = nQ_r$
$Q_r = PL$
$Q_p \geq 0$, and
m is integer and greater than one

Here, $Q_p = mx$ and $Q_p = nQ_r$ indicate that mx must be equal to nQ_r. In the other word Q_r is equal to the ratio of m and n multiplied by x that is $(m/n)x$. In this paper we assume that the time intervals for raw material receiving will be L time units. So we can write $Q_r = PL$, that means the raw material purchasing quantity is known and it is independent of Q_p. Substituting the relatioship $Q_r = PL$ in (3.6) and then simplifying, we get the total cost equation as follows:

$$TC_2 = \frac{D_p A_p}{Q_p} + \frac{f_r D_p A_r}{PL} + \frac{D_p f_r PLH_r}{2P} + m(C - \frac{xH_p}{2}) + Q_p(1 - \frac{D_p}{2P})H_p + \frac{xH_p}{2}, \qquad (3.7)$$

4 SOLUTION METHODOLOGY

In this section, the solution procedures for both the problems stated in the previous section are presented. The cost functions derived in (3.5) and (3.7) are nonconvex.

Efficient algorithms may be applied to solve the above two problems using an optimization technique. To solve a similar problem, Moinzadeh and Aggarwal(1990) proposed an algorithm for discrete optimization, Park and Yun(1984) developed a heuristic method, and Golhar and Sarker(1992) used a one directional search procedure. In this paper, we propose a heuristic method which is simple, easy to apply and computationally efficient. In this method, an initial solution for Q_p is determined by relaxing the integrality requirement for m. Then the corresponding value of m is computed. If the computed m is not integer, we then search for an integer m in the neighbouring points which gives a minimum TC.

4.1 Problem I

In order to find an optimal manufacturing quantity, it is necessary to differentiate relation (3.5) with respect to Q_p. But the function is nondifferential as it contains an integer decision variable m. However, it can be shown that TC is a convex function of Q_p for a given m. As we assumed in an earlier section, the integer m can be replaced by Q_p/x. So the modified cost equation is:

$$TC_1 = \frac{D_p(A_r + A_p)}{Q_p} + Q_p(\frac{D_p f_r H_r}{2P} - \frac{D_p H_p}{2P} + \frac{C}{x} + \frac{H_p}{2}) + \frac{x H_p}{2} \tag{4.1}$$

Now differentiating TC_1 with respect to Q_p and equating to zero, we get

$$Q_p^* = \sqrt{\frac{D_p(A_r + A_p)}{KK}} \tag{4.2}$$

where,

$$KK = \frac{D_p f_r H_r}{2P} - \frac{D_p H_p}{2P} + \frac{C}{x} + \frac{H_p}{2} \tag{4.3}$$

Now the calculated $m(= Q_p^*/x)$ may not be integer as we replaced m by continuous variable in (4.1). In case of non-integer m, we take the neighbouring integer value corresponding to which the total cost is minimum. The algorithm to solve the batch sizing problem for Case I is as follows:

Algorithm 1: *finding batch size*

Step 0. Initialize and store D_p, P, A_p, A_r, H_p and H_r
Step 1. Compute the batch size Q_p^* using (4.2)
Step 2. Compute $m(=Q_p^*/x)$. If m is an integer, then stop.
Step 3. Compute TC using (3.5) for $m = \lceil m \rceil$ and $\lfloor m \rfloor$. Choose the $m^* = m$ that gives minimum TC.
Stop.

4.2 *Problem II*

Substituting the integer variable m by Q_p/x, we can rewrite (3.7) as follows:

$$\begin{aligned} TC_2 &= \frac{D_pA_p}{Q_p} + \frac{D_rA_r}{PL} + \frac{D_pf_rPLH_r}{2P} + Q_p(\frac{C}{x} - \frac{H_p}{2}) \\ &+Q_p(1-\frac{D_p}{2P})H_p + \frac{xH_p}{2} \end{aligned} \tag{4.4}$$

Now, differentiating TC_2 with respect to Q_p and then equating to zero we get

$$Q_p^* = \sqrt{\frac{(D_pA_p)}{CC}} \tag{4.5}$$

where,

$$CC = \frac{H_p}{2}(1 - \frac{D_p}{P}) + \frac{C}{x} \tag{4.6}$$

Now the calculated $m(=Q_p^*/x)$ may not be integer as we replaced it by the continuous variable in (4.4). In case of non-integer m, we take the neighbouring integer value corresponding to which the total cost is minimum. The algorithm to solve the batch sizing problem for Case II is as follows:

Algorithm 2: *finding batch size*
Step 0. Initialize and store D_p, P, A_p, A_r, H_p and H_r
Step 1. Compute the batch size Q_p^* using (4.5)
Step 2. Compute $m(=Q_p^*/x)$. If m is an integer, then stop
Step 3. Compute TC using (3.7) for $m = \lceil m \rceil$ and $\lfloor m \rfloor$. Choose the $m^* = m$ that gives minimum TC.
Stop.

5 RESULTS AND DISCUSSIONS

The total cost functions with integer m are plotted in Figure 5.2 with various transportation costs (C). As an assumption made $m = Q_p/x$ in an earlier section, the total cost is computed for each paired m (integer) and Q_p data.

The data used, to generate these curves, are similar to Sarker and Parija (1994) and Jamal and Sarker (1993): $D_p = 2400$ units/year, $P = 3600$ units/year, $A_p = \$300$/set-up, $A_r = \$200$/order, $H_p = \$2$/unit/year, $H_r = \$1$/unit/year, $f_r = 1$ and $x = 100$ units/shipment. The behaviour of the cost function, in Figure 5.2, ensures a smoth and convex pattern.

If we restrict m as an integer for the problems stated in Sarker and Parija (1994), Jamal and Sarker (1993) and Golhar and Sarker (1992), then their cost equations become similar to our equation (3.5) with $C = 0$. With the above mentioned data, the optimum TC, m, Q_p with $C = 0, 50, 100, 150, 200, 250$ and 300 are shown in Table 5.1. With $C = 0$, the solution is $Q_p^* = 1300$, $m^* = 13$ and $TC = \$1889.744$.

As per the solution of Golhar and Sarker (1992), $Q_p = 1342$ units/batch which is incidentally similar to our Q_p solution in Step-1 for $C = 0$. The optimum Q_p is 1346 as determined by Sarker and Parija (1994) and Jamal and Sarker (1993). As we see in Figure 5.2 and Table 5.1, the total cost increases, and optimum m and Q_p decrease, with increase of C.

The results produced by the heuristic developed in this paper were also compared with the solutions of enumeration method and simulated annealing, a stochastic search algorithm. They produced exactly the same results for the test problems. It is not intended to present the detail of Simulated Annealing results in this paper.

The total cost plotted against raw material ordering cost for both problem 1 and problem 2 is shown in Figure 5.3. The data are similar to the above problem. Problem 2 is more sensitive than problem 1 to the ordering cost. From Figure 5.3, it is clear that problem 2 is more attractive for very low ordering cost. That means if the ordering cost is very low, which is a condition of JIT supply, problem 2 can be set for more savings.

Table 5.1 Optimum TC, m and Qp with different C; *optimum TC

Values of	Calc. Values		Upr Int.(m)		Lwr Int.(m)		Solution	
C	Q_p	m	m	TC	m	TC	m^*	Q_p^*
0	1342	13.42	14	1890.47	13	1889.74	13	1300
50	1014	10.14	11	2474.24	10	2466.66*	10	1000
100	849	8.49	9	2933.33*	8	2933.33*	8 or 9	900 or 800
150	744	7.44	8	3333.33	7	3330.95*	7	700
200	670	6.70	7	3680.95*	6	3700.00	7	700
250	616	6.16	7	4030.95	6	4000.00*	6	600
300	572	5.72	6	4300.00*	5	4333.33	6	600

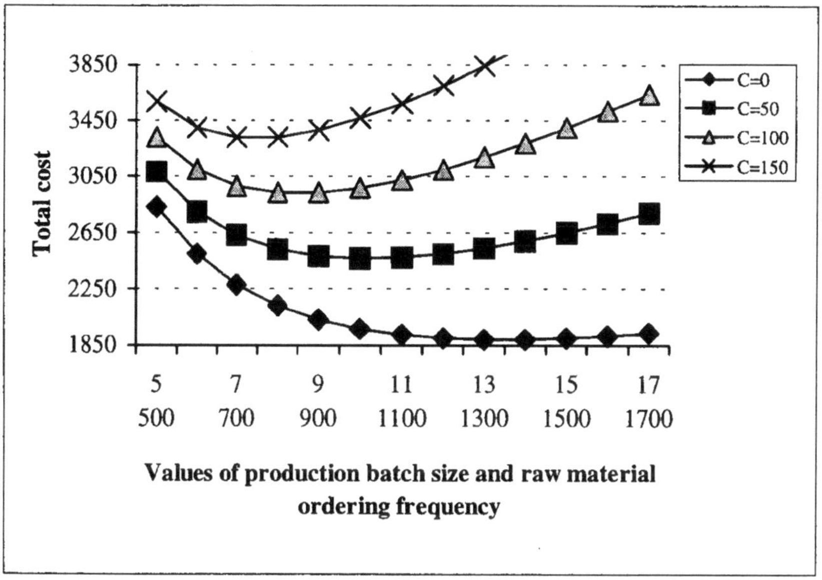

Figure 5.1 Total cost curve

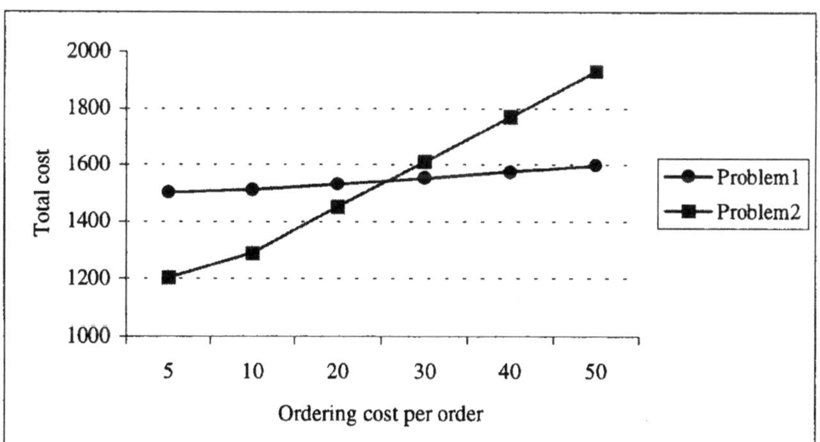

Figure 5.2 Sensitivity with raw material ordering cost

6 CONCLUSIONS

In a JIT manufacturing environment, a supplier is expected to frequently deliver in small lots. Ideally, a supplier to the JIT buyer is expected to synchronize his production volume with the buyer's demand so that the inventory in the supply pile line is reduced and eventually eliminated.

An inventory model is developed for perfect matching. The total cost function is non-differentiable with respect to manufacturing batch size. However, it can be shown that the total cost function is convex for a given m. A simple algorithm is developed to compute optimal manufacturing batch size and the associated raw material purchasing lot size. The quality of solution is discussed and sensitivity analysis is provided.

In problem II, it is assumed that $Q_p = nQ_r$, where $Q_r = PL$. To generalize the problem, the relation $Q_r = PL$ should be relaxed. Though the new problem will be much more complicated, our present research is moving towards that direction.

References

Aderohunmu, R., Mobolurin, A. and Bryson, N. (1995), Joint vendor-buyer policy in JIT manufacturing, *J. Opl. Res. Soc.*, Vol. 46, pp. 375-385.

Ansari, A. and Hechel, J. (1987), JIT purchasing: impact of freight and inventory costs, *J. Purchasing and Mat. Mgmt*, Vol. 23, pp. 24-28.

Ansari, A. and Modarress, B. (1987), The potential benefits of just-in-time purchasing for US manufacturing, *Prod. Inventory Mgmt*, Vol. 28, pp. 32-35.

Chakravarty, A. K. and Martin, G. E. (1991), Operational economies of a process positioning determinant, *Comps & Opns Res.*, Vol. 18, pp. 515-530.

Chapman, S. N. and Carter, P. L. (1990), Supplier/customer inventory relationships under just in time, *Decis. Sci.*, Vol. 21, pp. 35-51.

Golhar, D.Y., and Sarker, B. R. (1992), Economic manufacturing quantity in a just-in-time deliver system, *International Journal of Production Research*, Vol. 30 No.5, pp. 961- 972.

Jamal, A. M. M. and Sarker, B. R. (1993), An optimal batch size for a production system operating under a just-i-time delivery system, *Int. J. of Prod. Econ.*, Vol. 32, pp. 255-260.

Moinzadeh, K. and Aggarwal, P. (1990), Order expedition in multi-level production inventory system, *Paper presented at the TIMS/ORSA Joint National Meeting*, Las Vegas, NV, USA, May.

O'Neal, C. R. (1989), The buyer-seller linkage in a just-in-time environment, *J. Purchasing and Mat. Mgmt*, Vol. 25, pp. 34-40.

Park, K. S. and Yun, D. K. (1984), A stepwise partial enumeration algorithm for the economic lot scheduling problem, *IIE Trans.*, Vol. 16, pp. 363-370.

Sarker, B. R. and Golhar, D.Y. (1993), A reply to a note to economic manufacturing quantity in a just-in- time delivery system, *International Journal of Production Research*, Vol. 31, No. 11, pp. 27- 49.

Sarker, B. R., and Parija, G.R. (1996), Optimal batch size and raw material ordering policy for a production system with a fixed-interval, lumpy demand delivery system, *European Journal of Operational Research*, Vol. 89, pp. 593-608.

Sarker, B. R. and Parija, G. R. (1994), An optimal batch Size for a production system operating under a fixed-quantity, periodic delivery policy, *J. Opl Res. Soc.*, Vol. 45/8, pp. 891-900.

Sarker, R.A., Karim, A.N.M., and Azad, S. (1993), Integrated inventory system: cases of product-raw materials and producer-wholesalers, *37th Annual Convention of the Institution of Engineers Bangladesh*, Rajshahi.

Sarker, R.A., Karim, A.N.M., and Haque, A.F.M.A. (1995a), An optimal batch size for a production system operating under a continuous supply/demand, *International Journal of Industrial Engineering*, Vol. 2, No. 3, pp. 189-198.

Sarker, R.A., Karim, A.N.M., and Azad, S., (1995b) Two Cases of Integrated inventory, *Journal of the Institution of Engineers, Bangladesh*, Vol. 21, No. 4, pp. 45-52.

Sarker, R. A. and Uddin, J. (1995), Determination of batch size for a production system operating under a periodic delivery policy, *Internal Report*, Dept. of IPE, Bangladesh University of Engineering and Technology, Dhaka, pp. 1-80.

22 PARAMETER ESTIMATION IN A MATHEMATICAL MODEL FOR SUBSTRATE DIFFUSION IN A METABOLICALLY ACTIVE CUTANEOUS TISSUE

Klaus Schittkowski

Department of Mathematics,
University of Bayreuth,
95440 Bayreuth, Germany

Abstract: A numerical approach is described to determine parameters in a system of one-dimensional partial differential equations and coupled ordinary differential equations. The model allows arbitrary transition conditions between separate integration areas for functions and derivatives. The minimum least squares distance of the measured data from the solution of a system of differential equations at designated space values is computed.

A special application model is outlined in detail, that describes the diffusion of a substrate through cutaneous tissue. Metabolic reactions are included in form of Michaelis-Menten kinetics. The goal is to model transdermal drug delivery, where it is supposed that experimental data are available for substrate and metabolic fluxes. Some numerical results are included to show the efficiency of the implemented algorithms.

Key words: parameter estimation, data fitting, least squares optimization, partial differential equations, method of lines, substrate diffusion, cutaneous tissue, transdermal application

X. Yang et al. (eds.), Progress in Optimization, 329–342.

1 INTRODUCTION

The numerical identification of unknown parameters of a dynamical model is extremely important in many technical, natural sciences, and medical disciplines. In our case we proceed from a system of one-dimensional partial differential equations and want to minimize distances of some fitting criteria from experimental data. The underlying optimization problem consists of minimizing the sum of squared functions given in the form

$$F(p) = \sum_{i=1}^{l} f_i(p)^2 \ , \tag{1.1}$$

where all individual functions $f_i(p)$ are continuously differentiable w.r.t. p. In addition we allow at least constraints in form of upper and lower bounds for the parameters to be fitted, i.e.

$$p_l \leq p \leq p_u \ . \tag{1.2}$$

To formulate a data fitting or parameter estimation problem, respectively, we assume that a model function h is available depending on the unknown parameter vector p to be computed, and a time variable t. By adopting p the mathematical model is to be fitted to r measurement sets by a least squares estimate, where differences of model function values from given measurements at some experimental times are to be minimized. Thus we get

$$F(p) = \sum_{i=1}^{s} \sum_{k=1}^{r} w_i^k (h_k(t_i, p) - y_i^k)^2 \tag{1.3}$$

with experimental time values t_i, measured values y_i^k and additional weight factors w_i^k.

Moreover we assume that the fitting criterion depends on the solution of a dynamical system, in our case a one-dimensional partial differential equation with coupled ordinary differential equations. To specify time-dependent experimental data, we have to fix corresponding spatial variable values. The right-hand sides of the differential equations depend on the unknown parameter vector p.

The numerical solution is known as the method of lines. We discretize the space variable x and approximate the corresponding first and second derivatives w.r.t. x by polynomial interpolation or difference formulae, respectively. Initial, boundary, and transition conditions between different integration areas must be taken into account, and may depend on the parameters to be computed. The resulting large and eventually stiff system of ordinary differential equations can be solved by standard techniques, e.g., the ODE solvers implemented by Hairer et al (1993) and Hairer et al (1991) . For the iterative solution of the least squares problem, we propose any of the available Gauss-Newton and related methods, e.g., DN2GB of Dennis et al (1981), NLSNIP of Lindström (1983), or DFNLP of Schittkowski (1988).

A special model is investigated in detail, that describes the diffusion of a substrate (Ala-MNA) through cutaneous tissue. Two different layers are considered, the metabolically active tissue and a passive porous membrane. Metabolic reactions are taken into account by a Michaelis-Menten kinetics, leading to a system of two partial differential equations in two different integration areas. Non-continuous transition conditions between both areas are formulated.

The assumed laboraty experiment is to simulate transdermal drug delivery where experimental data are known for substrate and metabolic fluxes. To be able to compute the total amounts entering and leaving the diffusion area, ordinary differential equations are coupled to the partial ones at both boundary points. Some numerical results are included to show the efficiency of the implemented algorithms.

In Section 2 we outline the general structure of the dynamical system we are considering, i.e., the one-dimensional partial differential equation. The numerical solution method is described in Section 3, where we show how the method of lines is applied to get a system of ordinary differential equations. Section 4 contains the mathematical description of the the transdermal diffusion model, followed by some numerical results in Section 5.

2 THE DYNAMICAL SYSTEM

The fitting criteria depend on the solution of a system of one-dimensional partial differential equations and some coupled ordinary differential equations. To be able to define the model equations in different integration areas, we introduce n_a subintervals, where n_p partial differential equations can be defined separately:

Thus we introduce the notation

t	-	time coordinate,
T	-	final time,
x	-	space or spatial coordinate, respectively,
n_a	-	number of integration areas,
x_i^a	-	i-th boundary or transition point, $i = 0, \ldots, n_a$,
n_p	-	number of partial differential equations,
u	-	solution vector of partial differential equation,
v	-	solution vector of coupled ordinary differential equations.

The system of partial differential equations must be defined now separately for each subinterval in time-dependent form

$$u_t^i = f^i(x, t, v, u^i, u_x^i, u_{xx}^i, p) \tag{2.1}$$

for $i = 1, \ldots, n_a$ with initial values $u^i(x, 0, p) = u_0^i(x, p)$. Without loss of generality it is assumed that the initial time is zero. Thus the above equations are to be satisfied for $t \in (0, T]$ and $x \in (x_i^a, x_{i+1}^a)$, $i = 0, \ldots, n_a - 1$.

In addition we require boundaray conditions in Dirichlet or Neumann form

$$\begin{aligned} u^1(x_0^a, t, p) &= u^L(t, p) \ , \\ u^{n_a}(x_{n_a}^a, t, p) &= u^R(t, p) \ , \\ u_x^1(x_0^a, t, p) &= \hat{u}^L(t, u(x_0^a, t, p), p) \ , \\ u_x^{n_a}(x_{n_a}^a, t, p) &= \hat{u}^R(t, u(x_{n_a}^a, t, p), p) \end{aligned} \tag{2.2}$$

to be satisfied at x_0^a and $x_{n_a}^a$ for $t \in (0, T]$.

Transition conditions between different integration areas may depend in addition on the solution of the previous or subsequent integration area, and are formulated by alternative equations of the form

$$\begin{aligned} u^i(x_i^a, t, p) &= c_i^R(t, u^{i+1}(x_i^a, t, p), p) \ , \\ u^{i+1}(x_i^a, t, p) &= c_i^L(t, u^i(x_i^a, t, p), p) \ , \\ u_x^i(x_i^a, t, p) &= \hat{c}_i^R(t, u^{i+1}(x_i^a, t, p), u_x^{i+1}(x_i^a, t, p), p) \ , \\ u_x^{i+1}(x_i^a, t, p) &= \hat{c}_i^L(t, u^i(x_i^a, t, p), u_x^i(x_i^a, t, p), p) \end{aligned} \tag{2.3}$$

for $i = 1, \ldots, n_a - 1$.

Ordinary differntial equations can be coupled to the partial ones at arbitrary spatial variable values x_j^o leading to a set of additional equations of the form

$$\dot{v} = H(t, v, u(x_j^o, t, p), u_x(x_j^o, t, p), u_{xx}(x_j^o, t, p), p)$$

with initial conditions $v(0, p) = v_0(p)$.

It is essential to understand that the user has to select a suitable combination of boundary, transition, and coupled ordinary differential equations, so that the partial differntial equations are uniquely solvable. These conditions depend also on the structure of right-hand side of (2.1), e.g., whether second spatial derivatives are available or not. Note also that there are no restrictions concerning the right-hand side functions f_i, which may be highly nonlinear w.r.t. all parameters. Initial, boundary, and transition conditions may depend on the parameters to be fitted.

To compute the fitting criterion, the solution vectors $u(x, t, p)$ and $v(t, p)$ at predefined spatial variable values x_j^c are inserted into given functions

$$\overline{h}_k(t, x, u, u_x, u_{xx}, v, p) \ ,$$

$j = 1, \ldots, n_c$, $k = 1, \ldots, n_m$, i.e.,

$$f_k(p) = \overline{h}_j(t_i, x_j^c, u(x_j^c, t_i, p), u_x(x_j^c, t_i, p), u_{xx}(x_j^c, t_i, p), v(t_i, p), p) \tag{2.4}$$

where k runs from 1 to $n_m n_n s$ in any order. n_c is the number of spatial variable values at which measurements are available, and n_m represents the number of measurements taken at $t = t_i$ and $x = x_j^c$. It is assumed that $\overline{h}_j$ is differentiable w.r.t. the parameter vector p to be optimized.

More general dynamic systems are studied in Schittkowski (1998), that permit additional flux functions, algebraic partial differential equations, switching points w.r.t. t, and hyperbolic equations. But these options are not needed to describe the transdermal drug delivery model of Section 4. More details of numerical implementation and a set of 20 test examples are found in Schittkowski (1997b).

3 SOLUTION METHOD

Partial differential equations are discretized by the method of lines, see e.g. Schiesser (1991), which is outlined briefly. More details about the approach chosen are found in Schittkowski (1997b) and Schittkowski (1998).

For the i-th integration interval of the spatial variable, we denote the number of discretization points by n_i, $i = 1, \ldots, n_a$. We proceed from uniform grid points within each interval and get a discretization of the whole space interval from x_0^a to $x_{n_a}^a$. To approximate the first and second partial derivatives of $u(x, t, p)$ w.r.t. the spatial variable at a given point x, several different alternatives are available:

a) Polynomial approximation: We compute an interpolating polynomial subject to some neighbouring values by Newton's formula. The number of interpolation points depends on the polynomial degree selected. By differentiating the interpolation formulae, first and second order approximations are obtained. In case of Neumann boundary conditions, Hermite interpolation is used to exploit known derivative values.

b) Difference Formulae: First and second derivatives can be approximated by difference formulae, see for example Schiesser (1991). Difference formulae with 3 and 5 points for first derivatives can be applied recursively to get also second derivative approximations. Alternatively a 5-point difference formula for second derivatives can be implemented as well. The difference formulae are adapted at the boundary to accept given function and gradient values.

Whenever a boundary or transition condition is given in Dirichlet-form, we know the value of the boundary function and use it to interpolate or approximate the function $u(x, t, p)$ as decribed above. In other words the corresponding function value of the right-hand side of the discretized system is replaced by the value given. Alternatively a boundary condition may appear in Neumann-form. In this case the derivative values at the boundary are replaced by the given ones before evaluating the second order spatial derivative approximations.

Ordinary differential equations are added to the discretized system without any further modifications. Since arbitrary coupling points are allowed, they are rounded to the nearest line of the discretized system. The resulting large and eventually stiff system of ordinary differential equations can be integrated by any available standard technique, e.g., the implicit ODE solver RADAU5 of Hairer et al (1991) which is used for our numerical tests.

The resulting least squares problem can be solved iteratively by any of the existing codes based on Gauss-Newton or related methods, e.g., DN2GB of Dennis et al (1981), NLSNIP of Lindström (1983), or DFNLP of Schittkowski (1988), which is also used for the numerical tests of the subsequent sections. In this case the original problem is transformed into a general nonlinear programming problem by introducing additional variables and constraints. It can be shown, that typical features of a Gauss-Newton and quasi-Newton method are retained, see Schittkowski (1988). The resulting nonlinear programming problem is solved by a standard SQP code called NLPQL, cf. Schittkowski (1985). Numerical derivatives are approximated by forward differences.

4 A TRANSDERMAL DIFFUSION MODEL

We consider the permeation of a substrate through cutaneous tissue with simultaneous metabolism and through a subsequent passive porous membrane . Goal is the anlysis and numerical simulation of transdermal processes for developing new drugs and application devices, see e.g. Guy and Hadcraft (1988). Especially the affects of various parameters influencing diffusion and metabolism can be studied in detail, once a reliable model validated by experimental data is available.

A special laboratory experiment is to be modelled where a so-called donor contains a given volume and concentration of a substrate, e.g., Ala-MNA, that penetrates through a thin layer of a cutaneous tissue of given size and through an underlying membrane with known thickness and porosity. ¿From time to time the concentration of the substrate is measured on both sides of the membrane.

The situation is a bit more complicated in the sense that at the same time another substrate is generated, the so-called metabolite, because of enzymatic interactions, see e.g. Hotchkiss (1992), Hadgraft (1979), or Guy and Hadcraft (1988). Also the concentration of this new substrate is measured on both sides of the membrane. It is assumed that metabolism can be described by Michaelis-Menten kinetics, see Pratt (1990), that the distribution of metabollically active enzyms in the layer is homogeneous, and that the mass transport is one-dimensional along the x-axis.

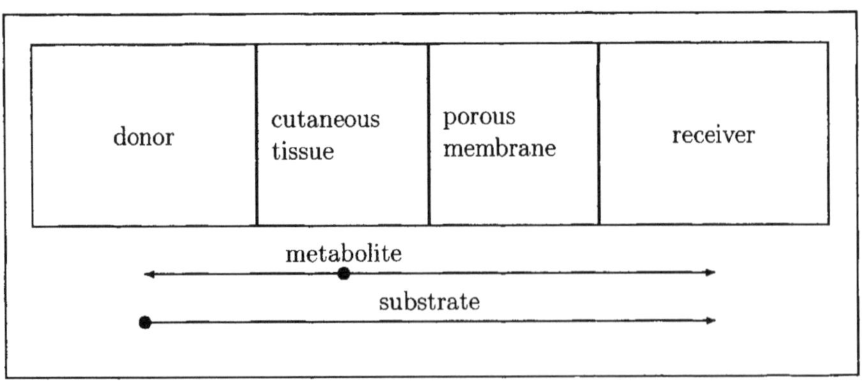

The dynamic variables we need to describe the process, are

$u^s(x,t)$ substrate concentration at x and t $[\frac{nmol}{mm^3}]$,
$u^m(x,t)$ metabolite concentration at x and t $[\frac{nmol}{mm^3}]$,
$v^s(t)$ substrate mass at donor, i.e. for $x=0$ $[nmol]$,
$w^s(t)$ substrate mass at receiver, i.e. for $x=l$ $[nmol]$,
$v^m(t)$ metabolite mass at donor, i.e. for $x=0$ $[nmol]$,
$w^m(t)$ metabolite mass at receiver, i.e. for $x=l$ $[nmol]$.

Two partial differential equations describe the diffusion through the skin and take the Michaelis-Menten effect into account, see also Wolf (1994), Steinsträsser (1994), or Boderke et al (1998). They are given by

$$\begin{aligned} u_t^s(x,t) &= D_T^s\, u_{xx}^s(x,t) - V_{\max} u^s(x,t)/(K_m + u^s(x,t)) \;, \\ u_t^m(x,t) &= D_T^m\, u_{xx}^m(x,t) + V_{\max} u^s(x,t)/(K_m + u^s(x,t)) \end{aligned} \tag{4.1}$$

for $0 < x < l_T$ and $0 < t \leq T$. The corresponding initial conditions are $u^s(x,0)=0$ and $u^m(x,0)=0$ for all $x>0$, and $u^s(0,0)=Y_0P^s/V_a$, $u^m(0,0)=0$. Diffusion of substrate and metabolite through membrane are described by the equations

$$\begin{aligned} u_t^s(x,t) &= D_M^s\, u_{xx}^s(x,t) \;, \\ u_t^m(x,t) &= D_M^m\, u_{xx}^m(x,t) \end{aligned} \tag{4.2}$$

defined for $l_T < x < l$ and $0 < t \leq T$. Corresponding initial values are $u^s(x,0)=0$ and $u^m(x,0)=0$ for all x.

Accummulation of mass fluxes at both sides of the two layers leads to the integrals

$$\begin{aligned} v^s(t) &= F_a D_T^s \int_0^t u_x^s(0,\tau)d\tau \;, \\ v^m(t) &= F_a D_T^m \int_0^t u_x^m(0,\tau)d\tau \;, \\ w^s(t) &= -F_a D_M^s \int_0^t u_x^s(l,\tau)d\tau \;, \\ w^m(t) &= -F_a D_M^m \int_0^t u_x^m(l,\tau)d\tau \;. \end{aligned} \tag{4.3}$$

¿From these equations we get four ordinary differential equations that are coupled to (4.1) and (4.2) to obtain a dynamic system in standard form, i.e.,

$$\begin{aligned} \dot{v}^s(t) &= F_a D_T^s\, u_x^s(0,t) \;, \\ \dot{v}^m(t) &= F_a D_T^m\, u_x^m(0,t) \;, \\ \dot{w}^s(t) &= -F_a D_M^s\, u_x^s(l,t) \;, \\ \dot{w}^m(t) &= -F_a D_M^m\, u_x^m(l,t) \end{aligned} \tag{4.4}$$

for $0 < t \leq T$. They describe the behaviour of substrate and metabolite at both end points, i.e. the mass fluxes into and out of tissue and membrane. Initial conditions are $v^s(0)=Y_0$, $v^m(0)=0$, $w^s(0)=0$, and $w^m(0)=0$.

Boundary conditions are formulated to couple the partial differential equations and the ordinary differential equations, i.e.,

$$\begin{aligned} u^s(0,t) &= P^s/V_a\, v^s(t) \ , \\ u^m(0,t) &= P^m/V_a\, v^m(t) \ , \\ u^s(l,t) &= P^s/V_a\, w^s(t) \ , \\ u^m(l,t) &= P^m/V_a\, w^m(t) \end{aligned} \tag{4.5}$$

for all t, $0 < t \leq T$.

Between both integration areas we consider non-continuous conditions for the transition of substrate and metabolite from tissue to membrane and vice versa, i.e.,

$$\begin{aligned} u^s(l_T^-,t) &= T^s\, u^s(l_T^+,t) \ , \\ u^m(l_T^-,t) &= T^m\, u^m(l_T^+,t) \ , \\ u_x^s(l_T^+,t) &= D_M^s/D_T^s\, u_x^s(l_T^-,t) \ , \\ u_x^m(l_T^+,t) &= D_M^m/D_T^m\, u_x^m(l_T^-,t) \ . \end{aligned} \tag{4.6}$$

These conditions are to be valid for all t with $0 < t \leq T$. Here $u^s(l_T^-,t)$ and $u^m(l_T^-,t)$ denote the concentrations of substrate and metabolite at the right tissue boundary and $u^s(l_T^+,t)$ or $u^m(l_T^+,t)$ the corresponding values at left membrane boundary.

To describe the mathematical model of the experiment as oulined above, we use the notation

l_T	-	thickness of tissue,
l_M	-	thickness of membrane,
l	-	total thickness of diffusion area, i.e., $l = l_T + l_M$,
K_m	-	Michaelis-Menten constant $[\frac{nmol}{mm^3}]$,
P^s	-	distribution coefficient of substrate,
P^m	-	distribution coefficient of metabolite,
V_a	-	volume of donor $[mm^3]$,
F_a	-	surface of the membrane $[mm^2]$,
Y_0	-	initial mass of substrate $[nmol]$,
T	-	maximum observation time $[min]$.

Parameters to be estimated, are

D_T^s	-	diffusion coefficient of substrate in tissue,
D_M^s	-	diffusion coefficient of substrate in membrane,
D_T^m	-	diffusion coefficient of metabolite in tissue,
D_M^m	-	diffusion coefficient of metabolite in membrane,
$V_{\max}$	-	maximum metabolization rate,
T^s	-	transition parameter of substrate,
T^m	-	transition parameter of s metabolite.

Table 5.1 Experimental Data

t_i	y_i^1	y_i^2	y_i^3	y_i^4
0.0	107	-	-	-
2.0	105	0.8	0.6	0.3
5.0	102	2.2	1.4	0.8
7.0	99	3.0	1.9	1.0
10.0	97	4.5	2.5	1.8
20.0	89	8.9	4.0	3.9
30.0	81	13.1	5.0	6.2

It is supposed that substrate and metabolite masses are measured at donor and receiver sides at different time values leading to the fitting functions

$$\begin{aligned} h_1(t) &= v^s(t) \ , \\ h_2(t) &= v^m(t) \ , \\ h_3(t) &= w^s(t) \ , \\ h_4(t) &= w^m(t) \ . \end{aligned} \tag{4.7}$$

5 NUMERICAL RESULTS

The mathematical model introduced in the previous section, is only one variant of a large number of different alternatives depending on the experimental design and the degree of refinement. To give an example, one could investigate alternative ways to describe the transition between different layers, see Wolf (1994).

The purpose of the numerical test is to show that the dynamic model yields reasonable and interpretable simulation results and that the numerical method can be implemented in an efficient way. Thus we consider only one experimental data set taken from Steinsträsser (1994), see Table 5.

The numerical tests were obtained by the Fortran code PDEFIT, see Schittkowski (1998), executed from an interactive system EASY-FIT, cf. Schittkowski (1997a) on a PC with a 200 MHz Pentium processor running under Windows NT. Model functions are provided in form of a special modelling language called PCOMP, see Dobmann et al (1994), based on a Fortran-similar syntax. Since model functions are interpreted during runtime, compilation and linking is prevented, and new model variants are tested rapidly.

The actual model uses the following constants according to the underlying experimental design:

Table 5.2 Parameter Values

p	p_0	p^*
D_M^s	0.0001	0.01095
D_M^m	1.0	0.1441
D_T^s	1.0	2773.2
D_T^m	1.0	0.55195
V_{max}	1.0	12.0571
T^s	1.0	0.2650
T^m	1.0	0.5634

$$
\begin{aligned}
l_T &= 0.005\, mm, \\
l_M &= 0.005\, mm, \\
K_m &= 7.0 \cdot 10^{-3}\, \frac{nmol}{mm^3}, \\
V_a &= 3.0 \cdot 10^3\, mm^3, \\
F_a &= 63.6\, mm^2, \\
P_s &= 0.106.3, \\
P_m &= 0.07, \\
Y_0 &= 105.0\, nmol, \\
T &= 30\, min.
\end{aligned}
$$

For each integration area we define 11 grid points, i.e. the resulting discretized ODE consists of 26 equations. Second spatial derivatives w.r.t. x are approximated by a 5-point-difference formula. For the numerical integration we apply the implicit Runge-Kutta method RADAU5, see Hairer et al (1991), with final error tolerances $ABSERR = RELERR = 10^{-6}$. The resulting least squares problem is solved by the Gauss-Newton-type algorithm DFNLP of Schittkowski (1988), with termination accuracy $ACC = 10^{-10}$.

DFNLP needed 109 iteration to reduce the residual scaled by the sum of squared measurement values, from 1.12 to 0.00062. The relatively large number of iterations is due to poor choice of initial parameter values listed in Table 5.1 together with the final ones. The fitting criteria and the experimental data are shown in Figures 5.2 to 5.3, the concentration profiles in Figures 5.4 and 5.5.

The underlying experiment allows to measure mass data only within a relatively short time interval of 30 *min*. Another interesting question is, whether the system achieves a reasonable and interpretable steady-state condition, i.e. a constant substrate and metabolite distribution at both sides of the two layers in equlilibrium. Therefore we insert the optimal parameters of Table 5.1 and perform only a simulation, i.e., we integrate the system equations from $t = 0$ to $t = 600$. The resulting mass functions are shown in Figures 5.6 to 5.9. We see that after 400 *min* the system is in steady-state with metabolite mass of 70 *nmol* at donor substrate mass 36 *nmol* at receiver side, aproximately. Substrate at donor and metabolite at receiver vanish completely.

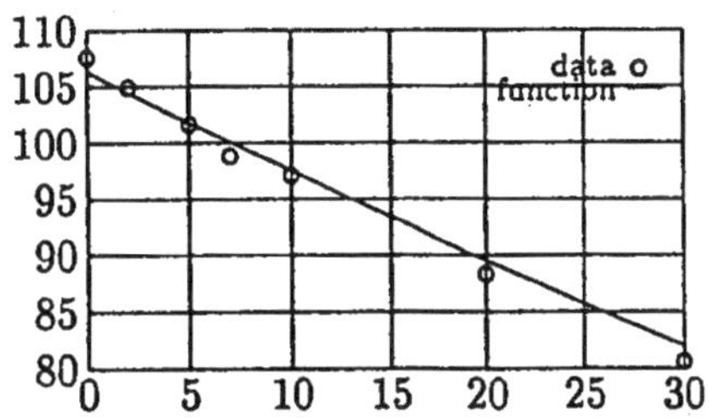

Figure 5.1 Substrate at Donor

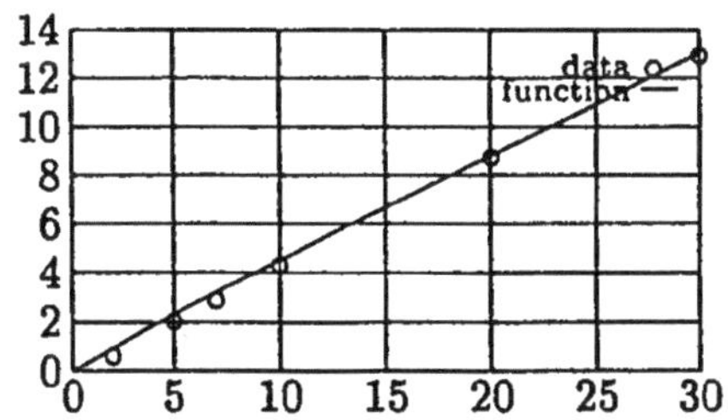

Figure 5.2 Metabolite at Donor

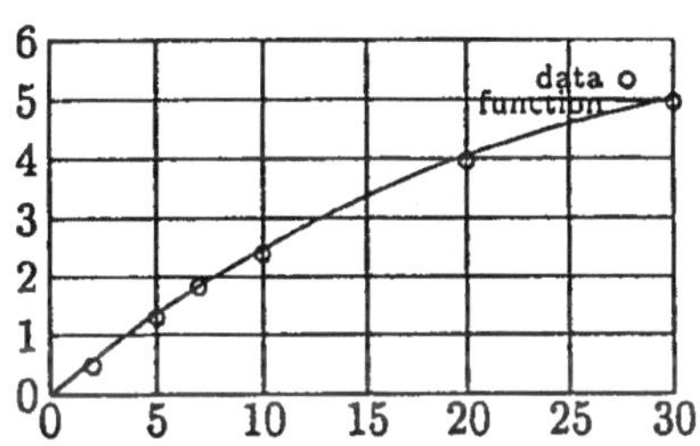

Figure 5.3 Substrate at Receiver

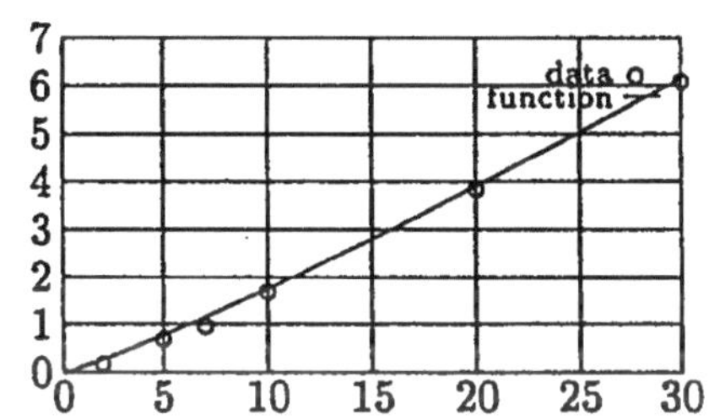

Figure 5.4 Metabolite at Receiver

A further possibility to check the validity of the model, is to compute the mass balance

$$
\begin{aligned}
b(t) &= v^s(t) + v^m(t) + w^s(t) + w^m(t) \\
&\quad + F_a \int_0^l u^s(x,t)dx + F_a \int_0^l u^m(x,t)dx - Y_0 \ .
\end{aligned} \tag{5.1}
$$

The mass distribution for 600 min is shown in Figure 5.10. The error is in the order of the discretization and numerical integration accuracy chosen.

6 CONCLUSIONS

We show an approach to compute unknown parameters in a dynamical model consisting of partial and coupled ordinary differential equations, by a least squares data fit. The numerical discretization is based on difference formulae for second derivatives w.r.t. the spatial variable. The resulting system of ordinary differential equations is integrated by a standard solver, and the least squares optimization problem is solved iteratively by a Gauss-Newton method. A practical application model is outlined in detail, to simulate the transdermal diffusion of drugs. Based on a laboratory experiment, partial differential equations describe diffusion of a substrate through a metabolically

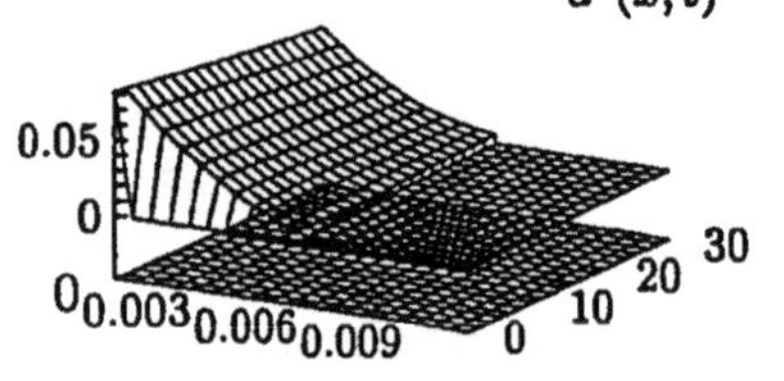

Figure 5.5 Substrate Concentration

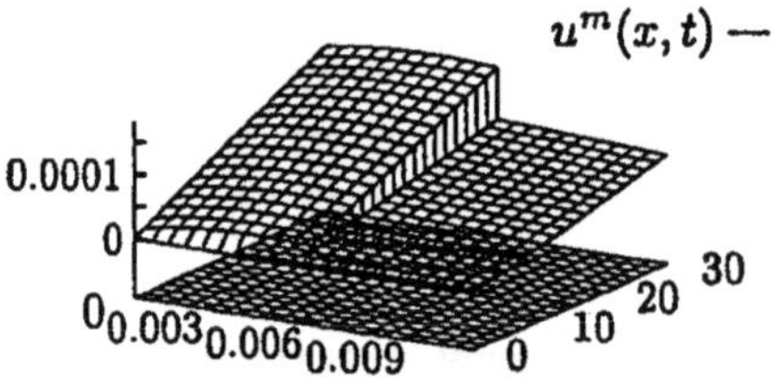

Figure 5.6 Metabolite Concentration

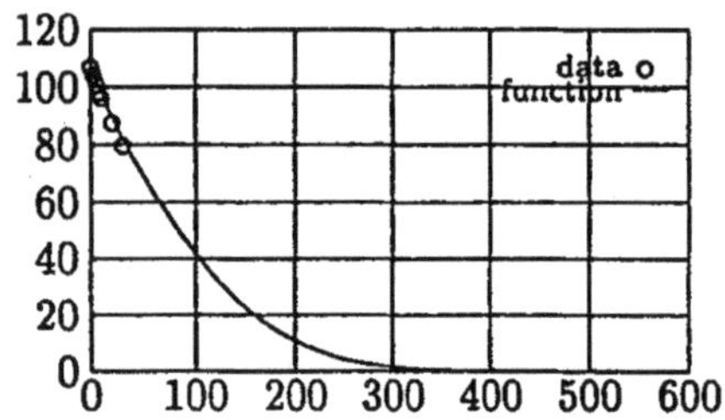

Figure 5.7 Substrate at Donor

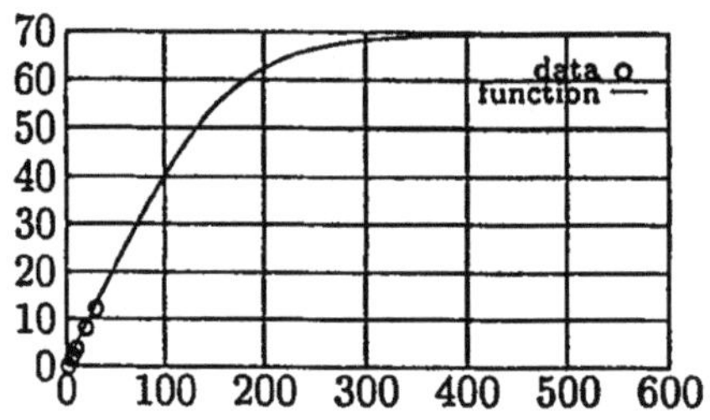

Figure 5.8 Metabolite at Donor

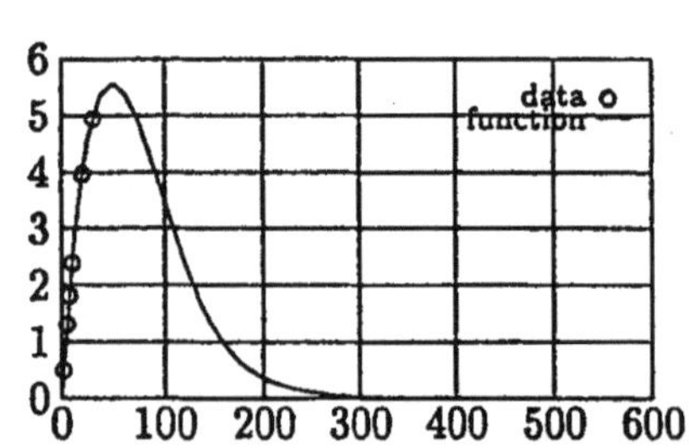

Figure 5.9 Substrate at Receiver

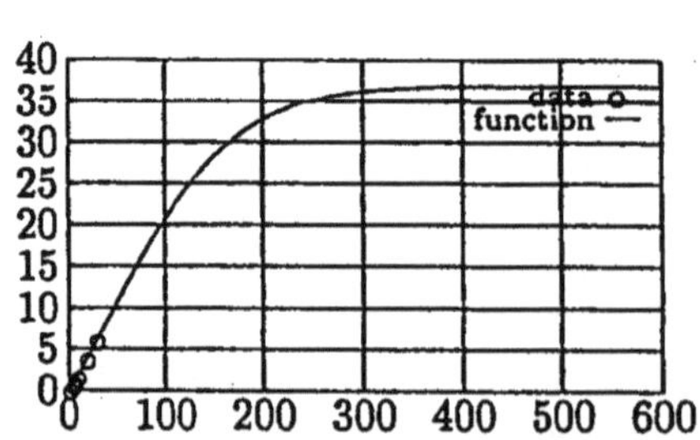

Figure 5.10 Metabolite at Receiver

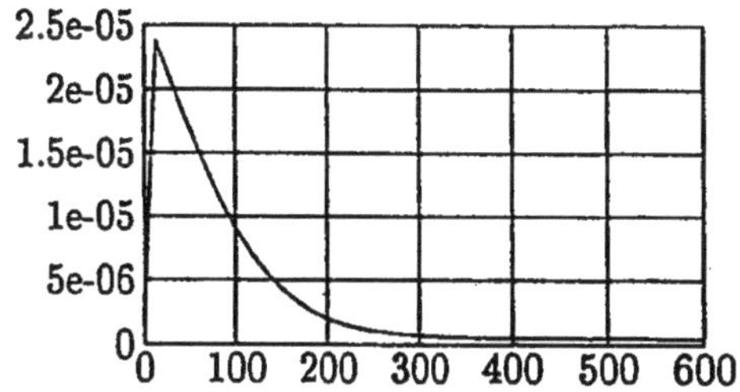

Figure 5.11 Mass Balance

active tissue, where generation of the metabolite is taken into account by the Michaelis-Menten effect. Non-continuous transitions between tissue and membrane are formulated. The mass fluxes at donor and reveiver sides are modelled by coupled ordinary differential equations. Numerical results based on some experimental data show the efficiency and feasibility of the proposed approach. Steady-state conditions and mass balance can be used to validate the model.

Acknowledgments

The author is grateful to Dr. M. Wolf from Boehringer Ingelheim Pharma KG, who developed the transdermal diffusion model. The paper was supported by a BMBF grant under contract number 03-SC7BAY-0.

References

P. Boderke, K. Schittkowski, M. Wolf, H.P. Merkle (1998), *A mathematical model for diffusion and concurrent metabolism in metabolically active tissue*, submitted for publication.

J.E. Dennis jr., D.M. Gay, R.E. Welsch (1993), Algorithm 573: NL2SOL-An adaptive nonlinear least-squares algorithm, *ACM Transactions on Mathematical Software*, Vol. 7, No. 3, pp. 369-383.

M. Dobmann, M. Liepelt, K. Schittkowski, Algorithm 746: PCOMP: A FORTRAN code for automatic differentiation, *ACM Transactions on Mathematical Software*, Vol. 21, No. 3, pp. 233-266.

R.H. Guy, J. Hadgraft (1988), Physicochemical aspects of percutaneous penetration and its enhancement, *Pharmaceutical Research*, Vol. 5, No. 12, pp. 753-758.

J. Hadgraft (1979), The epidermal reservoir, a theoretical approach, *International Journal of Pharmaceutics*, Vol. 2, pp. 265-274.

E. Hairer, S.P. Nørsett, G. Wanner (1993), Solving Ordinary Differential Equations I. Nonstiff Problems, Series Computational Mathematics, Springer.

E. Hairer, G. Wanner (1991), *Solving Ordinary Differential Equations II. Stiff and Differential-Algebraic Problems*, Series Computational Mathematics, Springer.

S.A.M. Hotchkiss (1992), Skin as a xenobiotic metabolizing organ, in: *Process in Drug Metabolism*, G.G. Gibson ed., Taylor and Francis Ltd., London, pp. 217-262.

P. Lindström (1983), *A general purpose algorithm for nonlinear least squares problems with nonlinear constraints*, Report UMINF-103.83, Institute of Information Processing, University of Umea, Umea, Sweden.

W.B. Pratt, P. Taylor (1990), *Principles of Drug Action*, Churchill Livingstone, New York.

W. E. Schiesser (1991), *The Numerical Method of Lines*, Academic Press, Inc., San Diego.

K. Schittkowski (1985), NLPQL: A FORTRAN subroutine solving constrained nonlinear programming problems, *Annals of Operations Research*, Vol. 5, pp. 485-500.

K. Schittkowski (1988), Solving nonlinear least squares problems by a general purpose SQP-method, in: *Trends in Mathematical Optimization*, K.-H. Hoffmann, J.-B. Hiriart-Urruty, C. Lemarechal, J. Zowe eds., International Series of Numerical Mathematics, Vol. 84, Birkhäuser, pp. 295-309.

K. Schittkowski (1997a), *EASY-FIT: Parameter estimation in dynamical systems - User's guide*, Report, Department of Mathematics, University of Bayreuth, Germany.

K. Schittkowski (1997b), Parameter estimation in one-dimensional time-dependent partial differential equations, *Optimization Methods and Software*, Vol. 7, No. 3-4, pp. 165-210.

K. Schittkowski (1998), PDEFIT: A FORTRAN subroutine for estimating parameters in systems of partial differential equations to appear: *Optimization Methods and Software.*

I. Steinsträsser (1994), *The organized HaCaT cell culture sheet: A model approach to study epidermal peptide drug metabolism*, Dissertation, Pharmaceutical Institute, ETH Zürich, Switzerland.

M. Wolf (1994), *Mathematisch-physikalische Berechnungs- und Simulationsmodelle zur Beschreibung und Entwicklung therapeutischer Systeme*, Habilitation, Faculty of Mathematics and Natural Sciences, University of Bonn, Germany.

Applied Optimization

1. D.-Z. Du and D.F. Hsu (eds.): *Combinatorial Network Theory.* 1996 ISBN 0-7923-3777-8
2. M.J. Panik: *Linear Programming: Mathematics, Theory and Algorithms.* 1996 ISBN 0-7923-3782-4
3. R.B. Kearfott and V. Kreinovich (eds.): *Applications of Interval Computations.* 1996 ISBN 0-7923-3847-2
4. N. Hritonenko and Y. Yatsenko: *Modeling and Optimimization of the Lifetime of Technology.* 1996 ISBN 0-7923-4014-0
5. T. Terlaky (ed.): *Interior Point Methods of Mathematical Programming.* 1996 ISBN 0-7923-4201-1
6. B. Jansen: *Interior Point Techniques in Optimization.* Complementarity, Sensitivity and Algorithms. 1997 ISBN 0-7923-4430-8
7. A. Migdalas, P.M. Pardalos and S. Storøy (eds.): *Parallel Computing in Optimization.* 1997 ISBN 0-7923-4583-5
8. F.A. Lootsma: *Fuzzy Logic for Planning and Decision Making.* 1997 ISBN 0-7923-4681-5
9. J.A. dos Santos Gromicho: *Quasiconvex Optimization and Location Theory.* 1998 ISBN 0-7923-4694-7
10. V. Kreinovich, A. Lakeyev, J. Rohn and P. Kahl: *Computational Complexity and Feasibility of Data Processing and Interval Computations.* 1998 ISBN 0-7923-4865-6
11. J. Gil-Aluja: *The Interactive Management of Human Resources in Uncertainty.* 1998 ISBN 0-7923-4886-9
12. C. Zopounidis and A.I. Dimitras: *Multicriteria Decision Aid Methods for the Prediction of Business Failure.* 1998 ISBN 0-7923-4900-8
13. F. Giannessi, S. Komlósi and T. Rapcsák (eds.): *New Trends in Mathematical Programming.* Homage to Steven Vajda. 1998 ISBN 0-7923-5036-7
14. Ya-xiang Yuan (ed.): *Advances in Nonlinear Programming.* Proceedings of the '96 International Conference on Nonlinear Programming. 1998 ISBN 0-7923-5053-7
15. W.W. Hager and P.M. Pardalos: *Optimal Control.* Theory, Algorithms, and Applications. 1998 ISBN 0-7923-5067-7
16. Gang Yu (ed.): *Industrial Applications of Combinatorial Optimization.* 1998 ISBN 0-7923-5073 1
17. D. Braha and O. Maimon (eds.): *A Mathematical Theory of Design: Foundations, Algorithms and Applications.* 1998 ISBN 0-7923-5079-0

18. O. Maimon, E. Khmelnitsky and K. Kogan: *Optimal Flow Control in Manufacturing.* Production Planning and Scheduling. 1998 ISBN 0-7923-5106-1
19. C. Zopounidis and P.M. Pardalos (eds.): *Managing in Uncertainty: Theory and Practice.* 1998 ISBN 0-7923-5110-X
20. A.S. Belenky: *Operations Research in Transportation Systems:* Ideas and Schemes of Optimization Methods for Strategic Planning and Operations Management. 1998 ISBN 0-7923-5157-6
21. J. Gil-Aluja: *Investment in Uncertainty.* 1999 ISBN 0-7923-5296-3
22. M. Fukushima and L. Qi (eds.): *Reformulation: Nonsmooth, Piecewise Smooth, Semismooth and Smooting Methods.* 1999 ISBN 0-7923-5320-X
23. M. Patriksson: *Nonlinear Programming and Variational Inequality Problems.* A Unified Approach. 1999 ISBN 0-7923-5455-9
24. R. De Leone, A. Murli, P.M. Pardalos and G. Toraldo (eds.): *High Performance Algorithms and Software in Nonlinear Optimization.* 1999 ISBN 0-7923-5483-4
25. A. Schöbel: *Locating Lines and Hyperplanes.* Theory and Algorithms. 1999 ISBN 0-7923-5559-8
26. R.B. Statnikov: *Multicriteria Design.* Optimization and Identification. 1999 ISBN 0-7923-5560-1
27. V. Tsurkov and A. Mironov: *Minimax under Transportation Constrains.* 1999 ISBN 0-7923-5609-8
28. V.I. Ivanov: *Model Development and Optimization.* 1999 ISBN 0-7923-5610-1
29. F.A. Lootsma: *Multi-Criteria Decision Analysis via Ratio and Difference Judgement.* 1999 ISBN 0-7923-5669-1
30. A. Eberhard, R. Hill, D. Ralph and B.M. Glover (eds.): *Progress in Optimization.* Contributions from Australasia. 1999 ISBN 0-7923-5733-7
31. T. Hürlimann: *Mathematical Modeling and Optimization.* An Essay for the Design of Computer-Based Modeling Tools. 1999 ISBN 0-7923-5927-5
32. J. Gil-Aluja: *Elements for a Theory of Decision in Uncertainty.* 1999 ISBN 0-7923-5987-9
33. H. Frenk, K. Roos, T. Terlaky and S. Zhang (eds.): *High Performance Optimization.* 1999 ISBN 0-7923-6013-3
34. N. Hritonenko and Y. Yatsenko: *Mathematical Modeling in Economics, Ecology and the Environment.* 1999 ISBN 0-7923-6015-X
35. J. Virant: *Design Considerations of Time in Fuzzy Systems.* 2000 ISBN 0-7923-6100-8

Applied Optimization

36. G. Di Pillo and F. Giannessi (eds.): *Nonlinear Optimization and Related Topics.* 2000
ISBN 0-7923-6109-1
37. V. Tsurkov: *Hierarchical Optimization and Mathematical Physics.* 2000
ISBN 0-7923-6175-X
38. C. Zopounidis and M. Doumpos: *Intelligent Decision Aiding Systems Based on Multiple Criteria for Financial Engineering.* 2000 ISBN 0-7923-6273-X
39. X. Yang, A.I. Mees, M. Fisher and L.Jennings (eds.): *Progress in Optimization.* Contributions from Australasia. 2000 ISBN 0-7923- 6286-1

KLUWER ACADEMIC PUBLISHERS – DORDRECHT / BOSTON / LONDON

MIX
Papier aus verantwortungsvollen Quellen
Paper from responsible sources
FSC® C105338

If you have any concerns about our products,
you can contact us on
ProductSafety@springernature.com

In case Publisher is established outside the EU,
the EU authorized representative is:
Springer Nature Customer Service Center GmbH
Europaplatz 3, 69115 Heidelberg, Germany

Printed by Libri Plureos GmbH
in Hamburg, Germany